C0-CCE-060

THE HISTORY OF GEOPHYSICS AND METEOROLOGY

BIBLIOGRAPHIES OF THE HISTORY
OF SCIENCE AND TECHNOLOGY
(Vol. 7)

GARLAND REFERENCE LIBRARY
OF THE HUMANITIES
(Vol. 421)

Bibliographies of the History of Science and Technology

Editors

Robert Multhauf, Smithsonian Institution, Washington, D.C.
Ellen Wells, Smithsonian Institution, Washington, D.C.

1. *The History of Modern Astronomy and Astrophysics*
   A Selected, Annotated Bibliography
   by David DeVorkin
2. *The History of Science and Technology in the United States*
   A Critical, Selective Bibliography
   by Marc Rothenberg
3. *The History of the Earth Sciences*
   An Annotated Bibliography
   by Roy Sydney Porter
4. *The History of Modern Physics*
   An International Bibliography
   by Stephen Brush and Lanfranco Belloni
5. *The History of Chemical Technology*
   An Annotated Bibliography
   by Robert P. Multhauf
6. *The History of Mathematics From Antiquity to the Present*
   A Selective Bibliography
   by Joseph Dauben, editor
7. *History of Geophysics and Meteorology*
   An Annotated Bibliography
   by Stephen G. Brush and Helmut E. Landsberg
8. *The History of Classical Physics*
   A Selected, Annotated Bibliography
   by R. W. Home, with assistance of Mark J. Gittins
9. *The History of Geography*
   An Annotated Bibliography
   by Gary S. Dunbar
10. *The History of the Health Care Sciences and Health Care, 1700–1980*
    A Selective Annotated Bibliography
    by Jonathon Erlen

THE HISTORY OF GEOPHYSICS AND METEOROLOGY
An Annotated Bibliography

Stephen G. Brush
Helmut E. Landsberg
with the assistance of Martin Collins

GARLAND PUBLISHING, INC. • NEW YORK & LONDON
1985

Library of Congress Cataloging in Publication Data

Brush, Stephen G., 1935–
The history of geophysics and meteorology.

(Bibliographies of the history of science and technology ; v. 7) (Garland reference library of the humanities ; vol. 421)
Includes bibliographical references and index.
1. Geophysics—History—Bibliography. 2. Meteorology—History—Bibliography. I. Landsberg, Helmut Erich, 1906– . II. Collins, Martin. III. Title.
IV. Series. V. Series: Garland reference library of the humanities ; v. 421.
Z6041.B78 1985 [QC804] 016.551 82-49292
ISBN 0-8240-9116-7 (alk. paper)

Printed on acid-free, 250-year-life paper
Manufactured in the United States of America

GENERAL INTRODUCTION

This bibliography is one of a series designed to guide the reader into the history of science and technology. Anyone interested in any of the components of this vast subject area is part of our intended audience, not only the student, but also the scientist interested in the history of his own field (or faced with the necessity of writing an "historical introduction") and the historian, amateur or professional. The latter will not find the bibliographies "exhaustive," although in some fields he may find them the only existing bibliographies. He will in any case not find one of those endless lists in which the important is lumped with the trivial, but rather a "critical" bibliography, largely annotated, and indexed to lead the reader quickly to the most important (or only existing) literature.

Inasmuch as everyone treasures bibliographies, it is surprising how few there are in this field. George Sarton's *Guide to the History of Science* (Waltham, Mass., 1952; 316 pp.), Eugene S. Ferguson's *Bibliography of the History of Technology* (Cambridge, Mass., 1968; 347 pp.), François Russo's *Histoire des Sciences et des Techniques. Bibliographie* (Paris, 2nd ed., 1969; 214 pp.) are justifiably treasured but they are, of necessity, limited in their coverage and need to be updated.

For various reasons, mostly bad, the average scholar prefers adding to the literature to sorting it out. The editors are indebted to the scholars represented in this series for their willingness to expend the time and effort required to pursue the latter objective. Our aim, and that of the publisher, has been to give the series enough uniformity to give some consistency to the series, but otherwise to leave the format and contents to the author/compiler. We have urged that introductions be used for essays on "the state of the field," and that selectivity be exercised to limit the length of each volume. Since the historical literature ranged from very large (e.g., medicine) to very small (e.g., chemical technology), some bibliographies will be limited to truly important writings while others will include modest "contributions" and even primary sources. The problem is intelligible guidance into a particular field—or subfield—and its solution is largely left to the author/compiler. In general, topical volumes (e.g., chemistry) will deal with the subject since about 1700, leaving earlier literature to area or chronological volumes (e.g., medieval science); but here, too, the volumes will vary according to the judgment of the author. The volumes are international (except for two, *Science and Technology in the United States* and *Science and Technology in Eastern Asia*) but the literature covered depends, of course, on the linguistic equipment of the author and his access to "exotic" literatures.

Robert Multhauf
Ellen Wells

Smithsonian Institution
Washington, D.C.

CONTENTS

INTRODUCTION

The purpose of a bibliography is twofold: first, to allow the researcher to find, quickly and accurately, specific information about a subject; second, to offer guidance to those users who are just beginning to look into the literature of the subject. The advanced researcher will demand that the literature be exhaustively listed and indexed and will have little need for an introduction (other than to learn how the bibliography is organized) or for evaluations of the works listed; the beginner will profit from the compiler's judgment in selecting the most appropriate items to read first. In this particular bibliography, we lean toward the second purpose, which would normally be accomplished by textbooks, scholarly treatises, and encyclopedias; unfortunately, very few such works give an overall view of the development of geophysics and meteorology or adequate references to historical information. We have therefore provided extensive annotations, and have called attention to some general works in this Introduction. On the other hand, we have omitted some of the more obscure publications which could not be obtained in a reasonable time in local libraries.

We have interpreted the term "geophysics" to include not only the study of the physics of the earth and its fluid envelopes, but also the origin and development of the solar system and the formation of the earth's surface features. The last topic obviously overlaps geology, and since another volume in this series is devoted to the literature of geology, we have included only a few of the more theoretical or fundamental works in this area.

There is no single comprehensive treatise dealing with the history of geophysics and meteorology, probably because, as pointed out by J. Tuzo Wilson (item A114), the earth sciences have until quite recently been fragmented into several non-communicating disciplines. General works on the history of geology include chapters on geophysical topics such as vulcanism, seismology, and the uniformitarian philosophy, but ignore oceanography, meteorology, and ionospheric research. Hall's *History of the Earth Sciences* (item A48) attempts an overview but says little about developments of the last 100 years. Schröder, in his *Disziplingeschichte* (item A94) and other works, discusses this problem from the standpoint of modern scholarship in the history and philosophy of science. There are some useful collections of case studies spanning the entire range of the earth sciences but most of these are limited to particular countries or time periods and do not attempt to offer an integrated account; see items A8, A13, A14, A17, A20, A98, A99, A103, A108, and other works cited in Section A.

In section A and in later sections devoted to specific topics, we include modern editions and translations of selected classic works such as Lyell's *Principles of Geology* (item A63), as well as anthologies of original research papers and extracts from important works. The reprint of Royal Institution Lectures on the *Earth Sciences* (item A86) is a fascinating collection of popular expositions presented by leading scientists from 1851 to 1937. In the annotations for such anthologies we have listed the authors of major contributions (more than a few pages), and have included these in the index at the end of the bibliography.

The biography section is divided into two parts: first, those items with the letter B followed by a number are works about several scientists, and are arranged by the author or editor of the work; the second part, items with the letter B followed by another letter and a number, includes works primarily about one scientist, arranged by the subject. Thus item BA3 would be a work about a scientist whose last name begins with A. But works which deal primarily with that scientist's contribution to a single topic would be found under that topic rather than in the biography section, so the index must be used in order to locate *all* items dealing with a particular scientist.

The best single source for information about scientists (other than those still alive) is the *Dictionary of Scientific Biography*. Since this valuable reference work is available in most libraries, we assume that it can be used in conjunction with this bibliography. Rather than give the *DSB* entry as a separate item we have simply indicated by the symbol "@" following the scientist's name that an entry may be found in the *DSB*; but if there are no other biographical sources we have included the *DSB* article in order to give an indication of the major contributions of that scientist.

Section C, "Social and Institutional History," consists largely of works on large-scale cooperative projects such as the International Geophysical Year. We found very few publications that could be called "social history" pertaining to geophysics, comparable to the substantial amount of scholarly research inspired by physics and biology. Items dealing with meteorological institutions will be found in section N.

Only one book, Jaki's *Planets and Planetarians* (item D33), attempts a comprehensive survey of all major theories of the origin of the solar system; two works by scientists (items D28 and D71) should be mentioned because they include more than the usual perfunctory review of earlier ideas. But none of these can be considered an objective history of cosmogony, nor do they attempt to relate ideas about the origin of the earth to theories of its early development.

This section (D) does include a number of good articles on individual theories, and the study by Philip Lawrence (item D44) shows how the most popular theory, the nebular hypothesis, interacted with physical ideas about the history of the earth.

Geochronology, insofar as it deals with the age of the earth, might seem to be part of the general topic of cosmogony, but we have assigned it a separate section (E) since most of the literature on this subject, until quite recently, has developed from a different standpoint. In geochronology we measure time backward from the present; in cosmogony we look forward from a hypothetical starting point in the past. Albritton's book (item E1) gives a good overview, including the connection of scientific estimates of the age of the earth to the current creation-evolution controversy. (Readers who wish to pursue that connection may be interested in the article by one of us, S. G. Brush, "Finding the Age of the Earth: By Physics or by Faith?" *Journal of Geological Education* 30 [1982]: 34–58, which is not listed in this bibliography because it is primarily polemical rather than historical.) The authoritative account of the 19th-century debate between Kelvin and the geologists is Burchfield's book (item E6), which may be supplemented by Dean's article (item E7) for the earlier background.

There seems to be no comprehensive account of researches on the physics of the earth's interior; section F includes a number of articles on specific aspects of this subject, or limited historical periods. (For a non-technical survey see S. G. Brush, "Inside the Earth," *Natural History* 93, no. 2 [Feb. 1984]: 26-34.) For geodesy and the shape of the earth there are several books but most of them do not cover the most recent period; the article by Fischer (item G27) is a good introduction which covers the two millennia from Greek antiquity to the early 1970s.

On the other hand there is a great abundance of works on the history of research and speculation relating to the formation of the earth's surface features (section H); we have included only a small selection of this literature since it is covered more comprehensively in the bibliography on geology. For a general introduction to the pre-20th-century theories, we recommend the books by Davies (item H17) and Greene (item H25). Carozzi's article (HA7) provides an authoritative account of the original formulation of the Ice Age theory, and the book by John and Katherine Imbrie (item HA16) reviews theories of the cause of ice ages in the 19th and 20th centuries. The history of ideas about continental drift and of the theory of plate tectonics is well treated in books by two scientists, Hallam (item HC18) and Marvin (item HC26), and a science journalist, Sullivan (item HC35); the anthology of original papers, with useful commentary by Cox (item HC7), is also quite valuable.

We could not find a recent comprehensive history of seismology; Davison's book (item J7) is a good introduction to the older researches, but technical reviews such as that by Stauder (item J35) must be consulted for information about 20th-century work. For the related area of applied geophysics, on the other hand, we can recommend the just-published book by Bates et al. (item JA2), which includes some aspects of seismology.

There is one book entirely devoted to the history of hydrology (Biswas, item K5); this is probably the best place to start exploring that subject.

The section on oceanography (L) includes both general surveys and works on physical aspects. There are two good comprehensive histories, the one by Deacon (item L17) covering the pre-20th-century periods, and the one by Schlee (item L48), which concentrates on the 19th and 20th centuries. The study of tides may be considered part of oceanography but has a special interest for some historians of science because it was one way in which Galileo attempted to prove the earth's motion (items LA3, LA5, LA10, LA12).

Two major works on the history of meteorology, by Khrgian (item M34) and by Schneider-Carius (item M56), provide extensive coverage of this topic. Middleton's book (item OA24) is the best single reference for the history of meteorological instruments. The collection of essays edited by McIntyre (item M39) touches on a number of interesting aspects of meteorology. More detailed treatments can be found in Boyer's monograph on the rainbow (item QC2) and Kutzbach's on cyclones (item R20).

The pre-instrumental views on weather and meteorology are covered in Henninger's *Handbook of Renaissance Meteorology* (item M28). Although it begins with meteorological thought of antiquity, the emerging science of meteorology is the topic of Frisinger's *History of Meteorology: To 1800* (item M19). The fascination of humans with rainfall since ancient times is documented in Middleton's *History of the Theories of Rain* (item PA25).

Writings on geomagnetism deal mainly with the early period, for example, the series of three articles by Mitchell (items U36–38), and the chapters in treatises by Chapman and Bartels (item U6) and Ludy and Howe (item U29). Twentieth-century discoveries are discussed in works on the history of plate tectonics, especially the books by Glen (item HC17) and by Takeuchi et al. (item U45).

Eather's book on the history of auroral research (item V17) demonstrates the spectacular visual appeal as well as the intellectual puzzles of one of the oldest branches of geophysics. Gillmor and Gran's article (item V27) discusses the growth of ionospheric studies in the early part of the 20th century, and Newell's book (item V37) surveys the research in "space science" made possible by rockets and satellites. This entire area of upper-atmosphere and ionospheric research is, of course, inseparable from that of solar-terrestrial relations, for which Meadows and

Kennedy provide a brief historical introduction (item W4). Histories of research on lunar-terrestrial relations, aside from those dealing specifically with tides (section LA), remain to be written.

Naturally the user of this bibliography will be most interested in items which we could not include, namely those published since 1982. To keep up with the latest historical research one should consult the annual "Critical Bibliography" published in *Isis*; a cumulative bibliography of the history of science covering most publications in the period 1913–65 has been compiled by Magda Whitrow, and a supplement for 1966–75 is now in the process of publication by John Neu. Technical reviews and biographical notes may be located in *Meteorological and Geoastrophysical Abstracts*, published by the American Meteorological Society, and in *Geophysical Abstracts* published by Geo Abstracts Ltd. These essential reference works should be available in every major scientific library.

ACKNOWLEDGMENT

Preparation of this bibliography was supported in part by a grant from the National Endowment for the Humanities.

The History of Geophysics and Meteorology

A. GENERAL HISTORIES

A1 Abbe, Cleveland. "A Plea for Terrestrial Physics." *Proceedings of the American Association for the Advancement of Science* 39 (1890): 65-79.

Complains that progress in this field is slow because "our best students find the grander terrestrial problems more difficult and less interesting than the minor problems." Presents a survey of outstanding problems.

A2 Adams, Frank Dawson. *The Birth and Development of the Geological Sciences*. Baltimore: Williams & Wilkins, 1938. Reprint, New York: Dover, 1954. 506 pp.

Includes chapters on "The Birth of Historical Geology with the Rise and Fall of the Neptunian Theory," "The Origin of Mountains" (with a long extract from *De Montium Origine* by Valerius Faventies, 1561), and "Earthquakes and the Nature of the Interior of the Earth," up to the middle of the 19th century.

A3 Albritton, Claude C., Jr. "Second Bibliography and Index for the Philosophy of Geology." *Journal of Graduate Research Center* (Southern Methodist University) 33 (1964): 73-114.

Supplement to the "bibliography and index" in *The Fabric of Geology* (item A5).

A4 Albritton, Claude C., Jr. "Third Bibliography and Index for the Philosophy of Geology." *Journal of Graduate Research Center* (Southern Methodist University) 35 (1966): 55-87.

A5 Albritton, Claude C., Jr., ed. *The Fabric of Geology*. Prepared under the direction of a Committee of The Geological Society of America, in commemoration of the Society's 75th Anniversary. San Francisco: Freeman, Cooper & Co., 1963, ix + 374 pp.

Includes: "James Hutton and the Philosophy of Geology" by D.B. McIntyre; "The Scientific Philosophy of G.K. Gilbert" by J. Gilluly; "Philosophy of Geology: A Selected Bibliography and Index" by C.C. Albritton, Jr.; essays on philosophical and methodological topics.

A6 Albritton, Claude C., Jr., ed. *Uniformity and Simplicity: A Symposium on the Principle of the Uniformity of Nature.* Geological Society of America, special paper No. 89. New York: Geological Society of America, 1967. [10] + 99 pp.

Includes: M. King Hubbert, "Critique of the Principle of Uniformity" (on Hutton, Lyell, Kelvin, Chamberlin, the cooling and age of the earth); Leonard G. Wilson, "The Origins of Charles Lyell's Uniformitarianism"; Nelson Goodman, "Uniformity and Simplicity."

A7 Albritton, Claude C., Jr., ed. *Philosophy of Geohistory, 1785-1970.* Benchmark Papers in Geology, vol. 13. Stroudsburg, Pa.: Dowden, Hutchison, Ross, 1975. xiii + 386 pp.

Includes extracts from philosophical-methodological writings of J. Hutton, J. Playfair, C. Lyell, W. Whewell, T.C. Chamberlin, G.K. Gilbert, and others, and historical writings of R. Hooykaas, D.B. Kitts, S. Toulmin, and J. Goodfield.

A8 Bartels, Julius, et al. *FIAT Review of German Science 1939-1946. Geophysics* (Parts I and II). Wiesbaden: Office of Military Government for Germany, Field Information Agencies Technical, 1948. 237 + 321 pp.

Contents: "Die Erde als Planet, Erdgestalt, Erdinneres, Schwere" by K. Jung; "Erdmagnetismus" by F. Errulat and J. Bartels; "Ionosphäre" by W. Dieminger; "Atmosphärische Elektrizität" by H. Israel; "Geologische Anwendungen der Geophysik" by A. Graf, H. Reich, and E. Bederke; "Seismik" by W. Hiller, G.-A. Schulze, O. Förtsch, R. Köhler, and H. Baule; "Ozeonographie" by G. Böhnecke, G. Neumann, W. Hansen, W. Horn, K. Kalle, and J. Joseph; "Hydrographie" by W. Friedrich; "Die Erde als geodätisches Messfeld für Lage- und Höhenbestimmungen" by F.R. Jung; "Das atmosphärische Ozon" by E. Regener.

A9 Batiushkova, Irina V. *Istoriia problemy proiskhozhdeniia materikov i okeanov.* Moscow: Nauka, 1975. 137 pp.

History of the problem of the origin of continents and oceans, since the 17th century; includes an account of

20th-century debates on mobility vs. permanence.

See also the book edited by Batiushkova, *Istoriia Geologii* (Moscow: Nauka, 1973. 389 pp.), which includes some sections on the history of geophysics and an extensive bibliography of sources.

A10 Becker, George F. "Present Problems of Geophysics." *Science* 20 (1904): 545-556.

An address at the International Congress of Arts and Sciences, St. Louis. The problems are: origin and physical constitution of the earth (e.g., whether crystalline solid or amorphous); former changes in the rate of rotation and their effects on the surface; seismology; the origin and mechanism of volcanoes; origin of petroleum and ore deposits; climatology and the cause of glaciation; magnetism of ancient lavas. Research on the properties of matter at high pressures and temperatures is needed.

A11 Bender, Paul A. "Resource Letter SEG-1: Solid-earth Geophysics." *American Journal of Physics* 44 (1976): 903-911.

Annotated bibliography of books and articles suitable for students and teachers, including selected items on gravity and geodesy, seismology, geomagnetism, geochronology and heat flow, and plate tectonics. Most of the items listed are not historical.

A12 Beringer, Carl Christoph. *Geschichte der Geologie und des geologischen Weltbildes*. Stuttgart: Ferdinand Enke Verlag, 1954. 158 pp.

"Surely no other natural science lent itself to philosophical speculation nor was it influenced by philosophical trends to the same extent as geology. Side by side with the growing consciousness of the existence of a history of mankind and of civilization, there was an interacting consciousness of a history of the earth. Western *faustische* man alone has come to appreciate that these are distinct. The growth of this historical feeling for the earth in the world of ideas and for its influence and interactions with science and philosophy are the principal concerns of this book. The path of conflicting ideas by which we have arrived at present-day cosmology was probably never so intricately embroidered as during the early nineteenth century, and Dr. Beringer has devoted much the largest and the best portion of this little book to this period and its principal theme--evolution....

In the history of geology, he finds an apt illustration of Spenglerian decline...."--C.J. Schneer, *Isis* 48 (1957): 358-359.

A13 Birett, H., K. Helbig, W. Kertz, and U. Schmucker, eds. *Zur Geschichte der Geophysik: Festschrift zur 50jährigen Wiederkehr der Gründung der Deutschen Geophysikalischen Gesellschaft*. Berlin: Springer-Verlag, 1974. xv + 288 pp.

Contents: M. Koenig, "Die Deutsche Geophysikalische Gesellschaft, 1922-1974"; H. Birett, "Zur Vorgeschichte der Newtonschen Theorie der Gezeiten"; W. Petri, "Geophysikalisch interessante Gedanken bei Copernicus und Kepler"; V. Bialas, "Modelle der Isostasie im 19. Jahrhundert"; G.G. Angenheister, "Geschichte des Samoa-Observatoriums von 1902-1921"; J. Meyer, "Künstliche Bodenerschütterungen mit der Mintrop-Kugel"; M. Toperczer, "Die Conrad-Diskontinuität"; F. Groten, "Die gravimetrischen Arbeiten Rudolf Tomascheks"; G.A. Schulze, "Anfänge der Krustenseismik"; R. Köhler, "Anfänge der Reflexionsseismik in Deutschland"; H. Closs, "Die geophysikalische Reichsaufnahme und ihr Vorgeschichte"; F. Errulat, "Die geophysikalische Wart Gross Raum der Universität Königsberg/Pr.--Ein Rückblick"; D. Voppel, "Hundert Jahre Erdmagnetischer Dienst in Norddeutschland"; W. Schröder, "Hermann Fritz und sein Wirken für die Polarlichtforschung"; R. Mühleisen and H.J. Fischer, "Erste luftelektrische Messungen der freien Atmosphere"; H.K. Paetzold, G. Pfotzer, E. Schopper, "Erich Regener als Wegbereiter der extraterrestrischen Physik"; B. Beckmann, "Die Entdeckungsgeschichte der Radiowellenausbreitung in den ersten fünfzig Jahren mit besonderer Berücksichtigung der deutschen Arbeiten"; N. Petersen, "Zur Geschichte der magnetischen Feldwagge"; U. Schmucker, H. Schmidlin, "Die Erforschung der remanenten Magnetisierung von Gesteinen bis zu den Arbeiten von J.G. Koenigsberger"; H. Birett, "Quellen zur Geschichte der Geophysik."

A14 Bowie, William, et al. "A Survey of Research Problems in Geophysics, Prepared by Chairmen of Sections of the American Geophysical Union." *Proceedings of the National Academy of Sciences* 6 (1920): 545-601.

Includes papers presented at the first annual meeting of AGU (1920), on geodesy by W. Bowie, on seismology by H.F. Reid, on meteorology by C.F. Marvin, on terrestrial magnetism and electricity by L.A. Bauer, on physical oceanography by G.W. Littlehales, on vulcanology by H.S. Washington, and on geophysical-chemical problems by R.B. Sosman.

A15 Bruce, Robert V. "A Statistical Profile of American Scientists, 1846-1876." *Nineteenth-Century American Science*. Edited by G.H. Daniels. Evanston, Ill.: Northwestern University Press, 1972, pp. 63-94.

The figures indicate a preference for earth science and natural history; the author interprets this as a preference for descriptive rather than laboratory or mathematical research.

A16 Caloi, P. "La geophysique dans le dernier demi-siècle." *Scientia* 92, Supplement (1957): 54-62.

On terrestrial magnetism, atmospheric electricity, and oceanography.

A17 Caloi, P. "Italian Pioneers in the Physics of the Universe. Second Part: Geophysics." *Cahiers d'Histoire Mondiale* 7 (1963): 453-469.

A survey of Italian contributions to atmospheric electricity, terrestrial magnetism, hydrography, physical oceanography, seismology, and vulcanology, since the second half of the 18th century. The most important scientists were L. Nobili and M. Melloni; also mentioned are G.B. Beccaria, L. Palmieri, C. Matteucci, T. Bertelli, L. dė Marchi, C. Somigliana, E. Oddone, A. Sella, and D. Pacini.

A18 Chao, Edward C.T. "Earth Sciences." *Science in Contemporary China* (item A75), pp. 189-212.

A survey of geology, seismology, etc. in the 1970s.

A19 Chapman, S. "Geophysics and Germany, Men and Enterprises." *Zeitschrift für Geophysik* (1970): 393-398.

On Schwabe, Bartels, Wiechert, and terrestrial magnetism.

A20 Chatterjee, S.P. *Fifty Years of Science in India; Progress of Geography*. Calcutta: Indian Science Congress Association, 1964. ix + 277 pp.

Includes brief surveys of glaciology, seismology (work of W.D. West, S.K. Ghosh, etc.), hydrology (P.K. Ghosh, S. Pranavananda, H.L. Chibber, J.M. Dutta, K. Bagchi, S.P. Chatterjee, A.K. Chakravarty), oceanography, climatology (S.K. Banerjee, B.N. Desai, C. Balasubrananian), geodesy (B.L. Gulatee).

A21 Cloud, Preston, ed. *Adventures in Earth History. Being a Volume of Significant Writings from Original Sources, on Cosmology, Geology, Climatology, Organic Evolution, and Related Topics of Interest to Students of Earth History, from the Time of Nicolaus Steno to the Present.* San Francisco: W.H. Freeman, 1970. xv + 992 pp.

Includes extracts from works on "ordering principles in earth history" by N. Steno (1669), J. Playfair (1822), G.K. Gilbert (1886), M.K. Hubbert (1967), S.J. Gould (1967), and K.B. Krauskopf (1968); on cosmogony by F.L. Whipple (1964), H.C. Urey (1952), J.H. Reynolds (1960), and others; on geochronology by C. Darwin (1859), A. Woodford (1965), C. Patterson (1956), W.F. Libby (1961), and S.K. Runcorn (1967); on "earth's air, water, and climate" by H. Spencer (1844), W.W. Rubey (1951), P.H. Abelson (1966), A.N. Strahler, (1963), E.N. Lorenz (1966), W. Munk (1955), H.E. Landsberg (1960); on "the differentiation of the solid earth" by K.E. Bullen (1955), G.P. Woollard (1960), P.J. Brancazio (1964), H.H. Hess (1962), A.E.J. Engel and C.G. Engel (1964), J. Gilluly (1964), P.M. Hurley (1968), and J.T. Wilson (1965); on reversals of the earth's magnetic field by A. Cox, G.B. Dalrymple, and R.R. Doell (1967); on atmospheric and hydrospheric evolution on the primitive earth by P. Cloud (1968); on tectonic evolution by C.E. Dutton (1882) and J. Gilluly (1963); other works on geology, paleontology, and evolution.

A22 Dauvillier, A. "La troisième confrontation: paleomagnetisme et paleoclimatologie." *Scientia* 102 (1967): 96-124.

The author recalls three confrontations between naturalists and astronomers or physicists: (1) age of the earth; (2) origin of meteorites; (3) polar migrations deduced from paleoclimatic or paleomagnetic facts. He concludes that theories pass but facts remain and constitute scientific progress.

A23 Davies, Gordon L. "The Eighteenth-Century Denudation Dilemma and the Huttonian Theory of the Earth." *Annals of Science* 22 (1966): 129-138.

The dilemma was the problem of reconciling belief in a tendency toward decay (attributed to the 17th-century Puritan conception of a wrathful deity who is punishing the world) with the tendency toward progress (resulting from a more optimistic theology). Hutton's resolution of the theological problem led to the geostrophic cycle of modern geology.

A24 Davies, Gordon L. "George Hoggart Toulmin and the Huttonian Theory of the Earth." *Geological Society of America Bulletin* 78 (1967): 121-124.

D.B. McIntyre (in item A5) suggested that James Hutton took some of his ideas from Toulmin, but contemporary sources indicate that Hutton formulated his theory more than a decade before the appearance of Toulmin's book. Textual similarities in the writings of Toulmin and Hutton may be the result of Toulmin's having plagiarized an early draft of the Huttonian theory.

A25 Ericson, David B., and Goesta Wollin. *The Deep and the Past.* New York: Knopf, 1964. xiv + 292 pp.

A personal account of research on cores of deep-sea sediment and the conclusions drawn concerning geological history and climatology.

A26 Ertel, Hans. *Entwicklungsphasen der Geophysik.* Vorträge und Schriften Deutsche Akademie der Wissenschaften, Heft 52. Berlin: Akademie-Verlag, 1953. 24 pp.

Emphasizes the contributions of Leibniz, and briefly mentions a number of 19th-century discoveries.

A27 Eyles, Victor A. "The History of Geology: Suggestions for Further Research." *History of Science* 5 (1966): 77-86.

Comments and elaborates on the review article by R. Rappaport (item A80). While noting the absence of any satisfactory comprehensive work on the history of geology since 1830, the author concentrates on the need for research in 18th-century topics, including general "theories of the earth," the origin of igneous rocks in relation to the idea that the earth has a hot interior, vulcanology, geological dynamics, and the age of the earth. He gives a brief discussion of sources.

A28 Fersman, A., et al. *The Pacific Russian Scientific Investigations.* Moscow: Academy of Sciences of the USSR, 1926. Reprint, New York: Greenwood Press, 1969. 191 pp.

Includes: P. Nikiforov, "Seismology," pp. 85-91; A. Belobrov, "Terrestrial Magnetism," pp. 92-94; V. Akhmatov, "Oceanography," pp. 95-112; W. Wiese, "Meteorology," pp. 113-120.

A29 Flint, W., and J.D.F. Gilchrist, eds. *Science in South Africa: A Handbook and Review.* Cape Town: Miller, 1905. x + 505 pp.

Includes: C.M. Stewart, "The Meteorology of South Africa," pp. 19-60; D. Gill, "Astronomy and Geodesy in South Africa," pp. 61-73; J.C. Beattie, "Earth Magnetism in South Africa," pp. 74-78.

A30 Frångsmyr, Tore. *Geologi och skapelsetro: Föreställningar om jordens historia från Hiärne till Bergman* [Geology and the Doctrine of Creation: Ideas on the History of the Earth from Hiärne to Bergman]. Lychnos-bibliotek, 26. Stockholm: Almqvist & Wiksell, 1969. 379 pp.

The author presents ideas of Swedish geologists on creation and the history of the earth, in the 17th and 18th centuries, in relation to those in other countries. From the summary in English:

"The Swedish pioneer in geology was without doubt Urban Hiärne (1641-1724), physician and chemist.... Hiärne described the inner structure of the earth according to an idea current in the seventeenth century. In this, the inside of the earth contained a system of channels through which the earth's waters passed.... Hiarne's explanation of earthquakes and volcanoes shows his chemical interest. He imagined great masses of sulphur and other combustible substances under the surface, the explosion of which would result in an earthquake and a fire which would produce an eruption.... Hiärne's most important achievement in geology, however, was his belief in the perpetual change in the face of the earth...."

"Christopher Polhem (1661-1751) ... was the technical genius of his time, best known for his pioneering achievements within the field of mining." He thought the earth is much older than the traditional 6000 years, and predicted its ultimate destruction due to "an internal imbalance which would eventually bring the earth to a stand-still.... [He] rejected the popular ideas of a central fire...."

"Emanuel Swedenborg (1688-1772) ... [argued] that the earth was losing speed in its orbit round the sun [and] thus steadily moving toward its own destruction.... In his second book Swedenborg gave his view on the origin of the mountains and other surface formations.... In the *Principia rerum naturalium* (1754) Swedenborg developed an extensive and somewhat Cartesian theory of the creation of the world."

"Carl von Linné (1707-1778) [discussed] the origin of the strata of the earth's crust...." He was willing to accept a much greater age of the earth than that inferred from the Bible.

"Johann Gottschalk Wallerius (1709-1785) ... maintained that the earth and all that originally existed on it had developed out of water...."

"Tobern Bergman (1735-1784) [wrote on] the origin of mountains. He seemed to be the first to distinguish between two processes of formation of mountains: one chemical and one mechanical.... He offered a mechanical view of the world which involved no biblical allusions of geological processes."

A31 Friedman, Robert Marc. "Nobel Physics Prize in Perspective." *Nature* 292 (1981): 783-789.

Traces the changes in tradition brought about in the award of the Nobel Prize in Physics, which, through prejudicial actions of members of the Swedish Nobel Committee, resulted in the rejection for eligibility of the famous Norwegian meteorologist Vilhelm Bjerknes. C.W. Oseen's "goal of promoting theoretical physics and of improving the international stature of Swedish physics motivated him to attempt restricting the definition of physics" so as to exclude geophysics, despite the absence of any such restriction in the original rules for the Prize.

A32 Fülöp, Jozsef. "Earth Sciences." *Science and Scholarship in Hungary*. Edited by Tibor Erdey-Grúz and Kálmán Kulcsár. Second Edition. [Budapest]: Corvina Press, 1975, pp. 81-97.

A review of Hungarian research in geography, geophysics, geochemistry, and meteorology, primarily in the 19th and 20th centuries.

A33 García Castellanos, Telasco. *Influencia de los conocimientos geológicos en la cultura europea del siglo XVIII*. Córdoba: Academia Nacional de Ciencias, 1972. 14 pp.

Influence of geology on Fontenelle, Montesquieu, Voltaire, Kant, Oliver Goldsmith, Condorcet, and Goethe.

A34 Garfield, Eugene. "*Science*: 101 Years of Publication of High Impact Science Journalism." *Current Contents, Physical, Chemical and Earth Sciences* 21, no. 39 (Sept. 28, 1981): 5-12.

Includes a list of the 20 physical sciences papers published in *Science* that were most cited during the period 1961-1980; 11 of them are in the area of geophysics or meteorology.

A35 Garfield, Eugene. "*Nature*: 112 Years of Continuous Publication of High Impact Research and Science Journalism." *Current Contents, Physical, Chemical and Earth Sciences* 21, no. 40 (Oct. 5, 1981): 5-12.

Includes a list of the 20 physical sciences papers published in *Nature* that were most cited during the period 1961-1980; 8 of them are in the area of geophysics or meteorology.

A36 George, T. Neville. "Charles Lyell: The Present is the Key to the Past." *Philosophical Journal: Transactions of the Royal Philosophical Society of Glasgow* 13 (1976): 3-24.

The Kelvin Lecture, Oct. 8, 1975, presenting an overview of the uniformitarian doctrine and its reception in the 19th century.

A37 Gillmor, C. Stewart. "The Place of the Geophysical Sciences in 19th-Century Natural Philosophy." *Eos* 56 (1975): 4-7.

Report on a workshop at Pittsburgh in 1974, with summaries of papers by N. Reingold, S.G. Brush, S. Goldberg, W.F. Cannon, J.D. Burchfield, J.L. Heilbron, E. Garber, and C.S. Gillmor. Problems of historiography and the relations between geophysics and other sciences were discussed.

A38 Gilluly, James. "The Role of Geological Concepts in Man's Intellectual Development." *Texas Quarterly* 11(2) (1968): 11-23.

The principal findings of geology are (1) the great antiquity and complex history of the earth; (2) that geologic history can be interpreted in terms of processes now acting, without postulating worldwide catastrophes; (3) that man is a very late comer on the geologic scene and is a product of evolution rather than special creation. These conclusions are far removed from the outlook of the Renaissance man. The author sketches how they came to be accepted by scientists, but when he says "we know" that they are valid he disregards the crucial point one might have expected on the basis of his title: the extent to which they have been accepted by modern thinkers who are not scientists.

A39 Gordeev, Demian I. "Les particularités de la géologie à la limité des XIXe et XXe siècles: Les symptômes de crise dans la théorie." *Actes, XIIe Congres International d'Histoire des Sciences, Paris 1968* (pub. 1971) 7: 35-38.

There was a crisis *circa* 1900 involving problems such as whole vs. part, irreversibility vs. cyclic development, randomness vs. determinism, etc.; this called for the application of dialectical materialism.

A40 Gould, Stephen Jay. "Is Uniformitarianism Necessary?" *American Journal of Science* 263 (1965): 223-228.

"Methodological uniformitarianism"--the claim that the laws of nature are invariant--was useful in the 19th century when Lyell and others were battling the doctrine that God intervenes in a geological history, but since that battle was won, it is now nothing more than the assertion that geology is a science. "Substantive uniformitarianism"--the hypothesis that rates or conditions of geological change are constant--has been refuted. Confusion still results from ignoring the distinction.

A41 Gould, Stephen Jay. "History versus Prophecy: Discussion with J.W. Harrington." *American Journal of Science* 268 (1970): 187-191.

Critique of Harrington's articles, "The First, First Principles of Geology," *ibid.*, 265 (1967): 449-461 and "The Prenatal Roots of Geology: A Study in the History of Ideas," *ibid.*, 267 (1969): 592-597. See Harrington's response, cited below (item A50).

A42 Gregory, Herbert E. "History of Geology." *Scientific Monthly* 12 (1921): 97-126.

Includes sections on origin of earth, making of mountains, glaciology, age of earth.

A43 Günther, Siegmund. *Lehrbuch der Geophysik und physikalischen Geographie*. 2 vols. Stuttgart: Enke, 1884-1885.

Includes numerous historical references.

A44 Günther, Siegmund. *Geschichte der anorganische Naturwissenschaften im Neunzehnten Jahrhundert*. Das Neunzehnte Jahrhundert in Deutschlands Entwicklung, Band V. Berlin: Bondi, 1901. xix + 984 pp.

Chapters 6 and 23 deal with geodesy and geophysics; chapter 4 is on Alexander v. Humboldt.

A45 Günther, S. *Geschichte der Erdkunde*. Wien: Deuticke, 1904.

A46 Guntau, Martin. "The Emergence of Geology as a Scientific Discipline." *History of Science* 16 (1978): 280-290.

Geology emerged at the end of the 18th century from the wreckage of religiously inspired "theories of the earth," stimulated by the needs of the mining industry and the general influence of the Enlightenment on scientific exploration of the natural world. Neptunism (stressing the role of water) provided a unifying theoretical conception for geology in many countries.

A47 Gunther, R.T. *Early Science in Oxford*, Volume 1. London: Dawsons of Pall Mall, 1967 (reprint of the 1921-23 edition). vii + vi + 407 pp.

Includes description of instruments and observations: magnetic compass, weather records, barometers, hygrometers and rain-gauges, surveying instruments.

A48 Hall, D.H. *History of the Earth Sciences During the Scientific and Industrial Revolutions, with Special Emphasis on the Physical Geosciences*. New York: American Elsevier, 1976. 297 pp.

The author emphasizes the links between scientific advances and social history in this brief, readable survey, concentrating on terrestrial magnetism and the figure of the earth before 1870. There is also a chapter on "Quantitative Study of the Growth of the Earth Sciences."

A49 Hall, D.H. "The Earth and Planetary Sciences in Science During the 20th Century." *Scientometrics* 3 (1981): 349-362.

A statistical study of articles in *Science* and *Nature*, 1900-1976. In both journals the relative attention given to earth science and planetary science dropped through the century to a low point about 1955. Thereafter the trend reversed, rising almost twice as rapidly as it had previously fallen.

A50 Harrington, John W. "Ontology of Geologic Reasoning with a Rationale for Evaluating Historical Contributions." *American Journal of Science* 269 (1970): 295-303.

In response to a criticism by S.J. Gould (item A41) of his earlier writings, the author attempts to provide a rationale for his use of "our modern intellectual framework as a basis for judging the insights of the ancients." It is assumed that there is a "singular state of nature toward which all estimates of reality converge," and that the development of science is cumulative. The author wants to break the "icy grip" of relativistic cultural historians (and to vindicate what is now known as the "Whig interpretation of the history of science").

A51 Hausen, Hans. *The History of Geology and Mineralogy in Finland. The History of Learning and Science in Finland, 1828-1918*. Helsinki: Societas Scientiarum Fennica, 1968. 147 pp.

Topics include glaciology and geophysical prospecting.

A52 Hawkes, Leonard. "Some Aspects of the Progress in Geology in the Last Fifty Years. II. Anniversary Address Delivered at the Annual General Meeting of the Society on 30th April, 1958." *Quarterly Journal of the Geological Society of London*, vol. 114, 455 (1958, pub. 1959): 395-410.

"The advances in our knowledge of the rocks in the hidden depths are reviewed. Through seismology three discoveries were made: The crust of the earth (J. Milne), its base (A. Mohorovicic) and the core of the earth (R.D. Oldham). The composition of the uppermost part of the mantel and the absence of the Mohorovicic discontinuity beneath certain seismic areas are discussed.... The significance of the fold mountain belts is discussed in the light of the evidence for gravitational sliding, and the hypothesis of a contracting globe in geological time is examined. The bearing of geomagnetic rock studies in the hypotheses of polar wandering and continental drift, and the pre-geological history of the earth's crust during which the continent-ocean separation took place are briefly considered" (Author's summary). "The subject of earth evolution in pre-geological time ... [is] outside our Quarterly Journal, and properly so. This society was founded to promote the discovery of facts, the discussion of their significance, and their integration into the main body of geological knowledge, and to eschew cosmological speculation concerning events the geologist has no knowledge of" (p. 406).

A53 Hazen, Robert M., ed. *North American Geology: Early Writings*. Benchmark Papers in Geology, Volume 51. New York: Academic Press, 1979. xvii + 356 pp.

Includes papers by T. Prince and J. Winthrop on earthquakes, B. Franklin on the formation of the earth and magnetism, N. Bowditch on oblateness of the earth.

A54 Hazen, Robert M., and Margaret Hindle Hazen, eds. *American Geological Literature, 1669 to 1850*. Stroudsburg, Pa.: Dowden, Hutchinson & Ross, 1980. xii + 431 pp.

List of 11,133 items, arranged by author, with subject index; subjects include earthquakes, volcanoes, glaciers, mountains.

A55 Herschel, John, ed. *Admiralty Manual of Scientific Enquiry*. New York: Science History Pubs., 1974 (reprint of the second edition, 1851, with a new introduction by David Knight). xxii + xi + 503 pp.

A collection of expository chapters intended for the use of seamen who might make scientific observations during expeditions. Includes chapters on magnetism by Sabine, hydrography by Beechey, tides by Whewell, earthquakes by R. Mallet, meteorology by J. Herschel, atmospheric waves by W.R. Birt.

A56 Hooykaas, R. *Natural Law and Divine Miracle: A Historical-Critical Study on the "Principle of Uniformity" in Geology, Biology, and Theology*. Leiden: Brill, 1959; reprinted, 1963, with new Preface. xiv + 237 pp.

A classic study which stimulated much of the recent writing on the history of uniformitarianism. In the first chapter, Hooykaas surveys the various ways in which the principle was formulated, discussed, and criticized by K.E.A. von Hoff (1822), C. Lyell (1830-33), J. Hutton (1785, 1788), J. Hall (1805, 1812), E. Kayser (1921), L. Cayeux (1941), R. Laffitte (1949), and M.G. Rutten (1949). He then shows how it was used in biology and how the views of Lamarck, Lyell, Charles Darwin, and some modern writers on organic evolution were related to uniformitarianism. A philosophical chapter argues that a strictly uniformitarian geology cannot be really "historical," yet in fact geology does have an historical character.

A57 Kelly, Suzanne. "Theories of the Earth in Renaissance Cosmologies." *Toward a History of Geology* (item A89), pp. 214-225.

A survey of ideas about the structure of the earth, origin of mountains, and other surface changes.

A58 Kitts, David B. *The Structure of Geology*. Dallas, Texas: Southern Methodist University Press, 1977. xix + 180 pp.

A collection of previously published essays on the history and philosophy of geology, including one on G.K. Gilbert, one on continental drift and Kuhn's theory of scientific revolutions, and one on geologic time.

A59 Krut, I[gor] V[asil'evich]. *Issledovanie osnovanii teoreticheskoi geologii* [Research on the Foundations of Theoretical Geology]. Moscow: Nauka, 1973. 207 pp.

A60 Launay, L. de. *La Science géologique: ses méthodes, ses résultats, ses problèmes, son histoire*. Second revised edition. Paris: Colin, 1913. 776 pp., 53 figs., 5 plates.

Includes chapters on the methods of physics and astronomy applied to geology (form, density, local variations of gravity, internal temperature, magnetism); cosmogony; tectonics; applied optics. The first edition appeared in 1905.

A61 Levere, Trevor H. *Poetry Realized in Nature: Samuel Taylor Coleridge and Early Nineteenth-Century Science*. New York: Cambridge University Press, 1981. ix + 271 pp.

Chapter 5, "The Construction of the World: Genesis, Cosmology and General Physics," pp. 122-158; Chapter 6, "Geology and Chemistry: The Inward Powers of Matter," pp. 159-200.

A62 Luck, James Murray. *Science in Switzerland*. New York: Columbia University Press, 1967. xvi + 419 pp.

Chapter 14, "Meteorology and Climatology"; Chapter 15, "Seismology and the Swiss Earthquake Service."

A63 Lyell, Charles. *Principles of Geology. An Attempt to Explain the Former Changes of the Earth's Surface, by Reference to Causes Now in Operation*. With an Introduction by Martin J.S. Rudwick. 3 vols. New York: Stechert-Hafner, 1970.

Facsimile reprint of the 1830-33 edition. In addition to being a major original source for modern ideas about geotectonics, climatology, vulcanology, and the uniformitarian philosophy, this book includes an interpretation of the earlier history of geology. The introduction by Rudwick is a useful guide.

A64 Mather, Kirtley F., and Shirley L. Mason. *A Source Book in Geology*. New York: McGraw-Hill, 1939. Reprint, New York: Hafner, 1964. xxii + 702 pp.

Short extracts from writings on several topics including origin of the earth, internal heat, volcanoes, glaciation, mountain building, earthquakes, crustal balance, etc.

A65 Mather, Kirtley F[letcher], ed. *Source Book in Geology, 1900-1950*. Source Books in the History of the Sciences. Cambridge, Mass.: Harvard University Press, 1967. 435 pp.

Includes extracts from works on diastrophism (including tectonophysics) by T.C. Chamberlin (1909), J. Barrell (1919), W.J. Mead (1925), W.H. Bucher (1940), F.A. Vening-Meinesz (1947), J.T. Wilson (1950), V.V. Belousov (1953); on seismology by R.D. Oldham (1906), H.F. Reid (1910), C.F. Richter (1935); on geochronology by E. Rutherford (1906) and W.F. Libby (1952); on isostasy by J. Barrell (1919); on internal structure of the earth by R.D. Oldham (1906) and B. Gutenberg (1941); on glaciology by W.D. Johnson (1904), W.M. Davis (1906), R.A. Daly (1929), G.F. Kay (1931), K. Bryan (1946), H.W. Ahlmann (1948); on geomagnetism by M. Matsuyama (1929), J.W. Graham (1949); on vulcanology by R.W. Goranson (1931) and F.M. Bullard (1947).

A66 Merriam, John C. "Earth Sciences as the Background of History." *Bulletin of the Geological Society of America* 31 (1920): 233-246; reprinted in *Scientific Monthly* 12, no. 1 (1921): 15-17.

A67 Merrill, George Perkins. *The First One Hundred Years of American Geology*. New York: Hafner, 1964 (facsimile reprint of 1924 edition). xxi + 773 pp., illus., maps, ports.

Includes chapters on glaciology and geochronology.

A68 Miller, Samuel. *A Brief Retrospect of the Eighteenth Century. Part First: in two volumes: Containing a*

Sketch of the Revolutions and Improvements in Science, Arts, and Literature, during that Period. 2 vols. Vol. 1, Ch. III. New York: T. & J. Swords, 1803. xvi + 544 pp.; vi + 510 pp. + errata page.

Chapter III, "Natural History," includes sections on geology, meteorology, and hydrology.

A69 Milton, Daniel J. "Astrogeology in the 19th Century." *Geotimes* 14(6) (1969): 22.

"The relation of geology to astronomy was apparently a burning question in Russian positivist circles in the 1860s and 1870s." V.V. Lesevich (1876) introduced the term "Astro-geology," which (quoting S. Meunier, 1874) "should extend to the whole of the visible universe the benefits of the methods employed in the study of the Earth, and conversely...." It was to be based primarily on the study of meteorites, secondarily on telescopic spectroscopy.

A70 Moore, Ruth. *The Earth We Live on: The Story of Geological Discovery*. New York: Knopf, 1956. xiv + 416 + x pp. Second edition, 1971. xiv + 437 + viii pp.

Includes chapters on L. Agassiz and the Ice Age, C. Lyell, C.E. Dutton on isostasy, T.C. Chamberlin and H.C. Urey on the origin of the earth, H. Jeffreys, E. Bullard, and S.K. Runcorn on interior of the earth, J.T. Wilson on growing continents, A.C. Waters on volcanoes, J.L. Kulp on age of the earth. The second edition has a new chapter on A. Wegener ("the greatest enigma is solved"); his name does not appear in the index of the first edition. Some of the chapters are based in part on interviews with the subjects, so the book has historical value despite being a "popular" account; the author "particularly" thanks H.C. Urey, J.T. Wilson, A.C. Waters, E. Cloos, B. Mason, and J.L. Kulp.

A71 Nairn, A.E.M. "Uniformitarianism and Environment." *Palaeogeography, Palaeoclimatology, Palaeoecology* 1 (1965): 5-11.

"The historical development of the law, principle, or doctrine of uniformitarianism is reviewed. It is suggested that the limitations proposed to the applicability of this principle are based on a misconception, a confusion of process and material environment. Confusion can be avoided by an explicit principle of the uniqueness of environment" (Author's summary).

A72 Neale, E.R.W., ed. *The Earth Sciences in Canada: A Centennial Appraisal and Forecast.* The Royal Society of Canada, Special Publications, No. 11. Toronto: University of Toronto Press, 1968. x + 259 pp.

Includes: "The Nature and Organization of Earth Sciences in Canada" by J.M. Harrison, D.C. Rose, and R.J. Uffen; "Trends in Geophysical Research in Canada" by G.D. Garland; "The Changing Role of Mining Geophysics in Canada" by H.O. Seigel.

A73 Nickles, Thomas, ed. *Scientific Discovery: Case Studies.* Boston: Reidel, 1980. xxv + 379 pp.

Papers presented at the Guy L. Leobard Memorial Conference in Philosophy, University of Nevada at Reno, October 1978. Includes items HC15, HC24, and PA30.

A74 Onoprienko, V.I., and A.S. Povarennykh. "Isotoriia geologii, metologii, i geologicheskaia teoiia" [History of Geology, Methodology and Theory of Geology]. *Geologicheskoe obrazovanie i istoriia geologii.* Edited by G.P. Gorshkov. Moscow: "Nauka," 1976, pp. 5-15.

In Russian with English abstract.

A75 Orleans, Leo A., ed. *Science in Contemporary China.* Stanford, Calif.: Stanford University Press, 1981. xxxii + 599 pp.

Primarily a review of developments in the 1970s with brief accounts of earlier history. Includes: N. Sivin, "Science in China's Past"; Edward C.T. Chao, "Earth Sciences" (item A18); Richard J. Reed, "Meteorology."

A76 Patten, Donald W. *The Biblical Flood and the Ice Epoch. A Study in Scientific History.* Seattle: Pacific Meridian Pub. Co., 1966. xvi + 336 pp.

"The central proposition of this book is to demonstrate the superiority of the theory of astral catastrophism over and against the uniformitarian view of earth history." ("Astral" refers to gravitational and magnetic forces of planetary magnitude.) The author presents a sympathetic review of ancient and modern catastrophic explanations of mountain formation, glaciogenesis, and the formation of planets.

A77 *Planet Earth. Readings from Scientific American.* With introductions by Frank Press and Raymond Siever. San Francisco: Freeman, 1974. viii + 303 pp.

A collection of popular articles on plate tectonics, the earth's interior, geomagnetism, tides and the earth-moon system, the atmosphere, and the ocean by authorities on those subjects, originally published 1969-73.

A78 Porter, Roy, and Kate Poulton. "Research in British Geology, 1660-1800: A Survey and Thematic Bibliography." *Annals of Science* 34 (1977): 33-42.

A79 Porter, Roy. "The Terraqueous Globe." *The Ferment of Knowledge*. Edited by G.S. Rousseau and Roy Porter. New York: Cambridge University Press, 1980, pp. 285-324.

A survey of 18th-century ideas about the earth.

A80 Rappaport, Rhoda. "Problems and Sources in the History of Geology." *History of Science* 3 (1964): 60-78.

Discusses the need for monographic studies of Buffon and Werner, and for research on what geologists thought about the time scale of the earth's history. The Neptunist-Vulcanist controversy needs more thorough investigation, as does late 18th-century thought about mountain-building. See item A27 for comments on this article.

A81 Ravikovich, A[leksandra] I[osifovna]. *Razvitie osnovnykh teoreticheskikh napravlenii v geologii XIX veka*. Akademiia Nauk SSSR, Geologicheskii Institut, trudy, 189. Moscow: Nauka, 1969. 248 pp.

Details development of the main theoretical tendencies in geology in the 19th century: catastrophism and uniformitarianism; and the idea of evolution in geology, before and after 1859. The monograph has a comprehensive bibliography and a section of biographical notes on about 100 scientists.

A82 [Ravikovich, A.I. (ed.)]. *Metodologiia i istoriia geologicheskikh nauk* [The Methodology and History of Geological Sciences]. Moscow: Nauka, 1977. 179 pp.

Contents in English (articles are in Russian) include: V.V. Tikhomirov, "About Periodization in the History of Geology"; V.E. Khain, "Actualism and Tectonics"; Ia.A. Kosigin, V.A. Soloviov, "The History and Methodology of Tectonic Classifications"; A.I. Ravikovich, "Probability: A Way of Thinking in the Natural Sciences of the 19th Century"; I.V. Krut, "From the History of Principles of Geonomy"; I.V. Batiushkova, "The Development of Acrogeological and Cosmo-geological Research"; T.P. Prolova, "Climate and Accumulation of Phosphates"; G.P. Chomizuri, "Strabo on the Movement of the Earth's Crust."

A83 Rees, Graham. "Francis Bacon on Verticity and the Bowels of the Earth." *Ambix* 26 (1979): 202-211.

Bacon accepted William Gilbert's principle of a universal rotational motion (verticity); though he did not infer the rotation of the earth itself, Bacon attributed winds and tides to the influence of cosmic motion, and seemed to believe that the earth's solid crust has some kind of verticity.

A84 Ronan, Colin A. *The Shorter Science and Civilisation in China*. An Abridgement of Joseph Needham's original text. Volume II. Volume III and a Section of Volume IV, Part I of the Major Series. New York: Cambridge University Press, 1981. xii + 459 pp.

See Chapter 5, "The Sciences of the Earth: (iii) Geology and Related Sciences," pp. 286-324; also Chapter 4 on geography and map-making.

A85 Rubey, William W. "Fifty Years of the Earth Sciences--A Renaissance." *Annual Review of Earth and Planetary Science* 2 (1974): 1-24.

A survey including six pages on geophysics.

A86 Runcorn, Stanley Keith, ed. *Earth Sciences* (The Royal Institution Library of Science, being the Friday Evening Discourses in Physical Sciences held at the Royal Institution: 1851-1939). 3 vols. London: Applied Science Publishers, 1971.

Volume 1 includes lectures on atmospheric magnetism by M. Faraday (1851), on the Cartesian barometer by W. Roxburgh (1854), on pendulum experiments for determining the mean density of the earth by G.B. Airy (1855), on glaciers by J. Tyndall (1857, 1858, 1859), on climate during the Permian epoch by A.C. Ramsay (1857), on meteorological observations during a balloon ascent by E. Vivian (1857), on earthquakes by J.P. Lacaita (1858), on volcanoes by C. Lyell (1859), on the earth's internal temperature and the thickness of its solid crust by W. Hopkins (1859), on atmospheric electricity by W. Thomson (1860), on glaciers by W. Hopkins (1862), on the glacial epoch by E. Frankland (1864), on volcanoes by D.T. Ansted (1866), on chemistry of the primeval earth by T. Sterry Hunt (1867), on meteorology by R.H. Scott (1869, 1873), on glaciers by H. Moseley (1870), on the "Challenger" expedition and geology by T.H. Huxley (1875), on tides by W. Thomson (1875), on geological measures of time by T. McK. Hughes (1876), and on land and sea in relation to geological time by W.B. Carpenter (1880).

Volume 2 includes lectures on geomagnetism by W.G. Adams (1881), on climate in town and country by E. Frankland (1882), on meteorology by R.H. Scott (1883), on rainbows by J. Tyndall (1884), on the building of the Alps by T.G. Bonney (1884), on sunlight and the earth's atmosphere by S.P. Langley (1885), on fogs, clouds, and lightning by S. Bidwell (1893), on atmospheric electricity by A. Schuster (1895), on physics of the upper atmosphere by A. Cornu (1895), on seismology by J. Milne (1897, 1908), on the atmosphere by J. Dewar (1902), on volcanoes by T. Anderson (1903), on the history of oceanic research by J.Y. Buchanan (1903), on weather forecasting by W.N. Shaw (1904), on the figure and constitution of the earth by A.E.H. Love (1908), and on radioactive changes in the earth by R.J. Strutt.

Volume 3 includes lectures on electrical and other properties of sand by C.E.S. Phillips (1910), on magnetic storms by C. Chree (1910), on winds in the free air by C.J.P. Cave (1913), on oceanography by W.S. Bruce (1915), on illusions of the upper air by N. Shaw (1916), on movements of the earth's pole by E.H. Hills (1916), on seismology by C.G. Knott (1919), on cloudland studies by J. Dewar (1921), on the age of the earth by J. Joly (1922), on water in the atmosphere by G.C. Simpson (1923), on the aurora and upper atmosphere by J.C. M'Lennan (1926), on thunderclouds by C.T.R. Wilson (1927), on atmospheric electricity by A.M. Tyndall (1928), on ozone in the upper atmosphere by G.M.B. Dobson (1931), on weather forecasting by G.C. Simpson (1932), on radio observations by E.V. Appleton (1933), on clouds by G. Walker (1935), on Northern Lights by A.S. Eve (1936), on the chemical exploration of the stratosphere by F.A. Paneth (1936), on the nature of snow by G. Seligman (1937), on volcanoes by G.P. Lenox-Conyngham (1937), and on ice ages by G. Simpson (1937).

A87 Sarton, George. "La synthèse geologique de 1775 à 1918." *Isis* 2 (1919): 357-394.

An essay inspired by the publication of E. de Margerie's French translation of Edward Suess' book on the face of the earth.

A88 Schmidt, Peter. *Zur Geschichte der Geologie, Geophysik, Mineralogie und Paläontologie: Bibliographie und Repertorium für die Deutsche Demokratische Republik.* Freiburg: Bibliothek der Bergakademie, 1970. 134 pp.

"A list is provided of the original works relating to the history of the geo-sciences as yet published within and outside the book trade of the German Democratic Republic, and a survey is given of the geological, geophysical, mineralogical and palaeontological bequests [Nachlässe] from the 18th to the 20th century extant in the GDR.... An index of subjects and persons conveys a conclusive indication of the contents of the bibliography and the repertory" (Author's summary).

A89 Schneer, Cecil J., ed. *Toward a History of Geology.* Proceedings of the New Hampshire Inter-Disciplinary Conference on the History of Geology, September 7-12, 1967. Cambridge, Mass.: MIT Press, 1969. vi + 469 pp.

Contains items F13, H8, H18, H34, H42, and HB10.

A90 Schneer, Cecil J., ed. *Two Hundred Years of Geology in America.* Hanover, N.H.: University Press of New England, 1979. 385 pp.

Contains items H36, HC14, and L5.

A91 Schröder, W. "Anregung zum Quellenstudium der Entwicklung von Meteorologie und Geophysik." *Wetter und Leben* 26 (1974): 42-47.

As an encouragement to study sources in the history of meteorology and geophysics, the author presents some case studies and accounts of the careers of scientists. He emphasizes the value of correspondence and mentions catalogs of observations of aurorae and noctilucent clouds.

A92 Schröder, W. "Einige Aspekte einer Geschichte der Meteorologie und Geophysik im Rahmen der Internationalen Union für Geophysik und Geodäsie (IUGG)." *Wetter und Leben* 28 (1976): 252-256.

The author urges systematic study of the history of meteorology and geophysics under the auspices of IUGG, with special symposia, special issues of international journals, etc. At present, historical articles are scattered widely in so many different publications that they are inaccessible to most historians and scientists.

A93 Schröder, Wilfried. "Why Research into the History of Geosciences?" *EOS* 62 (1981): 521-522.

A94 Schröder, W. *Disziplingeschichte als wissenschaftliche Selbstreflexion der historischen Wissenschaftsforschung.* Frankfurt am Main/Las Vegas: Verlag Peter D. Lang, 1981.

Case studies of the history of meteorology and geophysics--biographical reviews (e.g., Fritz, Wiechert, A. Wegener, O. Jesse, etc.)--auroral catalogs and observations and the solar-terrestrial relationships (e.g., Maunder-Minimum)--general history and philosophy of sciences with regard to the geosciences (and physics, astronomy, geography)--developmental phases of "exact" sciences and case studies of the philosophy of sciences (e.g., T.S. Kuhn, I. Lakatos)--history of meteorology and geophysics as "exact" sciences--research program for the history of geosciences (and relationship to geography, physics, astronomy).

A95 Schwarzbach, Martin. "Fortschritte der Geologie im letzten Jahrzehnt." *Naturwissenschaftliche Rundschau* 9 (1956): 291-296.

A review, briefly mentioning recent work in tectonics, seismology, vulcanology, geochronology, climatology, etc.

A96 Scientific American. *The Planet Earth.* New York: Simon & Schuster, 1957. viii + 168 pp.

Collection of articles originally published in *Scientific American* in the early 1950s, on the origin and structure of the earth, hydrosphere, and atmosphere; the book gives a good survey of how leading scientists presented their subjects to the public at the beginning of the Space Age. Authors included H.C. Urey, K.E. Bullen, S.K. Runcorn, W.H. Munk, H. Wexler, and H.E. Newell.

A97 Seneca, Lucius Annaeus. *Physical Science in the Time of Nero; being a translation of the Quaestiones Naturales of Seneca by John Clarke, with notes by Archibald Geikie.* London: Macmillan, 1910. liv + 368 pp.

Contents: I--Meteors, halo, rainbow, mock sun, etc.; II--Nature of air, thunder, and lightning; III--On forms of water; IV--Snow, hail, and rain; V--Winds, movement of the atmosphere; VI--Earthquakes; VII--Comets.

A98 Shcherbakov, D.I., et al. *The Interaction of Sciences in the Study of the Earth*. Translated from the Russian by Vladimir Talmy (originally *Vzaimodeistvie Nauk pri izuchenii Zemli*). Moscow: Progress Publishers, 1968. 323 pp.

Includes: D.I. Shcherbakov, "Trends of Development in Geosciences," pp. 7-10; V.V. Belousov, "Trends in Geoscience," pp. 11-24; Y.K. Fyodorov, "Some Problems of the Development of the Sciences of the Earth," pp. 25-54; B.Y. Levin, "The Interaction of Astronomy, Geophysics and Geology in the Study of the Earth," pp. 165-180; A.T. Donabedov, "On the Correlation of Geophysics with Geology and Physics," pp. 181-192; A.Y. Medunin, "Concerning the Historical Aspect of Certain Methods Employed by Geophysics in Studying the Earth....," pp. 193-209; V.V. Cherdyntsev, "The Geological Sciences and Nuclear Physics," pp. 210-219; V.I. Baranov, "Radioactivity and Geology," pp. 219-229; S.A. Djamalov, "Some Methodological Problems of Studying the Evolution of the Earth and the Value of Geothermal Researches at Small Depths," pp. 230-234.

A99 Simojoki, Heikki. *The History of Geophysics in Finland, 1828-1918*. Helsinki: Societas Scientiarum Fennica, 1978. 157 pp.

Contents: work of G.G. Hällström and J.J. Nervander; the University Magnetic Observatory; phenological and meteorological observations; Adolf Moberg; early history of the weather service; provisional decision in 1874 to establish a central meteorological institute; oceanography, J. Stjerncreutz; rise of geophysics in the 1880s--Selim Lemström; the International Polar Year and its follow-up in 1882-84; the Society of Sciences' Central Meteorological Institute (1881-1918)--directorships of N.K. Nordenskiöld (1880-89), Ernst Biese (1890-1907), and Gustav Melander (1908-18); the C.M.I. becomes a governmental body, 1918; meteorology at the University (1880-1918)--Sakari Levänen, A.F. Sundell, Th. Homén, Osc. V. Johansson, Hugo Karsten; the teaching of meteorology and other branches of geophysics; theses; oceanography (1880-1918)--water-level observations, study of ice conditions, hydrographical research; Rolf Witting; foundation of the Institute of Marine Research; study of the Inland Waters (1820-1918)--Edvard Blomqvist; geomagnetism, early history of the Sodankylä Observatory; seismology.

A100 Stokes, Evelyn. "Fifteenth Century Earth Science." *Earth Science Journal* 1 (1967): 130-148.

"The earth science content of two late medieval encyclopedias, the *Mirrour of the World* [translated from *Image du Monde* by Gossuin of Metz] and [Ranulph] Higden's *Polychronicon* ... is examined in relation to fifteenth century ideas about the physical nature of the earth and universe. Such topics as the four elements, the earth and the spheres, location of Hell and Paradise, the arrangement of continents and oceans, the unity of waters, earthquakes and volcanoes, erosion, fossils and mountain building, climatic zones and weather phenomena are summarized and reference made to the Biblical and classical Greek sources of these ideas" (Author's abstract).

* Takeuchi, H., S. Uyeda, and H. Kanamori. *Debate About the Earth. Approach to Geophysics Through Analysis of Continental Drift*. Cited below as item U45.

A101 Tardi, P. "Geophysics." *Science in the Twentieth Century*. Edited by René Taton. Translated by A.J. Pomerans from *La Science contemporaine II: Le XXe siècle* (1964). New York: Basic Books, 1966, pp. 286-295.

Includes brief sections on geodesy and gravimetry, seismology, geomagnetism, meteorology and aerology, physical oceanography.

A102 Taylor, E.G.R. "The Early Literature of Natural Calamities in Britain: with a Bibliography." *Scottish Geographical Magazine* 48 (1932): 83-89.

A survey of accounts of earthquakes and storms beginning in 1542.

A103 Thams, J[ohann] C[hristian], ed. *The Development of Geodesy and Geophysics in Switzerland*. Zurich: Berichthaus, 1967. 98 pp., illus.

Includes articles on geodesy by E. Huber and F. Kobold; seismology and physics of the earth's interior by F. Gassmann and M. Weber; nuclear studies in geophysics and geochemistry by J. Geiss, E. Jäger, and H. Oeschger; tectonophysics by N. Pavoni; meteorology and atmospheric physics by H.U. Dütsch and W. Kuhn; oceanography and limnology by H. Heberlein; glaciology by R. Haefelis; hydrology by P. Kasser and G. Schnitter; vulcanology by C. Burri.

A104 Tikhomirov, V.V., and V.E. Khain. *Kratkii ocherk istorii geologii*. Moscow: Gosudarstvennoe Nauchno-Tekhnicheskoe Izdatel'stvo Literatury po Geologii i okhrane Nedr, 1956. 260 pp.

Short history of geology, with extensive bibliography and many capsule biographies.

A105 Tikhomirov, V.V., and A.I. Ravikovich. "History of Geological Sciences in the USSR." *Geotimes* 13, no. 6 (July-August 1968): 19-22.

This is mainly a list of authors indicating the subjects on which they wrote but with no bibliography and little discussion of their conclusions. Geophysical topics are rarely mentioned explicitly.

A106 Tikhomirov, V.V., et al., eds. *Istoriia geologii*. Erevan: Izd-vo Akademii Nauk Armianskoi SSR, 1970. 362 pp.

Articles in Russian, with English summaries. Includes "Certain Regularities in the Evolution of Ideas on the Structure of the Earth" by I.V. Batiushkova (pp. 73-81).

A107 U. S. Coast and Geodetic Survey. *Centennial Celebration, April 5 and 6, 1916*. Washington, D.C.: U.S. Government Printing Office, 1916. 196 pp.

Includes addresses by L.A. Bauer on terrestrial magnetism, J.E. Pillsbury on ocean currents, W.H. Burger on geodesy, and C.L. Poor on tides.

A108 Vinogradov, A.P., ed. *Razvitie Nauk o Zemle v SSSR* (Akademiia Nauk SSSR, Otdelenie Nauk o Zemle. Institut Istorii Estestvoznaniia i tekhniki). Moscow: Nauka, 1967. 715 pp. English translation: *The Development of Earth Sciences in the USSR (Selected Articles)*. 3 vols. Wright-Patterson AFB, Ohio: Foreign Technology Division, Air Force Systems Command, 1969. Report no. FTD-MT; 24-291-68.

A collection of articles on the history of geology, geomagnetism, seismology, tectonics, hydrology, oceanography, meteorology, etc., in the USSR.

A109 Vitaliano, Dorothy B. *Legends of the Earth: Their Geologic Origins*. Bloomington: Indiana University Press, 1973. 305 pp. + illus., bibliography, and index.

A110 Vucinich, Alexander. *Science in Russian Culture 1861-1917*. Stanford, Calif.: Stanford University Press, 1970. xv + 575 pp.

Includes a chapter on "The Earth Sciences," pp. 397-423.

A111 Wagenbreth, Otfried. "Zur Dialektik in der Geologie und geologischen Forschung." *Zeitschrift für Geologische Wissenschaften* 5 (1977): 1215-1222.

"Difficulties arise in finding out true (causally connected) negations of negations, when investigating geological lines of development appearing phenomenologically as examples of objective dialectics in geology...."

A112 White, George W. "Reference Books for History of Geology." *Stechert-Hafner Book News* 18, no. 3 (1963): 29-30.

A113 Williams, Henry Smith. *A History of Science*. Volume III. *Modern Development of the Physical Sciences*. New York: Harper & Brothers, 1904. x + 309 pp.

Chapter IV, "The Origin and Development of Modern Geology" (from James Hutton); Chapter V, "The New Science of Meteorology" (including meteors and aurora).

A114 Wilson, J. Tuzo. "The Current Revolution in Earth Science." *Transactions of the Royal Society of Canada* series 4, 6 (1968): 273-281.

Considers the following questions: "Why is the study of the earth so fragmented? ... Is there a science of the earth at all? ... Is the trouble with earth sciences due to a neglect of mathematics, physics, and chemistry? ... Is earth science ripe for a scientific revolution [as described by T.S. Kuhn]? ... What has been the standard doctrine or paradigm?..."

A115 Young, Thomas. *A Course of Lectures on Natural Philosophy and the Mechanical Arts*. Volume II. London: Johnson, 1807. 738 pp. Reprint, New York: Johnson Reprint Corp., 1971.

The "Catalogue of Works Relating to Natural Philosophy and the Mechanical Arts" includes bibliographies of the following topics: Atmospheric Refraction, pp. 299-309; Rotation of the Earth and Planets, pp. 341-342; Theory of the Tides, pp. 342-343; Figure and Magnitude of the

Earth, pp. 358-364; Density of the Earth, p. 364; Terrestrial Magnetism, pp. 440-443; Meteorology, pp. 446-481; Atmospherical Electricity, pp. 481-486; Waterspouts, pp. 486-488; Aurora Borealis, pp. 488-490; Earthquakes and Agitations, pp. 490-493; Subterraneous Fires and Volcanos, pp. 493-494; History of Terrestrial Physics, pp. 519-520.

A116 Zharkov, V.N., and V.A. Magnitski."Evolution of Geophysics." *Akademiia Nauk SSSR. Izvestiya. Physics of the Solid Earth* no. 4 (1970): 217-221. Translated from Russian.

"Some trends in geophysics of the last few years are discussed. The following topics are considered: model of the earth; geophysical studies at high pressures; natural vibrations of the earth; convection; planetary geophysics (earth, moon, sun); planetary geophysics and cosmogonic hypothesis; and geophysics and geotectonic hypotheses" (Author's summary).

A117 Zittel, K.A. Ritter von. *Geschichte der Geologie und Paläontologie bis Ende des 19. Jahrhunderts*. München and Leipzig: Oldenbourg, 1889. Reprint, New York: Johnson Reprint, 1965. xi + 868 pp. English translation: *History of Geology and Palaeontology to the End of the Nineteenth Century*. Translated by Maria M. Ogilvie-Gordon, 1901; reprinted, New York: Hafner Publishing Company, 1962. xiii + 562 pp.

The original German edition provides a more comprehensive view of the subject than is found in recent works. It includes chapters on "cosmical geology," tectonics, glaciology, earthquakes, and volcanoes. The translation is "somewhat curtailed" with the approval of the author.

For additional information on this topic see works cited in the biography sections on the following scientists:

Fersman, A.E.
Guenther, A.W.S.
Humboldt, A. v.
Markham, C.R.
Nordenskiöld, A.E.
Walker, J.

B. BIOGRAPHIES

* Caloi, Pietro. "Italian Pioneers in the Physics of the Universe. Second Part: Geophysics." Cited above as item A17.

* Davison, Charles. *The Founders of Seismology.* Cited below as item J7.

B1 Fenton, Carroll, and Mildred Adams Fenton. *Giants of Geology.* Garden City, N.Y.: Doubleday, 1952. xvi + 333 pp.

Includes chapters on G.K. Gilbert, L. Agassiz, and T.C. Chamberlin.

B2 Guntau, M., ed. *Biographien bedeutender Geowissenschaftler der Sowjetunion.* Berlin: Akademie-Verlag, 1979. 199 pp.

Biographies of M.V. Lomonosov, V.M. Severghin, K.F. Rouillier, N.I. Koksharov, V.O. Kovalevsky, A.P. Karpinsky, E.S. Federov, A.P. Pavlov, B.B. Golitsin, V.I. Vernadsky, V.A. Obruchev, I.M. Gubkin. A.D. Arkhangelsky, A.E. Fersman, A.N. Kryshtofovich, S.S. Smirnov, N.S. Shatsky, A.P. Vinogradov, G.A. Gamburtsev. Summaries in English.

B3 Sarjeant, William A.S. *Geologists and the History of Geology. An International Bibliography from the Origins to 1978.* 5 vols. New York: Arno Press, 1980.

Volumes 2 and 3 list biographical and autobiographical works for several thousand geologists, including many who worked in meteorology and/or geophysics.

B4 Wells, John W. "A List of Books on the Personalities of Geology." *Ohio Journal of Science* 47 (1947): 192-200.

Approximately 150 items, with brief annotations.

B5 Wells, John W., and George W. White. "Biographies of Geologists." *Ohio Journal of Science* 58 (1958): 285-298.

Revised and expanded version of item B4; approximately 250 items with annotations.

Biographies, Autobiographies, Collected Works
(Works by the subject are listed first)

BA. Subjects: A

Abbe, C. @

BA1 Abbe, Truman. *Professor Abbe and the Isobars: The Story of Cleveland Abbe, America's First Weatherman.* New York: Vantage Press, 1955. 259 pp.

Biography of Cleveland Abbe (1838-1916).

BA2 Humphreys, W.J. "Biographical Memoir of Cleveland Abbe, 1838-1916." *National Academy of Sciences Biographical Memoirs* 8 (1919): 469-508.

A biography of Cleveland Abbe (1838-1916), who was the first meteorological scientist to work for the U.S. Weather Bureau.

Abbot, C.

BA3 Abbot, Charles Greeley. *Adventures in the World of Science.* Washington, D.C.: Public Affairs Press, 1958. 150 pp.

The erstwhile Secretary of the Smithsonian Institution reminisces about the work of that organization, including his studies on the solar constant and long-range weather forecasting.

Abich, O.H.W.

BA4 Tikhomirov, V.V. "Abich, Otto Hermann Wilhelm" (b. Berlin, Germany, 11 December 1806; d. Vienna, Austria, 1 July 1886). *Dictionary of Scientific Biography* 1 (1970): 19-21.

"Abich was a proponent of the volcanistic theory that assigned the decisive role in geologic process to hypogene forces and to magma ... in the Caucasus he distinguished four major directions of tectonic lines and, consequently, four stages of tectogenesis. Adhering to Buch's ideas, he considered all mountains to be elevation craters that originated from the intrusion of magmatic masses..."

Adams, F.D.

BA5 Clark, Thomas. "Adams, Frank Dawson" (b. Montreal, Canada, 17 September 1859; d. Montreal, 26 December 1942). *Dictionary of Scientific Biography* 1 (1970): 50-53.

Adams conducted experiments in which he subjected rocks to high pressures in an attempt to duplicate conditions within the earth's crust. This work contributed to the development of modern ideas on mountain building. His historical studies led to the writing of *Birth and Development of the Geological Sciences* (1938).

Agassiz, L. @

BA6 Agassiz, Louis. *His Life and Correspondence.* Edited by Elizabeth Cary Agassiz. 2 vols. Boston: Houghton Mifflin & Co., 1886. 794 pp.

Volume 1 includes correspondence and biographical details relating to the Ice Age theory.

BA7 Balmer, Heinz. "Louis Agassiz, 1807-1873." *Gesnerus* 31 (1974): 1-18.

BA8 Marcou, Jules. *Life, Letters and Work of Louis Agassiz.* New York: Macmillan, 1896; reprint, Westmead, Farnsborough, England: Gregg International Publishers, 1972. xxi + 318 pp.

Chapter V describes his conversion to the glacial theory of Jean de Charpentier and reprints his 1837 "Discours de Neuchatel" on the Ice Age. Chapters VI and VII describe the reception of this theory.

Albert I

BA9 Petit, Georges. "Albert I of Monaco (Honoré Charles Grimaldi)" (b. Paris, France, 13 November 1848; d. 26 June 1922). *Dictionary of Scientific Biography* 1 (1970): 92-93.

"Albert studied currents, especially the Gulf Stream (1885). He set out floating mines to study drift in both the North Atlantic and the Arctic. Using the Richard bottle, he took samples of water at various depths in order to determine the differences in temperature." His atlas with plates showing the bathymetry of all the oceans was a synthesis of all previous findings.

Anderson, E.M.

BA10 MacGregor, A.G. "Anderson, Ernest Masson" (b. Falkirk, Scotland, 9 August 1877; d. Edinburgh, Scotland, 8 August 1960). *Dictionary of Scientific Biography* 1 (1970): 153-155.

"His main scientific work ... concerned the dynamics of faulting and igneous intrusion, the lineation of schists, crustal heat and structure, and volcanism." In other geophysical work he studied heat flow in the crust and criticized the thermal earth-contraction theory.

Angström, A.

BA11 Taba, H. "The *Bulletin* Interviews: Dr. Anders K. Angström." *Bulletin of the World Meteorological Organization* 31 (1982): 77-86.

About a year before his death the editor of the *Bulletin of the World Meteorological Organization* interviewed Anders Angström (1888-1981), a scion of a famous family of physicists at the University of Uppsala. Like his father, Knut, he was interested in atmospheric radiation. With C.G. Abbot (1872-1973) he tried to determine the solar constant. He made important contributions toward the understanding of atmospheric turbidity. From 1949 to 1955, the year of his retirement, he served as director of the Swedish Hydrometeorological Service but continued his radiation research for many years thereafter.

Anuchin, D.N.

BA12 Esakov, V.A. "Anuchin, Dmitrii Nikolaevich" (b. St. Petersburg, Russia, 8 September 1843; d. Moscow, Union of Soviet Socialist Republics, 4 June 1923). *Dictionary of Scientific Biography* 1 (1970): 173-175.

"... he opposed the theory that mountains were formed as the result of the cooling of the earth. Anuchin considered the endogenic forces of the earth as the basic and decisive factor in the formation of mountains. He placed the origin of the internal energy of the earth 'in the pressure of strata and especially in radioactive substances, which are capable of giving off heat' (*Lektsii Po Fizicheskoj Geografii*, 1916, p. 33)." In his work in hydrology, he associated lake formation with the genesis of hollows in the watershed areas; he "is rightly acclaimed as one of the founders of limnology in Russia."

Appleton, E.V. @

BA13 Clark, Ronald. *Sir Edward Appleton.* Oxford: Pergamon Press, 1971. xiii + 240 pp.

Biography of Edward Victor Appleton (1892-1965) who received the 1947 Nobel Prize in Physics "for his investigation of the physics of the upper atmosphere, especially for the discovery of the so-called Appleton layers."

Arago, F. @

BA14 Danjou, A. "François Arago, 26 Fevrier 1786-2 Octobre 1853." *L'Astronomie* 67 (1953): 445-464.

This famous astronomer was also a major contributor to atmospheric optics, as pointed out in this biographical note on occasion of the centenary of his death.

Arctowski, H.

BA15 Kosiba, Aleksander, and Stanislaw Zych. "Henry K. Arctowski (July 15, 1871-Feb. 21, 1958)." *Przeglad Geofizyczny* 4 (1959): 83-90.

A biographical sketch with notes on his expeditionary work, his research on the solar constant, his teaching activity at Lwow, and his final association with C.G. Abbot at the Smithsonian Institution.

BA16 Kosiba, Aleksander. "Dzialalnosc naukowa Professora Henryka Arctowskiego" [Henryk Arctowski's Scientific Work]. *Acta Geophysica Polonica* 7 (1959): 250-281.

Arctowski (1871-1958) was one of the early explorers of Antarctic meteorology, but his major contributions were to the measurements of the solar constant and its influence on atmospheric parameters. See also obituary: Wlodzimierz Zinkiewicz, *Przeglad Geograficzny* 31 (1959): 198-209 (with portrait); bibliography: "Bibliografia prac naukovych Henryka Arctowskiego," *Acta Geophysica Polonica* 7 (1959): 282-295.

Argand, E.

BA17 Wegmann, C.E. "Argand, Emile" (b. Geneva, Switzerland, 6 January 1879; d. Neuchatel, Switzerland, 14 September 1940). *Dictionary of Scientific Biography* 1 (1970): 235-237.

"In 1915 Alfred Wegener published his fundamental paper containing his hypothesis of continental drift. Fascinated by the possibilities, Argand believed that continental drift offered a new motor for orogeny consistent with his evolutionary model. The Wegenerian hypothesis became the frame within which he created a new concept of Eurasian structural development."

Aristotle

BA18 Owen, G.E.L. "Aristotle" (b. Stagira in Chalcidice, 384 B.C.; d. Chalcis, 332 B.C.). *Dictionary of Scientific Biography* 1 (1970): 250-258.

The article presents a general discussion of Aristotle's views on physical phenomena but says little about his observations or "sublunary physics" in *Meteorologica*.

BB. Subjects: B

Babinet, J.

BB1 Frankel, Eugene. "Jacques Babinet" (b. Lusignan, France, 5 March 1794; d. Paris, France, 21 October 1872). *Dictionary of Scientific Biography* 1 (1970): 357-358.

"... the study of meteorology, particularly meteorological optics, occupied much of his career."

Bache, A.D. @

BB2 "Commemoration of the Life and Work of Alexander Dallas Bache and Symposium on Geomagnetism." *Proceedings of the American Philosophical Society* 84, no. 2 (1941): 125-351, port., fig.

Includes papers by Edwin G. Conklin, Henry Butler Allen, Edward P. Cheyney, Merle M. Odgers, Leo Otis Colbert, and Frank B. Jewett on his life and work. The symposium on geomagnetism contains semi-historical papers by O.H. Gish, J.A. Fleming, C.W. Gartlein, and Paul R. Heyl.

Bailey, E.B.

BB3 MacGregor, A.G. "Bailey, Edward Battersby" (b. Marden, England, 1 July 1881; d. London, England, 19 March 1965). *Dictionary of Scientific Biography* 1 (1970): 393-395.

"Bailey made notable contributions to tectonics and metamorphism, to igneous and general geology and to the history of the development of geological ideas."

Bailey, S.I.

BB4 Dieke, Sally H. "Bailey, Solon Irving" (b. Lisbon, New Hampshire, 29 December 1854; d. Norwell, Massachusetts, 5 June 1931). *Dictionary of Scientific Biography* 1 (1970): 397-398.

"Bailey traveled throughout Peru and Chile trying out various sites [for an observatory]. No information of weather patterns was available to make his task easier, so he established a chain of meteorological stations--from sea level up to 19,000 feet. He published the data that he accumulated over the following 37 years in his 'Peruvian Meteorology' (1889-1930)."

Bartels, J. @

BB5 Chapman, Sydney. "Julius Bartels [Obituary]." *Quarterly Journal of the Royal Astronomical Society* 6 (1965): 235-245.

Julius [August] Bartels (1899-1964) worked in geomagnetism and meteorology. This notice is based largely on personal recollections.

BB6 Fanselau, G. "Nachruf auf Julius Bartels." *Gerlands Beiträge zur Geophysik* 74 (1965): 1-6.

Bartels made his principal contributions in the field of geomagnetism, with emphasis on solar-terrestrial relations. His original methods for determining periodicities and coincidences are widely used also in other fields of geophysics.

BB7 Schröder, Wilfried. "Nachruf auf Julius Bartels." *Meteorologische Zeitschrift* 17 (1964): 129.

A brief obituary for Julius Bartels, professor of geophysics at the University of Göttingen, who made important contributions to the statistical analysis of time series to solar-terrestrial relations.

Barus, C. @

BB8 Lindsay, R.B. "Carl Barus." *Biographical Memoirs of the National Academy of Sciences* 22 (1943): 171-213.

Bauer, L.A. @

BB9 "Bauer Memorial Number." *Terrestrial Magnetism and Atmospheric Electricity* 37, no. 3 (Sept. 1932). 2 ports.

Includes articles on his work in terrestrial magnetism by A. Nippoldi, G.W. Littlehales, H.V. Sverdrup, and J. de Moidrey, and a list of his principal published papers.

Baur, F.

BB10 Amelung, Walther. "Prof. Dr. Phil. nat. Dr. agr. h.c. Franz Baur 85 Jahre alt." *Zeitschrift für Angewandte Bäder--und Klimaheilkunde* 19 (1972): 1-4.

Commemorative biography of Baur (1887-1977) on occasion of his 85th birthday. His merits for bioclimatic research are extolled and his pioneering efforts in the realm of long-range weather forecasts are appraised.

BB11 Schlaak, Paul. "Franz Baur--14 February 1887-20 November 1977." *Beilage zur Berliner Wetterkarte, Institut für Meteorologie, Freie Universität Berlin* (1 Dec. 1977). 2 pp.

Obituary for the indefatigable researcher, who against difficult odds and opposition worked from 1923 to the day of his death to produce long-range forecasts for 10 days, months, and seasons. No details on his work or his publications are given.

Beaufort, F.

BB12 Rathbone, Kenneth C. "Beaufort of the Beaufort Scale." *Nautical Magazine* 175 (1956): 223-224.

A brief biography of Admiral Sir Francis Beaufort (1774-1857), the developer of the wind scale.

Becke, F.J.K.

BB13 Baumgartel, Hans. "Becke, Friedrich Johann Karl" (b. Prague, Czechoslovakia, 31 December 1855; d. Vienna, Austria, 18 June 1931). *Dictionary of Scientific Biography* 1 (1970): 551-552.

In 1896 he proposed "the Becke volume rule, which states that--assuming isothermal conditions--with increased pressure, the formation of minerals with the smallest molecular volume (the greatest density) will be favored."

Becker, G.F.

BB14 Kendall, Martha B. "Becker, George Ferdinand" (b. New York, New York, 5 January 1847; d. Washington, D.C., 20 April 1919). *Dictionary of Scientific Biography* 1 (1970): 252-253.

"Several of his more general theoretical papers are devoted to geophysical problems, including the structure of the globe: 'An elementary proof of the earth's rigidity' (1890), 'The finite elastic stress-strain function' (1893), and 'Finite homogeneous strain, flow and rupture of rocks' (1893)." He was the first director of the Carnegie Geophysical Laboratory.

Bellinsgauzen, F.F.

BB15 Fedoseyev, Ivan A. "Bellinsgauzen, Faddei F." (b. Arensburg, on the island of Oesel, Russia [now Kingissepp, Sarema, Estonian Soviet Socialist Republic], 30 August 1779; d. Kronstadt, Russia, 25 January 1852). *Dictionary of Scientific Biography* 1 (1970): 594-595.

Bellinsgauzen determined the precise position of the South Magnetic Pole during an Antarctic expedition in 1819-21.

Benioff, V.H.

BB16 Press, Frank. "Victor Hugo Benioff. September 14, 1899-February 29, 1968." *Biographical Memoirs, National Academy of Sciences* 43 (1973): 27-40.

Benioff first specialized in the development of seismographs, then after 1950 began working on the general problem of earthquake mechanisms and introduced several of the concepts which form the basis for plate tectonics. He showed that great earthquakes form a global pattern of strain accumulation and release. His last major contribution was participation in one of the teams which detected free vibrations of the earth, a phenomenon which he had discussed as early as 1952.

Benndorf, H.

BB17 Israel, H., and M. Toperczer. "Hans Benndorf, †." *Archiv für Meteorologie, Geophysik und Bioklimatologie*, Serie A, 6 (1953): 111-113.

Obituary of the physics professor at the University of Graz, Austria (1870-1953), who made extensive contributions to knowledge of atmospheric electricity and invented a recording electrometer.

Bentley, W.

BB18 Blanchard, Duncan C. "Wilson Bentley, Pioneer in Snowflake Photomicrography." *Photographic Applications in Science, Technology and Medicine* 8, no. 3 (1973): 26-28, 39-41.

An illustrated account of Bentley's (1865-1931) painstaking and tireless work to photograph snow flakes and other icy crystals, starting his work in 1885--at home in Vermont. About half of his pictures, 25,000 of them, were published in a remarkable volume jointly with William J. Humphreys (1862-1949), entitled *Snow Crystals* (New York: McGraw-Hill Book Co., 1931, 227 pp.; 1962 reprint by Dover Publications).

BB19 Blanchard, Duncan C. "Wilson Bentley, the Snowflake Man." *Weatherwise* 23 (1970): 260-269.

An illustrated biographical sketch of Wilson Bentley (1865-1931), the Vermont amateur photographer, and his important contributions to the knowledge of hydrometeors, especially the forms of snow crystals, with a bibliography of his major scientific publications.

Benzenberg, J.F.

BB20 Sticker, Bernhard. "Benzenberg, Johann Fredrich" (b. Scholler, near Dusseldorf, Germany, 5 May 1777; d. Bilk, near Dusseldorf, 7 June 1846). *Dictionary of Scientific Biography* 1 (1970): 615-616.

Benzenberg corroborated Chladni's theory of the cosmic origin of meteorites and their relation to fireballs. His experiments on the deflection of falling bodies "demonstrated the revolution of the earth some fifty years before Foucault did."

Bergeron, H.P.

BB21 Weickmann, Helmut K. "For Harold Percival Bergeron." *Bulletin of the American Meteorological Society* 60 (1979): 406-414.

Bergeron (1891-1977) was one of the giants of modern meteorology. He was a prominent member of the Bergen meteorological school of Vilhelm Bjerknes (1862-1951). His revolutionary contribution on three-dimensional weather analysis (1922) completely reformed procedures in synoptic meteorology. His recognition of the role of ice nuclei and ice crystals for the formation of rain and the dissipation of clouds, although not immediately recognized, joined his name with that of another researcher in cloud thermodynamics in the eponym Bergeron-Findeisen process. His final studies on the influence of small terrain elevations on the distribution and amount of precipitation were another very important achievement.

Bergman, T.O.

BB22 Smeaton, W.A. "Bergman, Torbern Olof" (b. Katrineberg, Sweden, 9 March 1735; d. Medevi, Sweden, 8 July 1784). *Dictionary of Scientific Biography* 2 (1970): 4-8.

"Bergman's early contributions to physical science included studies of the rainbow twilight, and the aurora borealis, of which he estimated the height to be 460 miles.... Bergman's inaugural address to the Stockholm Academy in 1764 was 'The Possibility of Preventing the Harmful Effects of Lightning,' .... in 1766 Bergman contributed *Physical Description of the Earth*. This was an important treatise on physical geography...."

Bertrand, M.

BB23 Hansen, Bert. "Bertrand, Marcel-Alexandre" (b. Paris, France, 2 July 1847; d. Paris, 13 February 1906). *Dictionary of Scientific Biography* 2 (1970): 89-90.

"Bertrand was the first to conceive of the overthrust structure of the Alps, and by this theory of *grandes nappes* he attempted to connect the structure of the Pyrenees, Provence, and the Alps.... Bertrand developed an orogenic wave concept that he used to separate Earth history into natural divisions on the basis of successive periods of intense folding and orogeny...."

Berzelius, J.J. @

BB24 Frängsmyr, Tore. "The Geological Ideas of J.J. Berzelius." *British Journal for the History of Science* 9 (1976): 228-236.

The Swedish chemist Jöns Jacob Berzelius (1779-1848) saw the cooling of the earth as the fundamental process of geology and supported Elie de Beaumont's contraction theory; he rejected attempts to make geological periods conform to the 6000-year biblical time scale.

Bessel, F.W.

BB25 Fricke, Walter. "Bessel, Friedrich Wilhelm" (b. Minden, Germany, 22 July 1784; d. Konigsberg, Germany [now Kaliningrad, U.S.S.R.], 17 March 1846). *Dictionary of Scientific Biography* 2 (1970): 97-103.

"Among Bessel's works that contributed to geophysics were his investigations on the length of the simple seconds' pendulum (1826), the length of the seconds' pendulum for Berlin (1835), and the determination of the acceleration of gravity derived from observing the pendulum."

Bessemoulin, J.

BB26 [Taba, H.]. "The *Bulletin* Interviews: Mr. J. Bessemoulin." *World Meteorological Organization Bulletin* 31 (1982): 3-14.

Question-and-answer conversation with the retired head of the French weather service, covering his career in that service from 1933 to 1976. It is a mixture of biographical information, personal experiences, and developments in French meteorology.

Birge, E.A.

BB27 Lowell, E. Noland. "Birge, Edward Asahel" (b. Troy, New York, 7 September 1851; d. Madison, Wisconsin, 9 June 1950). *Dictionary of Scientific Biography* 2 (1970): 141-142.

"From a study of the Finger Lakes of New York and comparable investigations in Wisconsin, Birge worked out the principle of the 'heat budget' of lakes; he also studied the thermal exchange between lake water and air in considerable detail...."

Bjerknes, J.

BB28 Holmboe, Jörgen, Jerome Namias, and Morton G. Wurtele. "Jacob Bjerknes--1897-1975." *Bulletin of the American Meteorological Society* 56 (1975): 1089-1090.

An obituary of the discoverer of the polar front, whose contributions to synoptic meteorology and knowledge of the general circulation of the atmosphere revolutionized the field of meteorology. Born a Norwegian, he came to America in 1940 and became Professor at the University of California at Los Angeles.

BB29 Namias, Jerome. "Professor Jacob Aall Bonnevie Bjerknes." *Quarterly Journal of the Royal Meteorological Society* 102 (1976): 271.

A brief obituary of Bjerknes with an appraisal of his principal meteorological contributions, especially the life history of cyclones. Includes a list of his honors.

BB30 Schereschewsky, Philippe. "Jack Bjerknes" (Letter to the editor). *Bulletin of the American Meteorological Society* 56 (1975): 1105.

Personal reminiscence by the erstwhile director of the meteorological service of the French army of Jacob Bjerknes (1897-1975), with references to his antecedents in synoptic meteorology and a meeting of meteorological directors in Bergen, Norway, in 1921.

Bjerknes, V.

BB31 Bergeron, Tor, Olaf Devik, and Carl Ludwig Godske. "Vilhelm Bjerknes, March 14, 1862-April 19, 1951." *In Memory of Vilhelm Bjerknes on the 100th Anniversary of His Birth. Geofysiske Publikasjoner* 24 (1962): 7-25.

For publications of Vilhelm Bjerknes, see pp. 26-37.

BB32 D[ouglas], C.K.M. "Obituary. Professor V.F.K. Bjerknes." *Quarterly Journal of the Royal Meteorological Society* 77 (1951): 529-530.

Appraises the extraordinary contributions of V. Bjerknes (1862-1951) to dynamic meteorology.

BB33 Eliassen, Arnt. "Vilhelm Bjerknes and His Students." *Annual Review of Fluid Mechanics* 14 (1982): 1-11.

With photographs of V. Bjerknes, V.W. Ekman, H.U. Sverdrup, T. Hesselberg, H. Solberg, T. Bergeron, C.G. Rossby, and S. Rosseland. A brief appraisal of Bjerknes' contributions to dynamic meteorology.

BB34 Gold, Ernest. "Prof. V.F.K. Bjerknes. For. Mem. R.S." *Nature* 167 (1951): 838-839.

Obituary of Vilhelm Bjerknes.

BB35 Gold, Ernest. "V.F.K. Bjerknes." *Weather* 6 (1951): 216-217.

An obituary with excellent portrait of Vilhelm Bjerknes; an appraisal of his fundamental contributions to dynamic meteorology.

BB36 Groen, P. "In Memoriam Vilhelm Bjerknes." *Hemel en Dampkring* 49 (1951): 160.

Brief obituary of Bjerknes.

BB37 Neiburger, Morris. "Vilhelm Bjerknes: An Example of the Essential Ingredient." *Bulletin of the American Meteorological Society* 43 (1962): 301.

Commemoration of the 100th birthday of V. Bjerknes and his achievements in meteorology.

BB38 Robitzsch, Max. "V. Bjerknes †." *Zeitschrift für Meteorologie* 5 (1951): 257-258.

Appraisal of Bjerknes' contribution to dynamic meteorology.

BB39 Sverdrup, Harald Ulrik. "Vilhelm Bjerknes in Memoriam." *Tellus* 3 (1951): 217-221.

Obituary of V. Bjerknes, whose work revolutionized modern meteorology.

BB40 Sverdrup, Harald Ulrik. "A la memoire de Vilhelm Bjerknes (1862-1951)." *Association of Meteorology, International Union of Geodesy and Geophysics, Procès-Verbaux des Séances*, Bruxelles, 1953: 1-8.

A biographical note.

BB41 "Vilhelm F.K. Bjerknes, For. Mem. R.S." *Meteorological Magazine* 80 (1951): 181-184.

Biographical note on Bjerknes, his associations with Britain and a bibliography of his main contributions.

See also obituaries: Godske, C.L., "Vilhelm Bjerknes," *Naturen* 76 (1952): 34-48; Guillermo, Jans Huelin, "Professor V.F.K. Bjerknes," *Revista de Geofisica* 10 (1951): 257-258; Rossmann, F., *Naturwissenschaftliche Rundschau* 5 (1952): 218.

Blodgett, L.

BB42 McGuire, J.K. "The Father of American Climatology." *Weatherwise* 10 (1957): 92-94, 97.

A biographical note of Lorin Blodgett (1823-1901) on the hundredth anniversary of the publication of his important book *The Climate of the United States* (Philadelphia: J.B. Lippincott & Sons, 1857; 536 pp.).

Bonaventura, F.

BB43 Shmitt, Charles B. "Bonaventura, Federigo" (b. Ancona, Italy, 24 August 1555; d. Urbina [?], Italy, March 1602). *Dictionary of Scientific Biography* 2 (1970): 283.

"Bonaventura's most important scientific writings deal with meteorology.... They include *De causa ventorum motus* (1592), in which he argues, in opposition to many later interpreters, that there is no basic disagreement between the theory of winds of Aristotle and that of Theophrastus...."

Borelli, G.

BB44 Settle, Thomas B. "Borelli, Giovanni Alfonso" (b. Naples, Italy, 24 August 1555; d. Urbino [?], Italy, March 79). *Dictionary of Scientific Biography* 2 (1970): 306-314.

Borelli published a description and speculation on the causes and sources of the heat of the eruption of Etna in 1669.

Boscovich, R.J.

BB45 *Atti del Convegno internazionale celebrativo del 250 anniversario della nascita di R.G. Boscovich e del 200 anniversario della fondazione dell' Osservatorio di Brera. Milano-Nerate 6-8 Ottobre 1962.* Milano: Instituto per la storia della Tecnica, 1963. 297 pp.

Includes: L. Santomauro, "La meteorologia a Brera," pp. 83-93; N. Cubranic, "Il contributo di Ruggero Boskovic allo sviluppo della geodesia," pp. 103-113; A. Marussi, "Ruggero Boscovich e l'isostasia," pp. 121-128.

BB46 Markovic, Zeljko. "Boscovic, Rudjer J." (b. Dubrovnik, Yugoslavia, 18 May 1711; d. Milan, Italy, 13 February 1787). *Dictionary of Scientific Biography* 2 (1970): 326-332.

"It was his idea that mountains had originated from the undulation of rock strata under the influence of subterranean fires.... He also conceived the idea of a kind of gravimeter for measuring gravitation even in the ocean. At the same time, he proposed a method for determining the mean density of the earth by measuring the incremental attraction of masses of water at high tide by the deviation of a pendulum situation in the proximity." In his work in geodesy he improved existing measurements and developed better instruments.

BB47 Whyte, Lancelot Law, ed. *Roger Joseph Boscovich, S.J., F.R.S., 1711-1787. Studies of His Life and Work on the 250th Anniversary of His Birth*. New York: Fordham University Press, 1961. 230 pp.

Includes: Z. Kopal, "The Contribution of Boscovich to Astronomy and Geology," pp. 173-182; J.F. Scott, "Boscovich's Mathematics," pp. 183-192; C. Eisenhart, "Boscovich and the Combination of Observations," pp. 200-212 [the last two papers are also related to geology]; bibliography of works by and about Boscovich.

Bouguer, P.

BB48 Middleton, W.E. Knowles. "Bouguer, Pierre" (b. Croisic, France, 16 February 1698; d. Paris, France, 15 August 1758). *Dictionary of Scientific Biography* 2 (1970): 343-344.

Bouguer was one of the leaders of the expedition to Peru that measured the arc of the meridian near the equator, leading to a confirmation of the Newtonian theory of the shape of the earth. His work in photometry makes him a founder of atmospheric optics.

Boullanger, N.

BB49 Birembaut, Arthur. "Boullanger, Nicholas-Antoine" (b. Paris, France, 11 November 1722; d. Paris, 1 September 1759). *Dictionary of Scientific Biography* 2 (1970): 347-348.

"Instead of adhering to observation, he sought to develop a theory on the formation of the earth that invoked a universal deluge ... and attributed the deluge to a vast and sudden eruption of this [subterranean] water through springs.... The modern distribution of the oceans suggested a complementary hypothesis, that of the elasticity of the earth's stata, half of which had, by bending, caused the elevation of the other half."

Bourguet, L.

BB50 Ellenberger, François. "Bourguet, Louis" (b. Nimes, France, 23 April 1678; d. Neuchatel, Switzerland, 31 December 1742). *Dictionary of Scientific Biography* 15 (1978): 52-59.

Bourguet's theory of the earth "is at first close to that proposed by Woodward. But, influenced by Scheuchzer, Bourguet elaborated a genuine theory of orogeny (however warped to fit orthodox diluvianism) which he linked to the dynamics of terrestrial rotation."

Bowie, W.

BB51 Auser, Cortland. "Bowie, William" (b. Grassland, Anne Arundel County, Maryland, 6 May 1872; d. Washington, D.C., 28 August 1940). *Dictionary of Scientific Biography* 2 (1970): 372-373.

"Much of Bowie's importance as a scientist lies in his presentation of theories of isostasy.... His early studies had been of the relations between gravity anomalies and geologic formations.... His investigations in isostasy ... culminated in the publication of his book *Isostasy* (1927)."

BB52 Heck, Nicholas H. "Memorial to William Bowie." *Proceedings of the Geological Society of America for 1940*, 163-166.

Biography of W.B. (May 6, 1872-August 28, 1940), photograph, selective bibliography. His three main accomplishments were in geodetic surveys, mapping, and the theory of application of isostasy.

Brooks, C.E.P.

BB53 Rigby, Malcolm. "C.E.P. Brooks." *Bulletin of the American Meteorological Society* 39 (1958): 40-41.

An obituary reviewing the life, career, and works of Brooks (1888-1957) who wrote books on climatic change and meteorological statistics and was the author of notable meteorological bibliographies.

See also obituaries: J. Glasspoole, "Charles Ernest Pelham Brooks, I.S.O., D.Sc.," *Meteorological Magazine* 87 (1958): 97-98; A.H.R. Goldie, *Quarterly Journal of the Royal Meteorological Society* 84 (1958): 202-203.

Brooks, C.F.

BB54 Houghton, Henry G. "Charles Franklin Brooks, 1891-1958." *Bulletin of the American Meteorological Society* 39 (1958): 155.

Obituary of the Director of the Blue Hill Observatory and Harvard Professor of Meteorology who produced the International Cloud Atlas.

Other obituaries: *Nature* 181 (1958): 1440; *Geographical Review* 48 (1958): 443-444.

Brounov, P.I.

BB55 Maksimov, S.A. "Osnovatels skolchoziaistvennol meteorologii Petr Ivanovich Brounov" [The Founder of Agricultural Meteorology, Petr Ivanovich Brounov]. *Meteorologiya i Godrologiya* 7 (1952): 51-53.

Brounov (1852-1927), erstwhile Professor at Kiev, is regarded as the originator of agricultural meteorology in Russia. This biographical sketch assesses his contributions.

BB56 Maksimov, S.A. *P.I. Brounov, osnovopolozhnik selskokhoziaistvennoi meteorologii* [P.I. Brounov, Founder of Agricultural Meteorology]. Leningrad: Gidrometeoizdat, 2nd edition, 1952. 53 pp.

Gives biographical notes and lists important work of the pioneer Russian agrometeorologist, who organized an agrometeorological service for the Empire in 1897.

Buch, L. von

BB57 Guntau, M. "Leopold von Buch, 1774-1853: Kolloquium aus Anlasz der Wiederkehr seines Geburtstages nach 200 Jahren." *Zeitschrift für Geologische Wissenschaften* 2 (1974): 1363-1365.

BB58 Nieuwenkamp, W. "Buch, [Christian] Leopold von" (b. Stolpe, Germany, 25 April 1774; d. Berlin, Germany, 4 March 1853). *Dictionary of Scientific Biography* 2 (1970): 552-557.

Buch "theorized that the flanks of the volcanoes [in Auvergne] were formed by thin strata of the country rock that had been heaved up to form the tentlike roof of a large conical cavern filled with lava. Buch's theory of elevation crater was given considerable attention in the nineteenth century...."

Buchan, A.

BB59 Hawke, E.L. "Editorial." *Weather* 12 (1957): 138-139.

An appraisal of the work of the famous Scottish meteorologist Alexander Buchan (1829-1907).

Buchanan, J.Y. @

BB60 Stoddart, David. "Buchanan: The Forgotten Apostle." *Geographical Magazine* 44 (1972): 858-862.

John Young Buchanan was the chemist on board HMS *Challenger* (1872-76). "He was responsible for deep sea and surface temperature and salinity measurement and identified the first nodules." In 1886 he discovered the Congo submarine canyon. (The article includes a color reproduction of his map showing specific gravity of ocean water.)

Bucher, W.H. @

BB61 Kay, Marshall, "In Memoriam: Walter Hermann Bucher (1888-1965)." *Tectonophysics* 1 (1965): 455-458.

Bucher was known for his experiments related to fractures and faulting and for his book *Deformation of the Earth's Crust* (1933). This short obituary includes a list of his "principal publications."

Buffon, G.

BB62 Roger, Jacques. "Buffon, Georges-Louis LeClerc, Comte De" (b. Montbard, France, 7 September 1707; d. Paris, France, 16 April 1788). *Dictionary of Scientific Biography* 2 (1970): 576-582.

To explain the origin of the earth, Buffon "hypothesized that a comet, hitting the sun tangentially, had projected into space a mass of liquids and gases equal to 1/650 of the sun's mass." The material separated into spheres that cooled and became planets. From experiments on cooling, Buffon estimated the age of the earth to be 75,000 years; he did not publish his estimate of 3,000,000 years based on sedimentation phenomena.

Bugge, T.

BB63 Andersen, Einar. *Thomas Bugge: Et Mindeskrift i anledning af 150 arsdagen for Lnas død 15. Januar 1815.* Copenhagen: Geodaetisk Institut, 1968. 101 pp.

On T. Bugge (1740-1815) and his work in geodesy.

Bullen, K.E.

BB64 Jeffreys, Harold. "Keith Edward Bullen, 29 June 1906-23 September 1976." *Biographical Memoirs of Fellows of the Royal Society of London* 23 (1977): 19-39.

Bullen was an applied mathematician who used seismological and other data to develop models of the interior of the earth.

Burnet, T.

BB65 Kelly, Suzanne. "Burnet, Thomas" (b. Croft, Yorkshire, England, ca. 1635; d. London, England, 27 September 1715). *Dictionary of Scientific Biography* 2 (1970): 612-614.

"For more than a hundred years after Burnet, writers discussing the origin of and changes in the surface of the earth felt impelled to reconcile their theories with the account of Creation in Genesis." According to Burnet's *Sacred Theory of the Earth* (1681), the original surface of the earth, which covered the waters and a great abyss, "was smooth, regular, and uniform, without mountains or seas.... When the surface caved into the abyss, an event due to the continued drying action of the sun, and was no longer smooth, the fluctuations of the waters over this irregular earth caused the universal deluge."

Buys Ballot, C.H.D. @

BB66 Van Everdingen, Edwood. *C.H.D. Buys Ballot, 1817-1890.* 's-Gravenhage: D.N. Daamen, 1953. 222 pp.

A biography of the famous Dutch meteorologist; includes his organization of the Royal Netherlands Meteorological Institute, a bibliography of his 226 publications, and a list of the honors he received.

Byerly, P.

BB67 Bolt, Bruce A. "Memorial. Perry Byerly (1897-1978)." *Bulletin of the Seismological Society of America* 69 (1979): 928-945.

"He was one of the first professors to teach seismology in the United States as a separate discipline.... His greatest contribution to knowledge was probably his development of the method of determining fault parameters from the polarities of seismic waves recorded in seismographic stations remote from the wave source...." List of 89 papers.

BC. Subjects: C

Carlheim-Gyllensköld, V.

BC1 Ljungdahl, Gustaf S. "Vilhelm Carlheim-Gyllensköld, 1859-1934." *Terrestrial Magnetism and Electricity* 40 (1935): 107-110.

Obituary of Carlheim-Gyllensköld, known for his observations of the auroral spectrum and his analysis of the geomagnetic field variation since 1660. List of publications; portrait (facing p. 1).

Cassini, J.

BC2 Taton, Rene. "Cassini, Jacques (Cassini II)" (b. Paris, France, 18 February 1677; d. Thury, near Clermont, Oise, France, 15 April 1756). *Dictionary of Scientific Biography* 3 (1971): 104-106.

"In 1722 he published the important work *De la grandeur et de la figure de la terre*, in which he confirmed his support for the hypothesis of the elongation of the terrestrial ellipsoid, and opposed that of its flattening, which was defended by the supporters of Newton and Huygens.

Cassini de Thury, C.

BC3 Taton, Rene. "Cassini de Thury, Cesar-François (Cassini III)" (b. Thury, near Clermont, Oise, France, 17 June 1714; d. Paris, France, 4 September 1784). *Dictionary of Scientific Biography* 3 (1971): 107-109.

Like his father (Cassini II), and grandfather, he supported the elongation of the terrestrial spheroid (as opposed to its flattening at the poles), but abandoned this view by 1744.

Chamberlin, T.C. @

BC4 Chamberlin, Rollin T. "Thomas Chrowder Chamberlin." *Biographical Memoirs, National Academy of Sciences* 15 (1934): 307-407.

Chamberlin (1843-1928), a leading American geologist of the 19th century, became the leading American cosmogonist in the early 20th century when he proposed (with F.R. Moulton) his "planetesimal hypothesis." He also wrote one of the most widely cited papers on scientific method, the "method of multiple working hypotheses."

BC5 Collie, George L. "A Distinguished Son of Wisconsin: Thomas C. Chamberlin." *Wisconsin Magazine of History* 15 (1932): 263-281, 412-445.

BC6 *Journal of Geology* 37, no. 4 (May 1929).

Articles about T.C. Chamberlin by C.K. Leith, W.C. Alden, R.A.F. Penrose, Jr., C. Schuchert, W.D. MacMillan, B. Willis, F.R. Moulton; bibliography.

BC7 Nash, J.V. "Professor Chamberlin, Dean of American Scientists, on the Future of Man." *Open Court* 42 (1928): 513-527, frontis. photograph.

Based on an interview just before Chamberlin's death.

* Pyne, Stephen J. "Methodologies for Geology: G.K. Gilbert and T.C. Chamberlin." Cited below as item BG12.

BC8 Schultz, Susan. "Thomas C. Chamberlin: An Intellectual Biography of a Geologist and Educator." Ph.D. dissertation, University of Wisconsin, Madison, 1976. University Microfilms Order no. 76-20921. xiv + 448 pp.

For summary see *Dissertation Abstracts International* 37 (1976): 3863-A.

Chang Heng

BC9 Juang, Su-Shu. "A Biography of Chang Heng." *Stellar Physics*, by Hong-Yee Chiu. Waltham, Mass.: Blaisdell, 1968, vol. 1, pp. vi-xv.

Chang Heng (78-139) developed the first seismograph.

Chapman, S.

BC10 Akasofu, Syun-Ichi, Benson Fogle, and Bernhard Hauriwitz. *Sidney Chapman, Eighty--From His Friends.* No place or date given (Library of Congress No. 68-15590). xvi + 230 pp.

An appreciation volume in celebration of Sidney Chapman's (1888-1973) 80th birthday. Contains biographical notes; three autobiographical notes by Chapman; a list of his publications from 1910 to 1967; an appraisal of contributions to various topics in physics, meteorology, geomagnetism and solar physics, ionospheric physics and aeronomy (by 10 authors); and a large collection of letters with facsimile signatures of a large number of friends, associates, students, and acquaintances, recalling incidents and contacts with Chapman.

Charcot, J.-B. @

BC11 Hoisington, William A., Jr. "In the Service of the Third French Republic: Jean-Baptiste Charcot (1867-1936) and the Antarctic." *Proceedings of the American Philosophical Society* 119 (1975): 315-324.

Charcot was an organizer and leader of oceanographic expeditions. He gained his reputation by two voyages to the Antarctic (1903-5, 1908-10).

Charney, J.

BC12 "Jule Charney." *University Corporation for Atmospheric Research Newsletter* vol. 5, no. 6 (July 1981): 5.

A brief obituary of the dynamic meteorologist (1917-1981) who with John von Neumann (1903-1957) was the creator of computerized numerical weather prediction and made other basic contributions to the theory of atmospheric motions.

BC13 Phillips, Norman A. "Jule Charney's Influence on Meteorology." *Bulletin of the American Meteorological Society* 63 (1982): 492-498.

A eulogy praising the many fundamental contributions by Charney (1917-1981) to dynamic meteorology and particularly to numerical weather prediction.

Charpentier, J. de @

BC14 Balmer, Heinz. "Jean de Charpentier, 1786-1855." *Gesnerus* 26 (1969): 213-232.

Charpentier was the originator of the "ice age" hypothesis, subsequently popularized by Louis Agassiz.

Childrey, J.

BC15 North, J.D. "Childrey, Joshua" (b. Rochester, England, 1623; d. Upwey, Dorset-Shire, England, 26 August 1670). *Dictionary of Scientific Biography* 3 (1971): 248.

In a controversy with J. Wallis, Childrey pointed out the meteorological dependence of tides whose periods are multiples or submultiples of a tropical year; the tidal irregularity due to the inclination of the moon's orbit and its varying distance, etc.

Chromov, S.P.

BC16 Koncek, N. "S.P. Chromow †, Nachruf." *Zeitschrift für Meteorologie* 28 (1978): 1-2.

Szergie Petrovich Chromov (1904-1977), who introduced nephanalysis and modern synoptic meteorology in the Soviet Union, was responsible for the education of a large number of Russian meteorologists, first as Professor at the University of Leningrad, 1947-1953, and then at the University of Moscow until his death. His books on synoptic meteorology were widely acclaimed and translated into several languages.

Conrad, V.A.

BC17 Hader, F. "Victor A. Conrad, Aug. 26, 1876--April 22, 1962." *Wetter und Leben* 14 (1962): 93-94.

An obituary of Conrad who started his career in Austria and ended it at Harvard University. In addition to numerous papers, he published several large, seminal treatises on the analysis of climatological data.

See also obituary: Helmut E. Landsberg, *Bulletin of the American Meteorological Society* 43 (1962): 522-524.

Cordier, P.

BC18 Burke, John G. "Cordier, Pierre-Louis-Antoine" (b. Abbeville, France, 31 March 1777; d. Paris, France, 30 March 1861). *Dictionary of Scientific Biography* 3 (1971): 411-412.

"From accumulated temperature observations at various depths in mines, Cordier estimated that the earth's temperature increased one degree centigrade for each thirty to forty meters of depth. He concluded that the earth was fluid beyond a depth of 5,000 meters and that it was a cooled star. Cordier attributed the occurrence of volcanoes to the action of the earth's internal high-temperature fluid."

Cotte, L.

BC19 Dettwiller, J. "Un météorologiste oublie...." *La Météorologie* 6, no. 23 (1980).

A biography and appraisal of the work of Louis Cotte (1740-1815), whose most important work, *Traité de Météorologie* (Paris: Imprimerie Royale, 1774; 650 pp.), set the style of meteorological texts for half a century. It presented instruments in use, procedures, observational results, phenological and demographic information. The prevailing meteorological theories are discussed. Samples of the published tables are given along with excerpts from archival material.

BC20 Nicolas, Jean Paul. "L'Abbé Cotte et ses papiers." *Comptes rendus du 91e Congres National des Sociétés Savantes, Rennes, 1966*. Section des sciences, tome 1. Paris: Gauthier-Villars, 1967, pp. 255-263.

Summary of the papers of the French climatologist Abbé Louis Cotte (1740-1815), held in the Bibliothèque Municipale de Laon. Cotte compiled a large amount of material on the history of meteorology.

Croll, J.

BC21 Burstyn, Harold L. "Croll, James" (b. Cargill, Perthshire, Scotland, 2 January 1821; d. Perth, Scotland, 15 December 1890). *Dictionary of Scientific Biography* 3 (1971): 470-471.

In 1864 Croll suggested "that a change in the eccentricity of the earth's orbit had been responsible for the drastic changes in climate associated with glaciation."

Cuvier, G. @

BC22 Carozzi, Albert V. "Une nouvelle interpretation du soi-disant catastrophisme de Cuvier." *Archives des Sciences* 24 (1971): 367-377.

"The purpose of this paper is to show that most of the aspects of Cuvier's catastrophism have been misunderstood and that his contribution should be considered in a totally different light ... he did not visualize fundamentally different past and present causes whose succession in time would imply an impossible and absurd reversal of the fundamental physico-chemical laws, but a set of permanent causes whose action in the past has been characterized by short periods of higher intensity which he called revolutions. Hence, Cuvier's position corresponds to one of the modern aspects of the principle of uniformity."

BD. Subjects: D

Dalrymple, A.

BD1 Spray, W.A. "Alexander Dalrymple, Hydrographer." *American Neptune* 30 (1970): 200-216.

Dalrymple (1752-1808) "worked incessantly to provide reliable information for navigators. He was not only the first hydrographer to both the East India Company and the Admiralty, but was also the initiator of British plans for expansion into the Malay Archipelago.... Although Dalrymple was a vain and opinionated individual who was unable to divorce his theories from his own interest, nevertheless, he had known more about eastern navigation than any man living, and no one in the eighteenth century did more to promote geographical exploration and marine surveying of the Eastern Seas and elsewhere."

Dalton, J.

BD2 Thackray, Arnold. "Dalton, John" (b. Eaglesfield, Cumberland, England, 6 September 1766; d. Manchester, England, 27 July 1844). *Dictionary of Scientific Biography* 3 (1971): 537-547.

His *Meteorological Observations and Essays* (1793) provides tables of barometric pressure, etc. "The essays include a theory of trade winds, anticipated by George Hadley--a theory of the aurora borealis, similarly anticipated by Anders Celsius and by Edmund Halley; speculations about variations in barometric pressure, anticipated by Jean Deluc; and ideas on evaporation which include the germs of his own later chemical atomic theory." His continuing interest in the state of water vapor in the atmosphere appears in later papers on rain, dew, river-water runoff, and evaporation.

Daly, R.A.

BD3 Mather, Kirtley F. "Daly, Reginald Aldworth" (b. Napanee, Ontario, Canada, 18 March 1871; d. Cambridge, Massachusetts, 19 September 1957). *Dictionary of Scientific Biography* 3 (1971): 547-548.

"His long-sustained interest in isostasy ... led to his geophysical studies, concerned primarily with the strength of the earth's crust and the structure of the interior of the earth."

Dana, J.D. @

BD4 Prendergast, Michael L. "James Dwight Dana: The Life and Thought of an American Scientist." Ph.D. dissertation, University of California, Los Angeles, 1978. 640 pp.

For summary see *Dissertation Abstracts International* 39 (1979): 4451-A.

"... Dana's system of geology provided much of the theoretical structure within which American geology operated well into the twentieth century. Aspects of

Dana's geological work discussed include coral reef formation, mountain building, historical geology, glaciers, volcanoes, and uniformitarianism ..." (Author's summary).

Daniell, J.F.

BD5 Thackray, Arnold. "Daniell, John Frederic" (b. London, England, 12 March 1790; d. London, 13 March 1845). *Dictionary of Scientific Biography* 3 (1971): 556-558.

Daniell invented a dew-point hygrometer and a water barometer. His *Meteorological Essays* had a considerable success.

Daubeny, C.G.B.

BD6 Thackray, Arnold. "Daubeny, Charles Giles Bridle" (b. Stratton, Gloucestershire, England, 11 February 1795; d. Oxford, England, 13 December 1867). *Dictionary of Scientific Biography* 3 (1971): 585-586.

Daubeny "developed a chemical theory of volcanic action, which stated that such action results from penetration of water to the free alkali and alkaline earth metals supposed to exist beneath the earth's crust."

Davidson, G.

BD7 King, William Forrest. "George Davidson: Pacific Coast Scientist for the U.S. Coast and Geodetic Survey, 1845-1895." Ph.D. dissertation, Claremont Graduate School, 1973. 317 pp.

For summary see *Dissertation Abstracts International* 34 (1974): 4159-A.

"Davidson was the most important scientist to live and work on the West Coast in the nineteenth century.... [He invented] the Davidson Meridian Instrument for surveying.... He determined the width of the continental United States using the telegraphic method.... In order fully to explain the work which Davidson was engaged in, this study has introduced material on the history of governmental science in America in the nineteenth century. Information on the level of scientific training in the nation in the early part of the century has been included ..." (Author's summary).

Davis, W.M. @

BD8 Chorley, Richard J., Robert P. Beckinsale, and Anthony J. Dunn. *The History of the Study of Landforms, or the Development of Geomorphology*. Volume 2: *The Life and Work of William Morris Davis*. London: Methuen, 1973. xxii + 874 pp.

Volume 1, "Geomorphology before Davis," is cited below as item H12.

Day, A.L.

BD9 Abelson, Philip H. "Arthur Louis Day, October 30, 1869-March 2, 1950." *Biographical Memoirs, National Academy of Sciences* 47 (1975): 27-47.

His "scientific achievements spanned the fields of physics, geophysical chemistry, volcanology, seismology, and ceramic research." He studied the equilibria of mineral systems at high temperatures, and was director of the Carnegie Institution's Geophysical Laboratory from 1906 to 1936.

De Golyer, E.L.

BD10 Shor, Elizabeth Noble. "DeGolyer, Everette Lee" (b. Greensburg, Kansas, 9 October 1886; d. Dallas, Texas, 14 December 1956). *Dictionary of Scientific Biography* 4 (1971): 7-8.

De Golyer pioneered the use of geophysical methods--the balance and refraction seismographs--to find petroleum.

Delambre, J.J.

BD11 Cohen, I. Bernard. "Delambre, Jean-Baptiste Joseph" (b. Amiens, France, 19 September 1749; d. Paris, France, 19 August 1822). *Dictionary of Scientific Biography* 4 (1971): 14-18.

Delambre carried out geodetic measurements in order to establish the unit of length in the metric system.

De Luc, J.A. @

BD12 Tunbridge, Paul A. "Jean André de Luc, F.R.S. (1727-1817)." *Notes and Records of the Royal Society of London* 26 (1971): 15-23.

The article is primarily biographical, with brief remarks on de Luc's controversy with Joseph Black over the discovery of latent heat, and his research on the physics of the atmosphere.

Deryugin, K.M.

BD13 Plakhotnik, A.F. "Deryugin, Konstantin Mikhailovich" (b. St. Petersburg, Russia, 10 February 1878; d. Moscow, U.S.S.R., 27 December 1938). *Dictionary of Scientific Biography* 4 (1971): 42-43.

Deryugin studied the hydrology of the Barents Sea and the White Sea, and organized various projects of oceanographic research. His discovery of the variation of warm currents in the North Lape Stream "had a profound influence on the development of the fishing industry in the Barents Sea."

DesBarres, J.F.W.

BD14 Evans, Geraint N.D. *Uncommon Obdurate: The Several Public Careers of J.F.W. DesBarres*. Salem, Mass.: Peabody Museum of Salem; Toronto: University of Toronto Press, 1969. 130 pp., map.

Joseph F.W. DesBarres (1721-1824) produced a famous set of hydrographic charts, *The Atlantic Neptune*.

Descartes, R. @

BD15 Lafeur, Laurence J. "Descartes, Father of Modern Meteorology." *Bulletin of the American Meteorological Society* 31 (1950): 138-140.

Comments on the 10 discourses on meteorology by Descartes (1596-1650), dealing with a correct explanation of land-and-sea breezes and of rainbows. Others deal with clouds, hydrometeors, and thunderstorm phenomena.

Desmarest, N.

BD16 Taylor, Kenneth L. "Desmarest, Nicolas" (b. Soulaines-Dhuys, France, 16 September 1725; d. Paris, France, 28 September 1815). *Dictionary of Scientific Biography* 4 (1971): 70-73.

Although known primarily for "his assertion that the basalt columns found in Auvergne and elsewhere are volcanic in origin," Desmarest also "rejected forever the notion that volcanoes or an internal heat of the globe had been primarily responsible for producing the earth's essential features."

Dietrich von Freiberg

BD17 Wallace, William A. "Dietrich von Freiberg" (b. Freiberg, Germany, ca. 1250; d. ca. 1310). *Dictionary of Scientific Biography* 4 (1971): 92-95.

"Dietrich's place in the history of science is assured by his *De iride et radialibus impressionibus* ('On the Rainbow and "Radiant Impressions,"' i.e., phenomena produced in the upper atmosphere by radiation from the sun or other heavenly body), a treatise composed shortly after 1304.... In an age when scientific experimentation was practically unknown, Dietrich investigated thoroughly the paths of light rays through crystalline spheres and flasks filled with water; and he deduced therefrom the main elements of a theory of the rainbow that was to be perfected only centuries later by Descartes and Newton."

Dobson, G.M.B.

BD18 Houghton, J.T., and C.D. Walshaw. "Gordon Miller Bourne Dobson, 25 February 1889-11 March 1976." *Biographical Memoirs of Fellows of the Royal Society of London* 23 (1977): 41-57.

"Much of his life was devoted to the observation and study of atmospheric ozone, results from which were to be of great importance in leading to an understanding of the structure and circulation of the stratosphere." The article includes excerpts from a 1975 interview with Dobson.

Doelter, C.A.S.

BD19 Fischer, Walther. "Doelter (Cisterich y De La Torre), Cornelio August Severinus" (b. Arroyo, Guayama, Puerto Rico, 16 September 1850; d. Kolbnitz, Carinthia, Austria, 8 August 1930). *Dictionary of Scientific Biography* 4 (1971): 140-142.

Doelter determined the melting points of minerals and studied their behavior at high pressures and temperatures.

Dolomieu, D.

BD20 Taylor, Kenneth L. "Dolomieu, Dieudonné (called Deodat) de Gratet de" (b. Dolomieu, Dauphine, France, 23 June 1750; d. Chateauneuf, Saone-et-Loire, France, 28 [16? 29?] November 1801). *Dictionary of Scientific Biography* 4 (1971): 149-153.

Dolomieu proposed a catastrophic scheme of earth history. While he was an expert on volcanoes, he did not try to attribute most geological events to volcanism.

Dorno, C.

BD21 Freiesleben, H.C. "Dorno, Carl W.M." (b. Konigsberg, Prussia [now Kaliningrad, U.S.S.R.], 3 August 1865; d. Davos, Switzerland, 22 April 1942). *Dictionary of Scientific Biography* 4 (1971): 171.

In order to determine the factors that make a climate beneficial, Dorno established a private observatory where he measured solar radiation. "Dorno's investigations on the ultraviolet were so pioneering that the radiation between 2,900 and 3,200 angstrom units is called Dorno's radiation. He also deduced the concept of biological cooling and was thus the founder of bioclimatology."

Dove, H.W. @

BD22 "Heinrich Wilhelm Dove." *Proceedings of the American Academy of Arts and Sciences* (New Series) 7 (1880): 383-391.

A biographical memorial to a foreign fellow of the Academy who was a leading meteorologist in the 19th century.

BD23 Neumann, H. *Heinrich Wilhelm Dove, eine Naturforscher Biographie*. Liegnitz: H. Krumbhaar, 1925. 88 pp.

A brief biography of the second director of the Prussian Meteorological Institute, Dove (1803-1879), who was a major contributor to the theory of storms, compiler of monthly world isotherm charts, world climatography, and theory of monsoons. A key figure in 19th-century meteorology, foreign member of the National Academy of Sciences and the American Academy of Arts and Sciences, who published 234 meteorological papers.

BD24 Weickmann, Ludwig. "Zum 150. Geburtstag von Heinrich Wilhelm Dove." *Meteorologische Abhandlungen* 2 (1954): 9-19.

An appraisal of the contributions of Dove (1803-1879) to the "law of storms" and the study of world climate.

Duperrey, L.

BD25 Taylor, Kenneth L. "Duperrey, Louis-Isidore" (b. Paris, France, 21 [22?] October 1786; d. Paris, 25 August [10 September?] 1865). *Dictionary of Scientific Biography* 4 (1971): 256-257.

His 1822-25 circumnavigating expedition produced "new knowledge of the behavior of ocean currents in the Atlantic and Pacific Oceans" and of terrestrial magnetism, especially the determination of the earth's magnetic equator.

Du Toit, A.L.

BD26 Wilson, J.T. "Du Toit, Alexander Logie" (b. Rondebosch, near Cape Town, South Africa, 14 March 1878; d. Cape Town, 25 February 1948). *Dictionary of Scientific Biography* 4 (1971): 261-263.

"Toward the end of his life he supported the hypothesis of continental drift with arguments drawn from all parts of the world and considerations about the underlying mantle ... he was the first to realize that the Southern continents had at one time formed the supercontinent of Gondwanaland, which was distinct from the northern supercontinent of Laurasia."

Dutton, C.E.

BD27 Stegner, Wallace. "Dutton, Clarence Edward" (b. Wallingford, Connecticut, 15 May 1841; d. Englewood, New Jersey, 4 January 1912). *Dictionary of Scientific Biography* 4 (1971): 265-266.

Dutton helped to establish the theory of isostasy, to explain crustal movements. "His last major contribution was *Earthquake in the Light of the New Seismology*, which linked volcanism to radioactivity."

BE. Subjects: E

Easton, C.

BE1 Blaauw, A. "Easton, Cornelis" (b. Dordrecht, Netherlands, 10 September 1864; d. The Hague, Netherlands, 3 June 1929). *Dictionary of Scientific Biography* 4 (1971): 272-273.

In 1928 he published "a statistical-historical study of the climatological conditions in Western Europe that attempts to convert data as far back as the thirteenth century with modern ones and critically studies suggested periodic variations."

Eckart, C.H.

BE2 Munk, Walter H., and Rudolph W. Preisendorfer. "Carl Henry Eckart, May 14, 1902-October 23, 1973." *Biographical Memoirs, National Academy of Sciences* 48 (1976): 195-219.

"At the age of forty, with the advent of World War II, he turned his attention [from atomic physics] to underwater acoustics and related problems in geophysical hydrodynamics." He was director of the Scripps Institution of Oceanography for two years, and was a leader in the application of mathematical methods to the theory of propagation of waves in the ocean.

Ekman, V.W.

BE3 Welander, Pierre. "Ekman, Vagn Walfrid" (b. Stockholm, Sweden, 3 May 1874; d. Gostad, Stockaryd, Sweden, 9 March 1954). *Dictionary of Scientific Biography* 4 (1971): 344-345.

Ekman studied the influence of the earth's rotation on ocean currents and developed the theory of wind-driven circulation in an oceanic basin. He also conducted experiments on the physical properties of sea water and constructed oceanographic instruments.

Elie de Beaumont, J.

BE4 Birembaut, Arthur. "Élie de Beaumont, Jean-Baptiste-Armand-Louis-Léonce" (b. Canon, Calvados, France, 25 September 1798; d. Canon, 21 September 1874). *Dictionary of Scientific Biography* 4 (1971): 347-350.

This article presents a rather negative picture of Élie de Beaumont, whose theory of the formation of mountain chains by periodic collapse of the earth's crust on a cooling, shrinking nucleus was quite influential in the 19th century.

Ellis, W.

BE5 Laurie, P.S. "Ellis, William" (b. Greenwich, England, 20 February 1828; d. Greenwich, 11 December 1916). *Dictionary of Scientific Biography* 4 (1971): 354.

His 1880 paper "was accepted by most people as proof of the relationship between terrestrial magnetism and sun-spots suggested in 1852 by Sabine and others."

Elsasser, W.M.

BE6 Tuve, Merle. "Twenty-first Award of the William Bowie Medal. Citation of Walter Maurice Elsasser." *Transactions of the American Geophysical Union* 40 (1959): 91-92.

Eratosthenes

BE7 Dicks, D.R. "Eratosthenes" (b. Cyrene [now Shahhat, Libya], ca. 276 B.C.; d. Alexandria, ca. 195 B.C.). *Dictionary of Scientific Biography* 4 (1971): 388-393.

"His most enduring work was in geography (particularly notable is his measurement of the circumference of the earth)...."

Ertel, H.

BE8 Schröder, Wilfried. "Nachruf auf Hans Ertel." *Wetter und Leben* 23 (1971): 244-245. Also *Meteorologische Rundschau* 24 (1971): 160.

Two obituaries, closely resembling each other, of Ertel (1904-1971) who made a number of important contributions to theoretical hydrodynamics, many of which apply to meteorological problems. References to his major papers are given.

Escholt, M.P.

BE9 Spjeldnaes, Nils. "Escholt, Mikkel Pederson" (b. ca. 1610; d. Christiania [now Oslo], Norway, 1669). *Dictionary of Scientific Biography* 4 (1971): 406-407.

"Escholt demonstrated the rather unusual regularity of the earthquakes (two each century) in the Oslo region and was aware of the relationship of earthquakes to volcanism. He was the first to use the word 'geology' in the modern sense as the science of the earth."

Eskola, P.E.

BE10 Amstutz, G.C. "Eskola, Pentti Elias" (b. Lellainen, Honkilahti, Finland, 8 January 1883; d. Helsinki, Finland, 6 December 1964). *Dictionary of Scientific Biography* 4 (1971): 408-410.

"The modern concept that deeper layers of the earth do not necessarily differ in composition but rather in the density of their minerals largely originated in Eskola's high-pressure facies ideas."

Espy, J.P.

BE11 Reingold, Nathan. "Espy, James Pollard" (b. Washington County, Pennsylvania, 9 May 1785; d. Cincinnati, Ohio, 24 January 1860). *Dictionary of Scientific Biography* 4 (1971): 410-411.

Espy conducted experiments to study the behavior of clouds, and deduced the role of latent heat in cloud formation and rainfall. "As the concept of the saturated adiabatic expansion of rising air currents is basic to meteorology, Espy clearly merits recognition as an important pioneer." But until recently his excessive enthusiasm led other scientists to overlook his contributions.

Etzold, Fr.

BE12 Schmidt, Peter. "Franz Etzold: Observator der ehemaligen Erdbebenstation Leipzig." *Zeitschrift für Geologische Wissenschaften* 3 (1975): 513-521.

"A brief biography of Franz Etzold [1859-1928] is followed by a consideration of his share in looking after the seismic stations of Leipzig and Plaven, of his efforts to develop macroseismics, of his views on the causes of earthquakes in the Saxon Vogtland and Erzgebirge, as well as of his working programme published in 1919 on further investigations of these recent crustal movements" (Author's summary).

Evans, F.J.O.

BE13 Cunham, K.C. "Evans, Frederick John Owen" (b. London [?], England, 9 March 1815; d. London, 20 December 1885). *Dictionary of Scientific Biography* 4 (1971): 491-492.

"Evans' recognition as an outstanding scientist comes from his solution of the problems associated with compass navigation in iron and armor-plated ships and from his observations leading to the publication of a chart of curves of equal magnetic declination for the navigable world."

Ewing, W.M.

BE14 Bullard, Edward. "William Maurice Ewing, 12 May 1906-4 May 1974." *Biographical Memoirs of the Fellows of the Royal Society* 21 (1975): 269-311.

BE15 Wertenbaker, William. *The Floor of the Sea: Maurice Ewing and the Search to Understand the Earth.* Boston: Little, Brown, 1974. 275 pp.

Biography of Ewing (1906-1974) who, with his students, explored two-thirds of the earth's crust and provided much of the basis for plate tectonics.

Exner, F.M.

BE16 Brunt, David. "Prof. F.M. Exner." *Nature* 125 (1930): 419.

Obituary of the Austrian meteorologist who made important contributions to atmospheric dynamics.

BE17 Lauscher, Friedrich, and Georg Skoda. "Zum Gedenken an Felix M. Exner." *Wetter und Leben* 33 (1981): 94-102.

One of the most versatile meteorologists in the early part of the 20th century, Exner (1876-1930) distinguished himself in the field of atmospheric optics and by his fundamental contributions to dynamic meteorology. He was the first to design a prognostic weather chart and numerical predictions for grid points. This appraisal of his scientific contributions gives a complete listing of his 132 scientific books and papers.

BF. Subjects: F

Fassig, O.L.

BF1 Brooks, C.F. "Oliver Lanard Fassig, 1860-1936." *Bulletin of the American Meteorological Society* 18 (1937): 28-34.

Obituary of the first American meteorological Ph.D. (Johns Hopkins University, 1899). His major contributions to meteorology were an analysis of weather and climate of Baltimore, the climate of Puerto Rico, and an analysis of U.S. rainfall. He played a major role in inaugurating weather services in the Caribbean and in other international meteorological undertakings. A bibliography of his writings, 77 titles, is included.

Faye, H.

BF2 Kovalevsky, J. "Faye, Herve" (b. St. Benoit-du-Sault, France, 1 October 1814; d. Paris, France, 4 July 1902). *Dictionary of Scientific Biography* 4 (1971): 555.

Faye developed and improved Laplace's cosmogony. In his geodetic work he "introduced an idea close to isostasy, that the figure of the earth is almost an equilibrium figure, continents being lighter than the crust under the oceans."

Fedorov, E.

BF3 T[aba], H. "Eugenij Fedorov." *WMO Bulletin* 31 (1982): 285.

An obituary of Fedorov (1910-1981) who for many years was head of the hydrometeorological service of the U.S.S.R. His scientific contributions were made while a meteorologist on an arctic floating ice station. He was made an academician in the U.S.S.R. Science Academy in 1960, held the order of service six times, and was the 1976 recipient of the International Meteorological Organization prize.

Fenner, C.N.

BF4 Yoder, Hatten S. "Fenner, Clarence Norman" (b. near Clifton, New Jersey, 19 July 1870; d. near Clifton, 24 December 1949). *Dictionary of Scientific Biography* 4 (1971): 563-565.

Fenner conducted laboratory studies of the high-temperature transformations of silica, and investigated the production of gaseous emanations in volcanism.

Ferrel, W. @

BF5 "William Ferrel: Complete List of His Publications." *The Naturalist* 4 (5) (1889): 2-3.

Biographical note on Ferrel (1817-1891) who was one of the most important meteorologists in 19th-century America. Lists 50 of his contributions.

Fersman, A.E.

BF6 Meniailov, A. "Fersman, Aleksandr Evgenievich" (b. St. Petersburg, Russia, 8 November 1883; d. Sochi, U.S.S.R., 20 May 1945). *Dictionary of Scientific Biography* 4 (1971): 597-601.

In addition to his research in geochemistry (especially on the distribution of chemical elements), Fersman wrote on the history of earth science.

Fichot, L.

BF7 Biswas, Asit K., and Margaret R. Peitsch. "Fichot, Lazare-Eugene" (b. Le Creusot, Saone-et-Loire, France, 18 January 1867; d. Tabanac, Gironde, France, 17 July 1939). *Dictionary of Scientific Biography* 4 (1971): 614.

Fichot's *Les Marees* (1923) "contained a comprehensive synthesis of existing knowledge in tides as well as some of his own work on the subject."

Ficker, H. von

BF8 Koschmieder, H. "Biographical Sketch of Heinrich von Ficker, Nov. 22, 1881-April 29, 1957." *Beiträge zur Physik der Atmosphäre* 31 (1959): 141-146.

An appraisal of Ficker's contributions to meteorology with portrait.

BF9 Reuter, Heinz. "Prof. Dr. Heinrich Ficker." *Quarterly Journal of the Royal Meteorological Society* 83 (1957): 565.

Obituary of the famous Austrian synoptician Ficker (1881-1957).

BF10 Schneider-Carius, K. "Heinrich Ficker als Synoptiker und Aerologe." *Zeitschrift für Meteorologie* 12 (1958): 101-108.

Appraises the contributions of Ficker (1881-1957), in particular his theory of cyclones and his anticipation of the polar front theory of the Norwegian school. He also pioneered the concept of "steering" of weather systems.

BF11 Schröder, Wilfried. "Heinrich Ficker zum 100. Geburtstag." *Wetter und Leben* 33 (1981): 197-199.

A brief biographical sketch of the Austrian-German meteorologist Ficker (1881-1957) on the occasion of the hundredth anniversary of his birth. He anticipated in his early work many aspects of the cyclone model of the Norwegian school. A facsimile letter from him to Vilhelm Bjerknes (1862-1951) is reproduced.

Other obituaries: Milulas Koncek, *Meteorologiske Zpravy* 10 (1957): 85-86; H. Flohn, *Petermann's Geographische Mitteilungen* 101 (1957): 272; F. Desi, *Idöjaras* 61 (1957): 219-220; R. Bilanchini, *Rivista de Meteorologica Aeronautica* 18 (1958): 523; F. Steinhauser, *Archiv für Meteorologie, Geophysik und Bioklimatologie*, Serie A, 10 (1957): 257-264; Ludwig Weickmann, *Meteorologische Rundschau* 10 (1957): 81; Hans Ertel, *Zeitschrift für Meteorologie* 12 (1958): 99.

Fitzroy, R. @

BF12 Charnock, H. "Fitzroy--Meteorological Statist." *Proceedings of the Second International Congress on the History of Oceanography* (item L8), pp. 115-122.

"Robert Fitzroy is remembered mainly as the captain of H.M.S. *Beagle* who had Charles Darwin on board as naturalist from 1832 to 1836.... This essay is concerned with the background to Fitzroy's achievements at the Meteorological Department of the Board of Trade during the last eleven years of his life, attempting to see him in the context of nineteenth-century meteorology."

BF13 Mellersh, H.E.L. *FitzRoy of the Beagle*. London: Hart-Davis, 1968. 307 pp.

A biography of the captain of the *Beagle*, the ship on which Charles Darwin traveled as a naturalist. Robert Fitzroy (1805-1865), who eventually became a Vice Admiral, was in 1854 appointed to head the Meteorological

Department of the Board of Trade to institute storm warnings. This book devotes little space to his meteorological activities. It comments on his *Weather Book* (London: Longman, Green and Co., 1862) but government support of his office lagged. His work was attacked by scientists because it was "empirical" and by the public for missed "conjectures." Frustration and health problems led to suicide. Annotations in the meteorological part are quite inadequate.

Fleming, J.

BF14 Page, L.E. "Fleming, John" (b. Kirkroads, near Bathgate, Linlithgowshire, Scotland, 10 January 1785; d. Edinburgh, Scotland, 18 November 1857). *Dictionary of Scientific Biography* 5 (1972): 31-32.

Fleming was a participant in the uniformitarian/catastrophist debate, who stated uniformitarian principles in his 1822 work *Philosophy of Zoology*. He suggested that "world climate might have been affected by an increase in the amount of land during the course of geologic history," and rejected "the theory of an originally molten earth that has slowly cooled down."

Forbes, J.D.

BF15 Burke, John G. "Forbes, James David" (b. Edinburgh, Scotland, 20 April 1809; d. Clifton, Scotland, 31 December 1868). *Dictionary of Scientific Biography* 5 (1972): 68-69.

Forbes developed a theory of the motion of glaciers.

Foster, H.

BF16 Dostrovsky, Sigalia. "Foster, Henry" (b. Wood Plumpton, England, August 1797; d. Chagres River, Isthmus of Panama, 5 February 1831). *Dictionary of Scientific Biography* 5 (1972): 78-79.

"Henry Foster was involved with geophysical observations throughout his career in the British navy--he studied geomagnetism, the velocity of sound, atmospheric refraction, and the acceleration of gravity." The latter contributed to a determination of the ellipticity of the earth.

Fouqué, F.A.

BF17 Burke, John G. "Fouqué, Ferdinand Andre" (b. Mortain, Manche, France, 21 June 1828; b. Paris, France, 7 March 1904). *Dictionary of Scientific Biography* 5 (1972): 88-89.

Fouqué studied the chemical products of volcanic eruptions and undertook artificial synthesis of igneous rocks to determine the conditions surrounding their origin.

Franklin, B. @

BF18 Pomerantz, Martin A. "Benjamin Franklin--The Compleat Geophysicist." *EOS, Transactions of the American Geophysical Union* 57 (1976): 492-507.

Reproduction of a number of letters of Franklin (1706-1790), several of which deal with his observations of the motion of storms and the cause for whirlwinds and waterspouts, for which he proposes a convective hypothesis. He also speculated on the nature of the aurora.

Fritz, H.

BF19 Schröder, W. "Hermann Fritz und sein Wirken für die Polarlichtforschung." *Zur Geschichte der Geophysik.* Edited by H. Birrett, K. Helbig, W. Kertz, and U. Schmucker. Berlin: Springer Verlag, 1974, pp. 149-154.

Gives an appraisal of the important contribution made by Fritz (1830-1893) to aurora research, with a portrait and facsimile letter and a good list of references pertinent to early exploration of the aurora.

Fuess, R.

BF20 Kamphoff, Heinz. "R. Fuess, Erinnerungen an eine Persönlichkeit, die eine Firma von Weltruf in Berlin gründete." *Beilage zur Berliner Wetterkarte, Institut für Meteorologie, Freie Universität Berlin* 37/39, SO 8/79 (1979). 6 pp.

A brief biographical sketch of Rudolf Fuess, founder (1865) of the meteorological instrument firm of the same name, which until its merger with another company (1976) was one of the principal suppliers of meteorological instruments in the world.

BG. Subjects: G

Gagnebin, E.

BG1 Kuhn-Schnyder, Emil. "Gagnebin, Elie" (b. Liege, Belgium, 4 February 1891; d. Lausanne, Switzerland, 16 July 1949). *Dictionary of Scientific Biography* 5 (1972): 221-222.

In 1941 he proposed (with M. Lugeon) that the Alps were formed by continental drift followed by gravity sliding.

Gauss, C.F. @

BG2 Biermann, Kurt-R. *Briefwechsel zwischen Alexander von Humboldt und Carl Friedrich Gauss. Zum 200. Geburtstag von C.F. Gauss im Auftrage des Gauss-Komitees bei der Akademie der Wissenschaften der DDR.* Beiträge zur Alexander-von-Humboldt-Forschung, 4. Berlin: Akademie-Verlag, 1977. 202 pp.

Correspondence between Humboldt and Gauss on terrestrial magnetism, astronomy, the atmosphere, and related topics, in the period 1807-54. Although most of the letters have been published before, the new edition is much more complete and accurate; it is extensively annotated and indexed.

BG3 Buhler, W.K. *Gauss: A Biographical Study.* New York: Springer-Verlag, 1981. 208 pp.

Chapter 9, "Geodesy and Geometry"; Chapter 11, "Physics" (including terrestrial magnetism).

BG4 Dunnington, C. Waldo. *Carl Friedrich Gauss: Titan of Science. A Study of His Life and Work.* New York: Exposition Press, 1955. xvi + 479 pp.

"By his exhaustive study of one of the most outstanding figures of the past two centuries, Dr. G. Waldo Dunnington has put us greatly in his debt ... his account will be welcomed not only by the specialist but by the general reader.... The chapters dealing with Gauss's various contributions to science are somewhat unequal. The story of his alliance with Weber and the invention of the electromagnetic telegraph is told with the fullness it deserves. The same may well be claimed in respect of his contributions to other branches of science: magnetism, astronomy, optics, and above all, his study of non-Euclidean geometry. The somewhat lengthy chapter on geodesy, however, hardly reaches the same high standard as the others."--J.F. Scott, *Isis* 50 (1959): 285-286.

BG5 Hall, Tord. *Carl Friedrich Gauss: A Biography*. Translated from Swedish by Albert Froderberg. Cambridge, Mass.: MIT Press, 1970. viii + 173 pp.

The emphasis is on Gauss's work in pure mathematics, indicating how this was related to his geodetic measurements and research in terrestrial magnetism.

Gautier, H.

BG6 Ellenberger, François. "A l'aube de la geologie moderne: Henri Gautier (1600-1737). Premiere partie: Les antecedents historiques et la vie d'Henri Gautier." *Histoire et Nature: Cahiers de l'Association pour l'Histoire des Sciences de la Nature* 7 (1975): 3-58.

Geikie, A. @

BG7 Cutter, Eric. "Sir Archibald Geikie: A Bibliography." *Journal of the Society for the Bibliography of Natural History* 7 (1974): 1-18.

Geikie, J.

BG8 Challinor, John. "Geikie, James" (b. Edinburgh, Scotland, 23 August 1839; d. Edinburgh, 1 March 1915). *Dictionary of Scientific Biography* 5 (1972): 338-339.

In *The Great Ice Age* (1874) he suggested that the glacial period as a whole had been interrupted by mild episodes; his theory of interglacial periods was later supported by A. Penck and E. Bruckner (1909).

Gessner, J.

BG9 Pilet, P.E. "Gessner (Gesner), Johannes" (b. Zurich, Switzerland, 18 March 1709; d. Zurich, 6 May 1790). *Dictionary of Scientific Biography* 5 (1972): 379-380.

"A brilliant advocate of the principle of geochronological extrapolation, his work foreshadowed modern geophysics."

Gilbert, G.K. @

BG10 Kitts, David B. "Grove Karl Gilbert and the Concept of 'Hypothesis' in Late Nineteenth-Century Geology."

Foundations of Scientific Method: The Nineteenth Century. Edited by Ronald N. Giere and Richard S. Westfall. Bloomington: Indiana University Press, 1973, pp. 259-274. Reprinted in item A58.

"The tendency to see two distinct levels of inference is quite characteristic of historical disciplines. It manifests itself in discussions of historical explanation which consider how historical events are explained and do not consider how historical events are obtained in the first place.... I suggest that geologists find Gilbert's almost wholly particularized account of geological inference to be a satisfactory treatment of what might be called 'primary historical inference.' Geologists apparently do not regard this kind of inference as resting on theoretical, or even general principles. According to them, however, higher level historical inferences do."

BG11 Pyne, Stephen J. "Grove Karl Gilbert: A Biography of American Geology." Ph.D. dissertation, University of Texas at Austin, 1976. 652 pp.

For summary see *Dissertation Abstracts International* 37 (1977): 5308-A. A revised version has been published, cited below as item BG13.

BG12 Pyne, Stephen J. "Methodologies for Geology: G.K. Gilbert and T.C. Chamberlin." *Isis* 69 (1978): 413-424.

Gilbert saw earth processes as a mechanical equilibrium between force and resistance, with a cyclic conception of time; he felt that geology should imitate physics and its language should be mathematics. Chamberlin saw earth processes as part of a grand scheme of progressive evolution. For Gilbert the method of multiple hypotheses was conservative, to check egotism; for Chamberlin it encouraged individual growth. "Apart from Gilbert and Chamberlin it is curious that the earth sciences have not had so rich a methodological tradition as many of the other sciences.... Yet the essays of Gilbert and Chamberlin addressed methodological issues beyond the provincial needs of the earth sciences."

BG13 Pyne, Stephen J. *Grove Karl Gilbert, A Great Engine of Research.* Austin: University of Texas Press, 1980. viii + 148 pp.

Includes articles on his work in glacial geology by G.W. White, on his studies of faults, scarps, and earthquakes by R.E. Wallace, on gravity and isostasy by D.R.

Mabey, on the moon by Farouk El-Baz, on ground water A.F. Agnew, on geochronology by A.G. Fischer, and on his philosophy of science by D.B. Kitts. R.J. Chorley and R.P. Beckinsale note Gilbert's recent influence on the application of systems philosophy to environmental sciences.

BG14 Item deleted.

Glaisher, J. @

BG15 Manley, Gordon. "A Venturesome Victorian." *Weather* 17 (1962): 340-341.

Describes the daring meteorological balloon ascents of James Glaisher (1809-1903) in the 1860s.

Goethe, J.W.

BG16 Wells, George A. "Goethe, Johann Wolfgang Von" (b. Frankfurt am Main, Germany, 28 August 1749; d. Weimar, Germany, 22 March 1832). *Dictionary of Scientific Biography* 5 (1972): 442-446.

In his writings on geology, Goethe opposed catastrophism and attributed Swiss erratic blocks to transport by glaciers.

Götz, F.W.P.

BG17 Perl, G. "Prof. F.W. Paul Götz (1890-1954)." *Geofisica Pura e Applicata* 29 (1954): 236.

An obituary of the Swiss meteorologist whose research on atmospheric radiation and ozone brought him fame.

Gold, E.

BG18 Kirk, T.H. "Meteorology in the First World War." *Weather* 5 (1950): 301-304.

Based on notes of Ernest Gold (1881-1972), this article relates Gold's experience with the British Army Meteorological Service in France from 1915 to 1918 to applications of the science to gas warfare and to gunnery by predictions of ballistic winds.

BG19 Sutcliffe, R.C., and A.C. Best. "Ernest Gold, 24 July 1881-30 January 1976." *Biographical Memoirs of Fellows of the Royal Society of London* 23 (1977): 115-131.

Golitsyn, B.B.

BG20 Dorfmann, J.G. "Golitsyn, Boris Borisovich" (b. St. Petersburg [now Leningrad], Russia, 2 March 1862; d. Petrograd [now Leningrad], 16 May 1916). *Dictionary of Scientific Biography* 5 (1972): 461-462.

Golitsyn developed a seismograph with a galvanometric register; through his efforts, "Russian seismometry occupied a leading place in world science" in the early 20th century.

Gorczynski, W.

BG21 Stenz, Edward. "Professor Dr. Wladislaw Gorczynski." *Acta Geophysica Polonica* 2 (1954): 3-6.

An obituary for the notable Polish meteorologist, who did important research on solar radiation; gives biographical notes and a bibliography of his 245 publications.

Gould, B.A. @

BG22 Pitchford, Leslie Lee, Jr. "Benjamin Apthorp Gould and the Development of the Argentine Meteorological Service." Master's thesis, George Washington University, 1950. 112 pp.

Gould (1824-1896) played a major role in developing meteorological observations in Argentina and was director of the meteorological service from 1873 to 1884.

Grabau, A.M.

BG23 Kay, Marshall. "Grabau, Amadeus William" (b. Cedarburg, Wisconsin, 9 January 1870; d. Peking, China, 20 March 1946). *Dictionary of Scientific Biography* 5 (1972): 486-488.

Grabau suggested that "the crust moved relative to the poles, causing climatic changes such as glaciation, and that the continents once formed a single mass, Pangaiea." He thought mountains were rising at the fore of the shifting continental plates and volcanism was at their rear. "His pulsation theory attributed the distribution of the principal stratigraphic units to great rhythmic advances and regressions of the seas, which were in turn dependent on restriction and expansion of capacities of ocean basins: eustatic control."

Green, C.H.

BG24 Green, Cecil H. "Comment." *Discovery--Research and Scholarship at The University of Texas at Austin* 6, no. 1 (Autumn 1981): 2-5.

Autobiographical note; Green has been active in geophysical prospecting and in sponsoring geophysical research at the University of Texas.

Greenwood, I. @

BG25 Frisinger, H. Howard. "Isaac Greenwood: Pioneer American Meteorologist." *Bulletin of the American Meteorological Society* 48 (1967): 265-267.

Greenwood (1702-1745) gave some early lectures on meteorology at Harvard University. A 1727 letter to Dr. Jurin (Royal Society) is reproduced, indicating that Mr. Feveryear's Boston barometer observations had been forwarded (MS preserved at the Royal Society). Offers to send, if desired, annual accounts of the weather. He also proposed marine observations.

BG26 Simon, L.G. "Isaac Greenwood, First Hollis Professor." *Scripta Mathematica* 2 (1934): 117-124.

Biography of an early American scientist (1702-1745) interested in meteorology who was a Harvard professor.

Guenther, A.W.S.

BG27 Hofman, J.E. "Guenther, Adam Wilhelm Siegmund" (b. Nuremberg, Germany, 6 February 1848; d. Munich, Germany, 3 February 1923). *Dictionary of Scientific Biography* 5 (1972): 573-574.

Guenther published numerous compendia and historical surveys on geophysics. "During World War I he headed the Bavarian flying weather service, beginning in 1917."

Guericke, O. v. @

BG28 Schimank, Hans. "Otto von Guericke." *Naturwissenschaften* 40 (1953): 397-403.

Guericke (1602-1686), long-time mayor of Magdeburg, inventor and physicist, speculated about causes for clouds and the origin of winds. With a baroscope of his own construction he predicted a severe storm in 1660.

Gutenberg, B.

BG29 Shor, George G., Jr., and Elizabeth Noble Shor. "Gutenberg, Beno" (b. Darmstadt, Germany, 4 June 1889; d. Pasadena, California, 25 January 1960). *Dictionary of Scientific Biography* 5 (1972): 596-597.

At Göttingen, "an interest in weather forecasting and climatology led him to Emil Wiechert's course on instrumental observation of geophysical phenomena," and he soon became one of the world's leading seismologists, with a thesis on microseisms. His determination of the radius of the earth's core is still considered accurate. Continuing his early interest in meteorology, he studied the structure of the upper atmosphere and derived temperature curves for the ionosphere. With C.F. Richter at California Institute of Technology, he studied quantitative measures of earthquakes and their geographical distribution. His discovery of a low-velocity channel between 100 and 200 km depth is "essential to theories of crustal movements."

Guyot, A. @

BG30 Dana, J.D. "Biographical Memoir of Arnold Guyot." *National Academy of Sciences, Biographical Memoirs* 2 (1886): 309-347.

Guyot, a political refugee from Switzerland, became professor at Princeton and acted from 1849 to 1881 as meteorological adviser to the Secretary of the Smithsonian Institution, Joseph Henry. In that capacity he selected station locations, issued instructions for observers, and prepared the first edition of the renowned Smithsonian Meteorological Tables (1852).

BH. Subjects: H

Hall, J.

BH1 Eyles, V.A. "Hall, Sir James" (b. Dunglass, East Lothian, Scotland, 17 January 1761; d. Edinburgh, Scotland, 23 June 1832). *Dictionary of Scientific Biography* 6 (1972): 53-56.

To counter criticisms of James Hutton's *Theory of the Earth*, Hall conducted experiments on the properties of rocks at high temperatures and pressures. He suggested that certain geological features (now attributed to glacial action) had been produced by enormous tidal waves (tsunamis), resulting from submarine igneous activity. He is recognized as a pioneer in experimental testing of geological theories.

Halley, E. @

BH2 Armitage, Angus, and Colin A. Roan. "Edmond Halley 1656-1742: Papers to Commemorate the Tercentary of His Birth." *British Astronomical Association Memoirs* 37 (1956). 39 pp.

Halley, although principally known for his astronomical work, contributed to early scientific studies of meteorolo[gy] through his paper on trade winds.

BH3 Chapman, Sydney. "Edmond Halley as Physical Geographer, and the Story of His Charts." *Occasional Notes of the Royal Astronomical Society* 9 (1941): 122-134.

Includes a brief account of Halley's writings on trade winds and terrestrial magnetism; reproduces his 1688 chart of trade winds, his 1700 chart of magnetic declination, and (for comparison) a magnetic chart published in isogoni[c] lines for epoch 1700.

BH4 Cotter, Charles H. "Captain Edmond Halley, R.N., F.R.S." *Notes and Records of the Royal Society of London* 36 (1981): 61-77.

Reviews Halley's contributions to maritime science--variation of the compass, tidal science, and hydrographic surveying.

* Ronan, Colin A. "Edmond Halley and Early Geophysics." Cite[d] below as item U41.

BH5 Item deleted.

Hamberg, A.

BH6 Spjeldnaes, Nils. "Hamberg, Axel" (b. Stockholm, Sweden, 17 January 1863; d. Djursholm, Sweden, 28 June 1933). *Dictionary of Scientific Biography* 6 (1972): 78-79.

Hamberg contributed to hydrology and glaciology through observations, surveys, and the development of more accurat[e] methods of measurement.

Hamilton, W. @

BH7 Fothergill, Brian. *Sir William Hamilton, Envoy Extraordinary*. London: Faber and Faber, 1969; New York: Harcourt, Brace and World, 1969. 459 pp., port., plates, bibl., index.

Chapter VII includes his observations of the eruption of Vesuvius (1765-66); otherwise there is nothing on his scientific work.

BH8 Sleep, Mark C.W. "Sir William Hamilton (1730-1803): His Work and Influence in Geology." *Annals of Science* 25 (1969): 319-338.

Hamilton (1776) suggested the volcanic origin of basalt independently of Desmarest and advocated Vulcanism as a cause in geology. He may have been the first to produce an isoseismic map, based on Italian earthquake records (1783). His writings were frequently cited in later works on vulcanology.

Hann, J. @

BH9 Lauscher, Friedrich. "Julius Hann und die weltweite Entwicklung der Meteorologie." *Wetter und Leben* 23 (1971): 223-231.

An appraisal of Hann's (1839-1921) enormous influence on the development of meteorology through his research, his editorship of the influential *Meteorologische Zeitschrift*, and his textbooks on meteorology and climatology, which went through numerous editions.

Harker, A.

BH10 Challinor, John. "Harker, Alfred" (b. Kingston-upon-Hull, England, 19 February 1859; d. Cambridge, England, 28 July 1939). *Dictionary of Scientific Biography* 6 (1972): 116.

In his *Natural History of Igneous Rocks* (1909) he discussed "the whole range of phenomena in light of the general principles of mathematics, physics, and chemistry." His volcanic-plutonic-hypabysmal cycle is a "principle of the first importance."

Hassler, F.R.

BH11 Cajori, Florian. *The Chequered Career of Ferdinand Rudolph Hassler, First Superintendent of the United States Coast Survey. A Chapter in the History of Science in America.* Boston: Christopher Pub. House, 1929. 245 pp. Reprint, New York: Arno Press, 1980.

"... very useful because of the extensive research on which it is based. Indeed, some of the MSS cited apparentl are no longer extant. Its greatest weakness is its lack of any comparable research on Hassler's contemporaries and the often uncritical treatment accorded its subject's activities."--N. Reingold, *DSB* 6: 166. (next item)

BH12 Reingold, Nathan. "Hassler, Ferdinand Rudolph" (b. Aarau, Switzerland, 7 October 1770; d. Philadelphia, Pennsylvania, 20 November 1843). *Dictionary of Scientific Biography* 6 (1972): 165-166.

Hassler, who emigrated to America in 1805, revived and expanded the coast survey: he promoted measurements in geodesy, geomagnetism, and tides. According to Reingold, the biography by F. Cajori (*The Chequered Career ...*, 1929 [above, item BH11]) contains valuable information but is somewhat misleading on Hassler's relations with contemporaries.

Haug, G.E.

BH13 Carozzi, Albert V. "Haug, Gustave Emile" (b. Drussenheim, Alsace, France, 19 July 1861; d. Paris, France, 28 August 1927). *Dictionary of Scientific Biography* 6 (1972): 168-169.

Haug's rule in tectonics states that "when subsidence takes place in a geosyncline, a regression of the sea occurs over the adjacent epicontinental areas; conversely when compression and folding begin in a geosyncline, there is a massive transgression over the epicontinental areas."

Haüy, R.-J.

BH14 Hooykaas, R. "Haüy, René-Just" (b. St.-Just-en-Chaussee, Oise, France, 28 February 1743; d. Paris, France, 1 June 1822). *Dictionary of Scientific Biography* 6 (1972) 178-183.

Haüy is known for his researches on the physical and chemical properties of crystals, fundamental to mineralogy.

Hayford, J.F.

BH15 Reingold, Nathan. "Hayford, John Fillmore" (b. Rouses Point, New York, 19 May 1868; d. Evanston, Illinois, 10 March 1925). *Dictionary of Scientific Biography* 6 (1972): 188-189.

Hayford's principal achievements occurred as head of geodetic work for the Coast and Geodetic Survey from 1900 to 1909. "By introducing the use of the area method, rather than the arc method, Hayford ended an era in geodesy that dated from the seventeenth century and inaugurated the modern procedure in this field." He was "the first to take isostasy into consideration in arriving at the figure of the earth" and his work was "the first demonstration of the validity of the concept of isostasy." Contrary to T.C. Chamberlin, Hayford "contended that isostatic compensation had been increasing and was at its highest level at present." He "concluded that the earth was a failing, not a stable, structure."

Heaviside, O. @

BH16 Watson-Watt, Sir Robert. "Oliver Heaviside: 1850-1925." *Scientific Monthly* 61 (1950): 353-358.

An appraisal of the scientific contributions of Heaviside, the discoverer of the ionosphere, with reproduction of a portrait.

Heck, N.H.

BH17 Fleming, J.A. "Nicholas Hunter Heck, Geophysicist." *Terrestrial Magnetism and Atmospheric Electricity* 50 (1945): 141-143.

Heck, who recently retired from the Coast and Geodetic Survey, was known for his work in geomagnetism, seismology, and oceanography.

Heim, A.

BH18 Chorley, R.J. "Heim, Albert" (b. Zurich, Switzerland, 12 April 1849; d. Zurich, 30 August 1937). *Dictionary of Scientific Biography* 6 (1972): 227-228.

His *Mechanismus der Gebirgsbildung* (1878) became "the authoritative work on Alpine tectonics."

Hellmann, G.

BH19 Knoch, Karl. "Gustav Hellman zum Gedenken seines 100. Geburtstages." *Annalen der Meteorologie* 7 (1955/56): 1-2.

Hellmann (1854-1939), Professor of Meteorology at Berlin, in addition to producing major works on rainfall, was an ardent collector of historical meteorological material.

BH20 König, Willi. "Zur 100. Wiederkehr des Geburtstages von Gustav Hellmann." *Zeitschrift für Meteorologie* 8 (1954): 193.

An appraisal of Hellmann's (1854-1939) work.

Hellpach, W.

BH21 Flohn, Hermann. "Willy Hellpach." *Meteorologische Rundschau* 8 (1955): 73.

Hellpach (1877-1954) was a physiologist and psychologist who published extensively on weather influences on moods and behavior.

Helmert, F.R.

BH22 Fischer, Walther. "Helmert, Friedrich Robert" (b. Freiberg, Saxony, Germany, 31 July 1843; d. Potsdam, Germany, 15 June 1917). *Dictionary of Scientific Biography* 6 (1972): 239-241.

Helmert developed the mathematical and physical theories of geodesy, and conducted numerous studies of the deviation of plumb lines and the variation of gravity on the earth's surface.

Helmholtz, H. von @

BH23 *Gedanken von Helmholtz über schöpferische Impulse und über das Zusammenwirken verschiedener Wissenschaftszweige*. Sitzungsberichte des Plenums und der Klassen der Akademie der Wissenschaften der DDR, 1972, 1. Berlin: Akademie-Verlag, 1972. 78 pp.

Includes "Zur Bedeutung der Arbeiten von H. v. Helmholtz für die geophysikalische Hydrodynamik und für die Physik des Erdinnern" by Heinz Stiller.

Hogg, A.R.

BH24 Gascoigne, S.C.B. "Arthur Robert Hogg." *Records of the Australian Academy of Science* 1 (1968): 58-70.

Hogg (1903-1966) worked in atmospheric electricity and astronomy.

Holmes, A.

BH25 Dunham, Kingsley. "Holmes, Arthur" (b. Hebburn on Tyne, England, 14 January 1890; d. London, England, 20 September 1965). *Dictionary of Scientific Biography* 6 (1972): 474-476.

Holmes pioneered radiometric dating of rocks and showed that the earth is billions of years old. Along with his work in geochronology he considered the effect of radioactive heat on the thermal history of the earth, and proposed "cyclical expansion alternating with contraction as a means of explaining tectonic movements in the crust." He also became a strong supporter of A.L. Wegener's hypothesis of drifting continents and was the first advocate of convection currents "in the mantle." His *Principles of Physical Geology* (1944, 1964) is an important synthesis of geophysical interpretation.

BH26 T[omkeieff], S.I. [Obituary Notice of Arthur Holmes]. *Proceedings of the Geologists' Association* 78 (1967): 374-377.

Arthur Holmes, born 14 January 1890, died 20 September 1965. His main interests were in geochronology, the petrology of igneous rocks, and in the radioactivity of the earth.

Hooke, R. @

BH27 Davies, Gordon L. "Robert Hooke and His Conception of Earth History." *Proceedings of the Geologists' Association* 75 (1964): 493-498.

"The geological writings of Robert Hooke (1635-1703) have never received the study they deserve. Attention is here drawn to Hooke's *Discourses of Earthquakes* of 1668 in which he displayed both an understanding of unconformities and a remarkable grasp of the cyclic nature of events at the earth's surface" (Author's summary). "Although he favoured the view that the earth's relief had a seismic origin, Hooke did not overlook the important role of denudation in moulding topography" (pp. 494-495).

Hopkins, W.

BH28 Beckinsale, Robert P. "Hopkins, William" (b. Kingston-on-Soar, Derbyshire, England, 2 February 1793; d. Cambridge, England, 13 October 1866). *Dictionary of Scientific Biography* 6 (1972): 502-504.

Hopkins applied mathematics to problems such as crustal elevation and its effect on surface fracturing, the natur of the earth's interior, and the causes of climatic change. His views on the transport of erratic boulders are now considered wrong, but led to a result in hydrauli now known as [G.K.] "Gilbert's sixth-power law." He refuted the current idea that the earth's interior is mostly liquid by computing the effect of such an interior on precession and nutation. Beckinsale concludes that "Hopkins's effect on contemporary geology was frequently retrogressive rather than progressive," and gives him credit only for quantifying ideas proposed by others.

Hough, G.W.

BH29 Dieke, Sally H. "Hough, George Washington" (b. Tribes Hill, New York, 24 October 1836; d. Evanston, Illinois, 1 January 1909). *Dictionary of Scientific Biography* 6 (1972): 522-523.

Hough devised several meteorological instruments such as a self-registering mercury barometer and a recording anemometer.

Howard, L.

BH30 Blench, B.J.R. "Luke Howard and His Contribution to Meteorology." *Weather* 18 (1963): 83-92.

Howard (1772-1864), a chemist by training, was one of the most productive British meteorologists in the early and mid-19th century. His work on cloud classification (1804) is still the basis for current cloud-form reporting. It so excited the German poet J.W. von Goethe (1749-1832) that he celebrated the work by a poem and had an artist sketch the various patterns. This led to some correspondence between the two. Another famous work was his *Climate of London* (2 vols., Y. and A. Arch, Cornhill; Longman and Co., London, 1818), which is the first description of the effect of an urban area on climate. He attempted a theory of rain formation in 1833, in which he suggested the importance of electric influence. The essay has a bibliography of Howard's most important publications.

Hulbert, E.O.

BH31 "Dr. Edward Hulburt, 91, Navy Researcher, Dies." *Washington Post* (October 12, 1982), p. B4.

Hulburt died October 11, 1982; he was the senior scientist for the U.S. National Committee for the IGY, and was an authority on the optics of the sea, the physics of the upper atmosphere, and terrestrial magnetism.

Humboldt, A. v. @

* Biermann, Kurt-R. *Briefwechsel zwischen Alexander von Humboldt und Carl Friedrich Gauss*. Cited above as item BG2.

BH32 Botting, Douglas. *Humboldt and the Cosmos*. New York: Harper and Row, 1973. 295 pp.

A profusely illustrated account of Humboldt's life and explorations, with very little information about his scientific work.

BH33 Gerasimov, I.P., et al. *Alexander von Humboldt, 1769-1859*. Acta Historica Leopoldina, 6. Leipzig: Barth, 1971. 59 pp.

A collection of essays on Humboldt's influence on science.

BH34 Hammacher, Klaus, ed. *Universalismus und Wissenschaft im Werk und Wirken der Brüder Humboldt*. With an appendix, "Wilhelm von Humboldts Briefe an John Pickering," edited by Kurt Müller-Vollmer. Studien zur Philosophie und Literatur des 19. Jahrhunderts, 31. Frankfurt am Main: Klostermann, 1976. 342 pp.

Papers presented at meetings in 1972 and 1974, including a few on the worldview and career of Alexander von Humboldt.

BH35 Kalesnik, S.V. "Zhisn i tvorcheskii put Aleksandra Humboldta: K stoletiya so dnia smerti" [Life and Works of Alexander Humboldt: Centenary of His Death]. *Istvestiya Vsesoiuznoe Geograficheskoe Obshchestvo* 91 (1959): 313-323.

A brief biography of Humboldt (1769-1859) and an appraisal of his scientific contributions, which included important work in climatology. At his urging, regular meteorological observations were started in Russia.

BH36 Keil, Karl. "Ein Brief Alexander von Humboldts an den Geh. Sanitätsrat Dr. med. Friedrich W. Stohlmann in Gütersloh." *Meteorologische Rundschau* 3 (1950): 221-222.

Reproduces a letter of von Humboldt (1769-1859) setting forth a plan for a Prussian meteorological institute.

BH37 Lange, Fritz G. *Alexander von Humboldt: Eine Bibliographie der in der Deutschen Demokratischen Republik erschienenen Literatur*. Beiträge zur Alexander-von-Humboldt-Forschung 3. Berlin: Akademie-Verlag, 1974. 98 pp.

Contains a survey of the owners of Humboldt manuscripts in the German Democratic Republic.

BH38 Melderis, G.A. "Materialy ob A. Gumbol'dte v TsGIA Latviiskoi SSR" [Material on A. von Humboldt in the Central State Historical Archives of the Latvian SSR]. *Iz Istorii Estestvoznaniia i Tekhniki Pribaltiki* 2 (1970): 307-319.

BH39 Minquet, Charles. *Alexandre de Humboldt, historien et géographe de l'Amerique espagnole (1799-1804)*. Paris: Ed. La Decouverte, Fr. Maspero. 1969; Paris: Institute des Hautes Études de l'Amerique Latine, 1969. 694 pp.

Chapter VI includes brief sections in climatology, meteorology, terrestrial magnetism, and "Le courant Humboldt."

BH40 Pfeiffer, Heinrich, ed. *Alexander von Humboldt. Werk und Weltgeltung*. München: Piper, 1969. 505 pp.

Includes essays by W. Heisenberg, M. Bunge, C. Hentschel and others.

BH41 Schneider-Carius, Karl. "Alexander von Humboldt in seinen Beziehungen zur Meteorologie." *Alexander von Humboldt.*

Edited by Johannes Gellert. Berlin: VEB Deutscher Verlag der Wissenschaften, 1960, pp. 17-24.

Cites 14 of Humboldt's (1769-1859) meteorological publications.

BH42 Wilhelmy, Herbert, Gerhard Engelmann, and Gerhard Hard. *Alexander von Humboldt: Eigene und neue Wertungen der Reisen, Arbeit und Gedankenwelt*. Geographische Zeitschrift. Beihefte: Erdkundliches Wissen, 23. Wiesbaden: Steiner, 1970. 73 pp.

See also: L. Kellner. *Alexander von Humboldt*. London: Oxford University Press, 1963. 247 pp.

Humphreys, W.J.

BH43 Humphreys, William Jackson. *Of Me*. Washington, D.C.: By the Author, 1947. 375 pp.

Autobiography of Humphreys (1862-1949). Contains much technical and political information on developments during his 30-year service in the Weather Bureau, especially the scientific work on the Mount Weather Observatory and the congressional investigation of the alleged misuse of the installation. A list of the many books and scientific papers of the author is given, with his evaluation of their relative importance.

Huntington, E.

BH44 Martin, Geoffrey J. *Ellsworth Huntington: His Life and Thought*. Hamden, Conn.: Archon, 1973. xx + 315 pp., illus., bibl.

Huntington (1876-1947) was an American geographer who wrote on climatology.

Hutton, J. @

BH45 Hutton, James. *System of the Earth, 1785; Theory of the Earth, 1788; Observations on Granite, 1794. Together with Playfair's Biography of Hutton*. Introduction by Victor A. Eyles. Foreword by George W. White. Contributions to the History of Geology, 5. Darien, Conn.: Hafner, 1970. xxiii + 203 pp.

BH46 Dean, Dennis R. "James Hutton and His Public, 1785-1802." *Annals of Science* 30 (1973): 89-105.

J.A. DeLuc (1796) attacked Hutton's views on the formation of continents and the extent of geological time. Other critics gave careful attention to his view, contrary to the later opinion that it was ignored until the publication of John Playfair's *Illustrations* in 1802.

BH47 Galbraith, Winslow H. "James Hutton: An Analytic and Historical Study." Ph.D. dissertation, University of Pittsburgh, 1974. 301 pp.

For summary see *Dissertation Abstracts International* 36 (1975): 1054-A.

"Hutton's theology, 'metaphysics,' physics, and geology are examined. In each domain Hutton is an innovator or heretic.... In physics, Hutton rejected the Newtonian concepts of action at a distance, atoms, the void, and what he deems the mistaken generalization of gravity to all matter.... Hutton founded modern geology by conceiving a set of cycles which, utilizing the powers discussed in his physics, operated steadily through uncountable ages to produce a world which would be conducive to the happiness of man ..." (Author's summary).

BI. Subjects: I

Ibáñez e Ibáñez de Ibero, C.

BI1 Azcona, J.M. Lopez de. "Ibáñez e Ibáñez de Ibero, Carlos" (b. Barcelona, Spain, 14 April 1825; d. Nice, France, 28 January 1891). *Dictionary of Scientific Biography* 7 (1973): 1-2.

Ibañez was the first director of the Spanish corps of Geodesists; he improved the accuracy of geodesic measurements and accomplished the geodesic union of Europe with Africa.

Imamura, A.

BI2 Bullen, K.E. "Imamura, Akitune" (b. Kagoshima, Japan, 14 June 1870; d. Tokyo, Japan, 1 January 1948). *Dictionary of Scientific Biography* 7 (1973): 9-10.

Imamura contributed to the development of the seismograph; he "was among the first to show systematic connections between ground tilting and earthquake occurrence ... his mission in life was to mitigate disastrous earthquake effects."

BJ. Subjects: J

Jaggar, T.A., Jr.

BJ1 Bothner, Wallace A. "Jaggar, Thomas Augustus, Jr." (b. Philadelphia, Pennsylvania, 24 January 1871; d. Honolulu, Hawaii, 17 January 1953). *Dictionary of Scientific Biography* 7 (1973): 62-63.

Jaggar contributed to volcanology, primarily through qualitative observations and experiments.

Jeans, J.H. @

BJ2 Chandrasekhar, S. "James Hopwood Jeans 1877-1946." *Science* 105 (1947): 224-226.

Mostly on his contributions to cosmogony.

Jeffreys, H.

BJ3 Jeffreys, Harold. *Collected Papers of Sir Harold Jeffreys on Geophysics and Other Sciences*. New York: Gordon & Breach, 1971- (multi-volume work in progress).

An indispensable reference work which reprints most of Jeffreys' research papers with occasional annotations and comments on later work. The "General Introduction" in Volume 1 gives a personal account of his career.

BJ4 Jeffreys, Harold. "Developments in Geophysics." *Annual Review of Earth and Planetary Sciences* 1 (1973): 1-14.

Jesse, O.

BJ5 Schröder, W. "Otto Jesse und die Erforschung der leuchtenden Nachtwolken." *Gerlands Beiträge zur Geophysik* 81 (1972): 423-432.

Discusses the merits of Otto Jesse (1838-1901) for the exploration of noctilucent clouds on the basis of many original sources.

Johnson, N.

BJ6 Gold, E. "Sir Nelson Johnson." *Weather* 17 (1954): 754-755.

Obituary of Johnson (1892-1954), who made major contributions to boundary layer meteorology in connection with warfare and later became director of the British Meteorological Office.

Joly, J.

BJ7 Eyles, V.A. "Joly, John" (b. Holywood, King's County [now Offaly], Ireland, 1 November 1857; d. Dublin, Ireland, 8 December 1933). *Dictionary of Scientific Biography* 7 (1973): 160-161.

In 1899 Joly estimated the age of the earth as 80-90 million years, based on the rate of increase of sodium content in oceans. Later he contributed to radiometric geochronology by studying pleochroic halos.

BK. Subjects: K

Karazin, V.N.

BK1 Budyko, M.I. "Vydainschchiisya russki meteorolog V.N. Karazin" [The Famous Russian Meteorologist V.N. Karazin]. *Meteorologiya i Gidrologiya* 4 (1950): 3-9.

Karazin (1773-1842) proposed in 1810 to establish a network of stations in Russia, and in 1818 advanced the idea of a State Meteorological Committee to supervise observations.

Kármán, T. von

BK2 Dryden, Hugh L. "Theodore von Kármán, May 11, 1881-May 7, 1963." *National Academy of Sciences Biographical Memoirs* 38 (1965): 348-384.

The work of von Kármán had a major influence not only on aerodynamics but also on boundary layer meteorology.

BK3 Roy, Maurice. "Theodore von Kármán." *Comptes Rendues, Academie des Sciences, Paris* 225 (1963): 4545-4551.

An obituary of von Kármán (1881-1963), whose aerodynamic work had a profound influence on studies of the atmospheric boundary layer and atmospheric turbulence.

See also obituary: J. Ackeret, *Zeitschrift für Angewandte Mathematik und Physik* 14 (1963): 389.

Kassner, C.

BK4 Hoffmeister, J. "Nachruf auf Carl Kassner." *Zeitschrift für Meteorologie* 4 (1950): 7-8.

An obituary of the German climatologist Kassner (1864-1950), who contributed about 300 papers.

Kater, H.

BK5 Dorn, Harold. "Kater, Henry" (b. Bristol, England, 16 April 1777; d. London, England, 26 April 1835). *Dictionary of Scientific Biography* 7 (1973): 262-263.

Kater devised a pendulum for use in determining the acceleration due to gravity, thus improving the accuracy of geodetic surveys.

Kavraysky, V.V.

BK6 Kulikovsky, P.G. "Kavraysky, Vladimir Vladimirovich" (b. Zherebyatnikovo, Simbirsk province [now Ulyanovsk oblast], Russia, 22 April 1884; d. Leningrad, U.S.S.R., 26 February 1954). *Dictionary of Scientific Biography* 7 (1973): 265-266.

Kavraysky invented instruments for geodetic and astronomical measurements, and developed their mathematical theory.

Keil, K.

BK6a Flohn, H. "Karl Keil: 23 Jahre Herausgeber der Meteorologischen Rundschau." *Meteorologische Rundschau* 24 (1971): 33.

A tribute to the founder and long-time editor of the principal German-language meteorological journal of the post-World War II era, with comments on defunct antecedents in Germany and the role of this journal in the vastly expanded post-war literature.

Kendrew, W.G.

BK7 Beckinsale, R.P. "Wilfrid George Kendrew." *Nature* 194 (1962): 817.

Kendrew (1885-1962) was geographer at Oxford University and published several books on climatology. His *The Climates of the Continents* went through many editions.

Kepler, J.

BK8 Pejml, Karel. "Johann Kepler (1571-1630) a ceška meteorologie v jeho dobe" [Johann Kepler and the Meteorology of the Period in Bohemia]. *Meteorologicke Zpravy* (Prague) 24 (1971): 133-134.

Meteorology was Kepler's hobby. He tried to make forecasts by astrological methods and compared his results with actual occurrences. He also contributed a work on the origin and structure of solid hydrometeors.

Khanevskii, V.A.

BK9 Pinus, N.A., and I.V. Khanevskaya. "Forming and Development of the Theoretical Investigations of the Atmosphere in the Studies of V.A. Khanevskii (on the Hundredth Anniversary of His Birth)." [Translated from the Russian title.] *Meteorologiya i Gidrologiya* 11 (1980): 118-121.

A eulogy for Vladimir Andreyevich Khanevskii (1880-1943) who made important contributions to aerology including a theory of multi-layer tropopauses and who played a major role for meteorological education in the U.S.S.R.--as author of textbooks and professor at Moscow University. His work was largely overlooked in Western countries.

King, C.R.

BK10 Aldrich, Michele L. "King, Clarence Rivers" (b. Newport, Rhode Island, 6 January 1842; d. Phoenix, Arizona, 24 December 1901). *Dictionary of Scientific Biography* 7 (1973): 370-371.

King estimated the age of the earth's crust at 24 million years based on Kelvin's methods, and developed a catastrophist theory of geological change to explain the American West. He extended this theory to explain volcanic lava in terms of subcrusted meeting induced by lowered pressure.

Kirwan, R.

BK11 Scott, E.L. "Kirwan, Richard" (b. Cloughballymore, County Galway, Ireland, 1733 [?]; d. Dublin, Ireland, 1 June 1812). *Dictionary of Scientific Biography* 7 (1973): 387-390.

Kirwan is known as a critic of James Hutton's geomorphological theory. His pioneering work in meterology, especially in forecasting Irish weather, has only recently claimed attention.

Klein, H.J.

BK12 Dieke, Sally H. "Klein, Hermann Joseph" (b. Cologne, Germany, 14 September 1844; d. Cologne, 1 July 1914). *Dictionary of Scientific Biography* 7 (1973): 400.

Klein was "one of the foremost popularizers of meteorology and astronomy of his day"; he studied cirrus clouds and lunar craters.

Knoch, K.

BK13 Amelung, Walther. "Professor Dr. phil. Karl Knoch." *Zeitschrift für angewandte Bader und Klimaheilkunde* 19 (1972).

An obituary of Karl Knoch (1883-1972) stressing in particular his contributions to bioclimatology and his knowledge of the climates of spas, also the large number of papers on general climatology and his administrative efforts to reestablish the German climatological service after World War II.

BK14 Flohn, H. "Karl Knoch (1883-1972) zum Gedenken." *Meteorologische Rundschau* 25 (1972): 131-132.

An obituary appraising the many contributions of Knoch to the science of climatology, with especially noteworthy contributions to bio-, agro-, and micro-climatology; also his appraisal of climatic classifications.

Knott, C.G.

BK15 Bullen, K.E. "Knott, Cargill Gilston" (b. Penicuik, Scotland, 30 June 1856; d. Edinburgh, Scotland, 26 October 1922). *Dictionary of Scientific Biography* 7 (1973): 413.

Knott, one of the group of British and Japanese physicists who "inaugurated the modern era in earthquake study," contributed to mathematical seismology. By tracing the connection between travel times and distance he provided the basis for later researches on the interior of the earth. Knott also organized the first magnetic survey of Japan.

Knudsen, M.H.C.

BK16 Pihl, Mogens. "Knudsen, Martin Hans Christian" (b. Hansmark, Denmark, 15 February 1871; d. Copenhagen,

Denmark, 27 May 1949). *Dictionary of Scientific Biography* 7 (1973): 416-417.

Knudsen developed methods to define the properties of seawater and edited the *Bulletin Hydrographique*.

Köhler, H.

BK17 Schröder, W. "Hilding Kohler zum 80. Geburtstag." *Gerlands Beiträge zur Geophysik* 77 (1968): 425-426.

A tribute, on his 80th birthday, to the Professor Emeritus and former director of the Meteorological Institute at the University of Uppsala. He was one of the important early contributors to cloud physics. His 1919 paper on hoarfrost deposits from fog at the mountain station Pårtetjåkko and a series of monographs on cloud elements, condensation nuclei, coalescence, and cloud formation (1925-1936) were seminal for the understanding of the colloidal systems called clouds.

Koeppen, W.P.

BK18 Loewe, F. "Prof. Wladimir Koeppen." *Quarterly Journal of the Royal Meteorological Society* 67 (1941): 389-391.

Obituary of the climatologist Wladimir Peter Koeppen (1846-1940).

BK19 Wegener-Köppen, Else. *Wladimir Köppen: Ein Gelehrtenleben*. Stuttgart: Wissenschaftliche Verlagsgesellschaft, 1955. 195 pp.

This biography of Köppen (1846-1940) was based on his own notes and on material made available by Professor Erich Kuhlbrodt of the Deutsche Seewarte, where Köppen spent most of his professional life. The author is Köppen's daughter, who married Alfred Wegener (1880-1930) with whom Koppen collaborated on paleoclimate.

Koschmieder, H.

BK20 Wippermann, F. "Harald Koschmieder." *Meteorologische Rundschau* 20 (1967): 1-2.

Obituary of Koschmieder (1897-1966), who made major contributions to the theory of visibility, gustiness, air pollution studies, structure of land and sea breeze.

See also: Leonhard Foitzik, "Harald Koschmieder, Sept. 19, 1897-Aug. 10, 1966," *Beiträge zur Physik der Atmosphäre* 39 (1966): 63-66.

Koto, B.

BK21 Kobayashi, Hideo. "Koto, Bunjiro" (b. Tsuwano, Iwami [now Shimane prefecture], Japan, 4 March 1856; d. Tokyo, Japan, 8 March 1935). *Dictionary of Scientific Biography* 7 (1973): 473-474.

Koto studied earthquakes, volcanoes, and tectonic phenomena and pointed out the role of fault lines in all of them.

Krasovsky, T.N.

BK22 Sheynin, O.B. "Krasovsky, Theodosy Nicolaevich" (b. Galich, Kostroma guberniya, Russia, 26 September 1878; d. Moscow, U.S.S.R., 1 October 1948). *Dictionary of Scientific Biography* 7 (1973): 496-497.

Krasovsky contributed to the mathematical theory of geodesy; he emphasized gravimetry rather than isostatic theory.

Kratzer, A.

BK23 Abt, Edelbert, and Konvent von Ettal. *P. Dr. Albert Kratzer OSB.* Flyer issued by the Monastery of Ettal, dated 14 April 1975.

An obituary of the Benedictine monk Albert Kratzer (1905-1975), who contributed to meteorology by his book on urban climate (*Das Stadtklima*, Braunschweig: Friedr. Vieweg & Sohn, 1st ed. 1937, 2nd ed. 1956).

Krayenhoff, C.R.T.

BK24 Nieuwenkamp, W. "Krayenhoff, Cornelis Rudolphus Theodorus" (b. Nijmegen, Netherlands, 2 June 1758; d. Nijmegen, 24 November 1840). *Dictionary of Scientific Biography* 7 (1973): 499-501.

Krayenhoff was a leader in Dutch geodetic surveys but his reputation was demolished because of an unfair criticism by Gauss.

Krick, I.

BK25 Boesen, Victor. *Storm: Irving Krick vs. the U.S. Weather Bureaucracy*. New York: G.P. Putnam's Sons, 1978. 156 pp.

A biography of Krick, the highly controversial personality, who has a Ph.D. from the California Institute of Technology. His claims in the field of long-range forecasting and in weather modification have never been subjected to the usual scientific scrutiny because he has not published his methods.

Kropotkin, P.A.

BK26 Naumov, G.V. "Kropotkin, Petr Alekseevich" (b. Moscow, Russia, 9 December 1842; d. Dmitrov, Moscow oblast, U.S.S.R., 8 February 1921). *Dictionary of Scientific Biography* 7 (1973): 510-512.

"Kropotkin's major scientific work was the corroboration of the theory of ancient continental glaciation."

Kulik, M.S.

BK27 "Obituary of Maksim Savvich Kulik (1907-1980)." Translated from the Russian." *Meteorologiya i Gidrologiya* 3 (March 1981): 128.

The research and teaching activities of this leading Ukrainian agrometeorologist are eulogized.

Kupfer, A.I.

BK28 Nevskaia, N.I. "Akademik A.Ia. Kupfer i ego Trudy po Geofizike." *Iz Istorii Estestvoznaniia i Tekhniki Pribaltiki* 3 (1971): 47-57.

On Adolf Izklovlevich Kupfer (1799-1865) and his researches in geophysics.

BL. Subjects: L

Lacaille, N. de

BL1 Gingerich, Owen. "Lacaille, Nicolas-Louis de" (b. Rumigny, near Rheims, France, 15 March 1713; d. Paris, France, 21 March 1762). *Dictionary of Scientific Biography* 7 (1973): 542-545.

Lacaille was a leader in geodetic measurements ca. 1740 before he turned his attention to astronomy. His results supported the Newtonian theory of the shape of the earth.

Lacroix, A.

BL2 Hooker, Marjorie. "Lacroix, Alfred" (b. Macon, France, 4 February 1863; d. Paris, France, 16 March 1948). *Dictionary of Scientific Biography* 7 (1973): 548-549.

Lacroix specialized in volcanology and volcanic rocks, and was one of the first to recognize the *nuée ardente* type of eruption.

Lamarck, J. de

BL3 Burlingame, Leslie J. "Lamarck, Jean Baptiste Pierre Antoine de Monet de" (b. Bazentin-le-Petit, Picardy, France, 1 August 1744; d. Paris, France, 28 December 1829). *Dictionary of Scientific Biography* 7 (1973): 584-594.

"Lamarck's meteorology was devoted to a search for those laws of nature which regulate climatic change ... he was concerned with a natural classification of meteorological phenomena." His project for daily reporting of weather observations to a central data bank, begun in 1880, was terminated by Napoleon in 1810. In his writings on geology he stressed the role of water acting gradually over millions of years.

Lamb, H.

BL4 Bullen, K.E. "Lamb, Horace" (b. Stockport, England, 29 November 1849; d. Manchester, England, 4 December 1934). *Dictionary of Scientific Biography* 7 (1973): 594-595.

Lamb applied mathematics to the propagation of seismic waves, the modes of oscillation of an elastic sphere (relevant to free vibration of the earth produced by major earthquakes), tides, and terrestrial magnetism.

Lamplugh, G.W.

BL5 Penny, L.F. "George William Lamplugh (1859-1926)." *Journal of Glaciology* 6 (1966): 307-309.

Langmuir, I. @

BL6 Rosenfeld, Albert. *The Quintessence of Irving Langmuir.* Oxford: Pergamon Press, 1965. 369 pp.

This review of Langmuir's work, with portrait and a complete list of his publications, includes his work on cloud physics.

BL7 Schaefer, Vincent J. "Irving Langmuir, Versatile Scientist." *Bulletin of the American Meteorological Society* 38 (1957): 483-484.

Obituary of Langmuir (1881-1957) with emphasis on his contributions to cloud physics.

A similar note by the same author appeared under the title "Irving Langmuir, Man of Many Interests" in *Science* 127 (1958): 1227-1229.

Langsdorf, G.I.

BL8 Komissarov, B.N. *Grigorii Ivanovich Langsdorf, 1774-1852.* Leningrad: Nauka, 1975. 123 pp., illus., bibl.

Laplace, P.

BL9 Gillispie, Charles Coulston, Robert Fox, and Ivor Grattan-Guiness. "Laplace, Pierre-Simon, Marquis de" (b. Beaumont-en-Auge, Normandy, France, 23 March 1749; d. Paris, France, 5 March 1827). *Dictionary of Scientific Biography* 15 (1978): 273-403.

This magnificent article serves as a grand finale to the *Dictionary of Scientific Biography*, edited with infinite pains and perseverance by Gillispie. Like the *DSB* itself, the article is a cooperative effort but is guided by the editor's firm hand. It gives due notice to Laplace's mathematical contributions to geophysics: the shape of the earth and Newtonian gravitation potential theory, the theory of tides, and the secular acceleration of the moon. His theory of the origin of the solar system is characterized not as a "nebular hypothesis" introducing an evolutionary picture of the development of the world, but as an "atmospheric hypothesis" employing the concept of "probability of cause."

Larsen, E.S., Jr.

BL10 Auser, Cortland P. "Larsen, Esper Signius, Jr." (b. Astoria, Oregon, 14 March 1879; d. Washington, D.C., 8 March 1961). *Dictionary of Scientific Biography* 8 (1973): 41-43.

Larsen "developed a method of determining the age of igneous rocks using the lead in accessory minerals" (the lead having been formed by alpha decay of uranium).

Lawson, A.C.

BL11 Vaughan, Francis E. *Andrew C. Lawson, Scientist, Teacher, Philosopher*. Glendale, Calif.: The Arthur H. Clark Co., 1970. 474 pp., ports.

Andrew Cowper Lawson (1861-1952) directed the official investigation of the 1900 California earthquake and promoted research in seismology. His geological research included work on isostasy.

BL12 Vaughan, Francis E. "Lawson, Andrew Cowper" (b. Anstruther, Scotland, 25 July 1861; d. San Leandro, California, 16 June 1952). *Dictionary of Scientific Biography* 8 (1973): 98-100.

"Lawson's studies of isostasy over the three decades 1920-1950 have drawn special attention to its importance in diastrophism ... he developed the logical consequences of isostatic adjustment as an important factor in orogenesis." He supervised the official report on the California earthquake of 18 April 1906, and "his continued activity in this field brought about the organization of the Seismological Society of America" and the establishment of a network of seismographic stations in California.

Lazarev, P.P.

BL13 Dorfman, J.G. "Lazarev, Petr Petrovich" (b. Moscow, Russia, 14 April 1878; d. Alma-Ata, U.S.S.R., 24 April 1942). *Dictionary of Scientific Biography* 8 (1973): 101-103.

Lazarev compiled geomagnetic maps and established geophysical prospecting surveys in the U.S.S.R.

Lehmann, J.G.

BL14 Freyberg, Bruno von. "Lehmann, Johann Gottlob" (b. Langenhennersdorf, near Pirna, Germany, 4 August 1719; d. St. Petersburg, Russia [now Leningrad, U.S.S.R.], 22 January 1767). *Dictionary of Scientific Biography* 8 (1973): 146-148.

Lehmann proposed a theory of the propagation of earthquakes as being dependent on the "inner structure of the surface of the earth."

Leibniz, G.W. v.

BL15 Keil, Karl. "Wilhelm von Leibniz und die Meteorologie." *Meteorologische Rundschau* 1 (1948): 321-322.

A note, based on a number of older sources, indicates that von Leibniz (1646-1716) made the first instrumental observations of temperature and pressure in Germany in 1677. He recommended to Peter the Great in 1700 the establishment of a meteorological network in Russia.

BL16 Schmöger, Friedrich. *Leibniz in seiner Stellung zur tellurischen Physik. Beitrag zur Würdigung von Leibniz in geophysikalischer Hinsicht.* Münchener Geographische Studien, elftes Stück. München: Theodor Ackermann, 1901. vi + 83 pp.

Contents: I. Leibniz' Aetherwellentheorie als Ursache der Erdbewegung. II. Das Wesen der Schwerkraft. III. Licht. IV. Der Magnetismus kosmischen Ursprungs. V. Atmosphäre--Barometrische Apparate. VI. Gründe für die Notwendigkeit der Kugelstalt der Erde. VIII. Ursprung der Gewässer. IX. Schichtung der Erdrinke. Schichten störungen. X. Vulkanismus. Erdinneres. XI. Gebirge, Höhlen, Thäler. XII. Hydrologie.

Leist, E.

BL17 Busygin, I.A. *Ernest Egorovich Leist.* Moscow: Nauka, 1969. 71 pp.

Biography of Leist (1852-1918), meteorologist and geomagnetician.

Lenz, E.K.

BL18 Lezhneva, Olga A. "Lenz, Emil Khristianovich (Heinrich Friedrich Emil)" (b. Dorpat, Russia [now Tartu,

Estonian S.S.R.], 24 February 1804; d. Rome, Italy, 10 February 1865). *Dictionary of Scientific Biography* 8 (1973): 187-189.

Lenz determined the specific gravity and temperature of seawater to the depth of about two kilometers; he also discovered and explained variations of salinity in the oceans. "He noted that at definite latitudes, water at the ocean's surface is warmer than the air above it; and he found two maximums and two minimums of barometric pressure in the tropics."

Le Roy, C.

BL19 Gough, J.B. "Le Roy, Charles" (b. Paris, France, 12 January 1726; d. Paris, 10 December 1779). *Dictionary of Scientific Biography* 8 (1973): 225-256.

Le Roy proposed a theory of evaporation; the use of the word "precipitation" in meteorology is a remnant of his once-popular theory.

Le Verrier, U.J.J.

BL20 Levy, Jacques R. "Le Verrier, Urbain Jean Joseph" (b. Saint-Lo, France, 11 March 1811; d. Paris, France, 23 September 1877). *Dictionary of Scientific Biography* 8 (1973): 276-279.

Le Verrier, best known as the discoverer of Neptune, was also "one of the founders of modern meteorology." He organized networks to report and warn of storms, later incorporated into the Bureau Central Météorologique de France.

Levinson-Lessing, F.

BL21 Meniailov, A.A. "Levinson-Lessing, Franz Yulevich" (b. St. Petersburg, Russia, 9 March 1861; d. Leningrad, U.S.S.R., 25 October 1939). *Dictionary of Scientific Biography* 8 (1973): 285-287.

He "advanced the doctrine of two ancestral magmas, granite and basalt, that had played a primary role in the creation of the rock of the earth's core ... [he] concluded that different parts of the earth's crust rise and fall simultaneously. Mountains were formed by these movements, and plastic or liquid magma was transferred at depth, serving as the source of magmatic and volcanic phenomena."

Leybenzon, L.S.

BL22 Grigorian, A.T. "Leybenzon, Leonid Samuilovich" (b. Kharkov, Russia, 26 June 1879; d. Moscow, U.S.S.R., 15 March 1951). *Dictionary of Scientific Biography* 8 (1973): 300-301.

"A large part of his work was in geophysics, particularly the applications of elasticity theory to the study of the structure of the earth."

Linke, F.

BL23 Amelung, W. "Nachruf auf F. Linke." *Balneologie* 11 (1944): 46.

Obituary of the German radiation researcher who introduced the atmospheric turbidity factor.

Lisboa, J.D.

BL24 Albuquerque, Luis de. "Lisboa, Joao de" (b. Portugal; d. Indian Ocean, before 1526). *Dictionary of Scientific Biography* 8 (1973): 390.

Lisboa published in 1514 "the first ... presentation of a method for measuring the magnetic declination ... he assumes that the declination of the compass undergoes variations proportioned to the longitude."

Loidis, A.P.

BL25 Loidis, Lidiia A. *Aleksandr Platonovich Loidis*. Leningrad: Gidrometeoizdat, 1975. 72 pp., illus.

Biography of the Russian meteorologist A.P. Loidis (1873-1947).

Lokhtin, V.M.

BL26 Fedoseyev, I.A. "Lokhtin, Vladimir Mikhaylovich" (b. St. Petersburg, Russia, 1849; d. Petrograd, Russia, 1919). *Dictionary of Scientific Biography* 8 (1973): 465-467.

"Lokhtin was one of the founders of the hydrology of rivers."

Lomonosov, M.V. @

BL27 "M.V. Lomonosov i meteorologiya." *Meteorologiya i Gidrologiya* No. 11 (1961): 3-6.

On occasion of the 250th birthday of Lomonosov (1711-1765), his contributions to knowledge of air currents, temperature distribution, and atmospheric electricity are praised.

Loomis, E. @

BL28 Miller, Eric R. "The Pioneer Meteorological Work of Elias Loomis at Western Reserve College, Hudson, Ohio, 1837-1844." *Monthly Weather Review* 59 (1931): 194-195.

Discusses Loomis' publication in 1841 and 1842 of synoptic weather maps on the basis of 24 barometer readings in North America. Although the method was first applied by H.W. Brandes (1777-1834) in 1820, these are the first such maps found in the literature. His papers also contain clear allusions to the existence of a cold front, with a diagram closely resembling modern frontal cross sections.

BL29 Newton, H.A. "A Memorial of Professor Elias Loomis." *American Meteorological Journal* 7 (1890): 97-117.

Obituary of Loomis (1811-1889) who was a very important contributor to the development of scientific meteorology in the United States in the 19th century.

Love, A.E.H.

BL30 Bullen, K.E. "Love, Augustus Edward Hough" (b. Weston-super-Mare, England, 17 April 1863; d. Oxford, England, 5 June 1940). *Dictionary of Scientific Biography* 8 (1973): 516-518.

Love developed elasticity theory and applied mathematical analysis to isostasy, tides of the solid earth, variation of latitude, effects of compressibility, gravitational instability, and the vibrations of a compressible planet. "Many of his contributions are basic in current geophysical research, especially Love waves and Love's numbers, the latter being key numbers in tidal theory." His analysis of surface seismic waves also led to a method for estimating crustal thicknesses in various geographical regions of the earth.

Lugeon, M.

BL31 Wegmann, Eugene. "Lugeon, Maurice" (b. Poissy, France, 10 July 1870; d. Lausanne, Switzerland, 23 October 1953). *Dictionary of Scientific Biography* 8 (1973): 543-545.

Lugeon developed theories of geotectonics, and "was responsible for both the concepts and terminology of such tectonic staples as autothy, allochthony, window, and involution."

Luneland, H.W.

BL32 Keränen, J. "Harald Wilhelm Luneland." *Geophysica (Meteorology Series) Helsinki* 4 (1950): 1-3.

Obituary of Lunelund (1882-1950), Professor of Physics at the University of Helsinki, who made noteworthy contributions to the study of solar radiation.

Lyell, C. @

BL33 Cannon, W. Faye. "Charles Lyell, Radical Actualism, and Theory." *British Journal for the History of Science* 9 (1976): 104-120.

Argues that one should discuss the scientific aspects of Lyell's non-progressionist principle apart from its alleged origin in his religious or philosophical views. Lyell's goal was to "erect a comprehensive theoretical system of hypotheses and deductions for all of geology." He was guided by "absolute actualism," a criterion of allowable forces ("causes") in geology, to be distinguished from "radical actualism" as practiced for example by William Buckland, who would make non-actualist concessions. Lyell made an exception for the creation of man, but that did not diminish the absoluteness of his actualism within the domain of geology. By contemporary standards of scientific method, i.e., those of John Herschel, Lyell's approach was generally satisfactory though perhaps too extreme.

BL34 Laudan, Rachel. "The Role of Methodology in Lyell's Science." *Studies in History and Philosophy of Science* 13 (1982): 215-249.

Lyell "attempted to base his geology on the *vera causa* tradition, which he believed was the best way to avoid both the unfounded conjectures involved in the method of hypothesis, and the theoretical crippling incurred by too strict an adherence to an inductive methodology.... To think of Lyell's main aim as being that of establishing a 'principle of uniformity' or 'uniformitarianism' in geology is to introduce categories that were only applied retrospectively to his work by Whewell."

BL35 Murphy, Anthony C. "Lyell's *Principles of Geology*: Explanation and Metaphor in Early 19th-Century Geology." Ph.D. dissertation, University of Notre Dame, 1976. 392 pp.

For summary see *Dissertation Abstracts International* 37 (1976): 1612-A.

"Charles Lyell ... attempted to extend a decidedly mechanical methodology to the earth sciences.... he describes all geological phenomena as periodic, i.e., characterized by minor fluctuations around a constant mean.... His efforts to extend the mechanical tradition led ultimately to an imaginative grasp of the immensity of geological time. Lyell's opponents, as represented by Whewell and Sedgwick, pictured all geological phenomena in terms of a developmental and teleological trend in earth history ..." (Author's summary).

BL36 Porter, Roy. "Charles Lyell and the Principles of the History of Geology." *British Journal for the History of Science* 9 (1976): 91-103.

BL37 Porter, Roy. "Charles Lyell: The Public and Private Faces of Science." *Janus* 69 (1982): 29-50.

BL38 Ruse, Michael. "Charles Lyell and the Philosophers of Science." *British Journal for the History of Science* 9 (1976): 121-131.

"Two of the most influential evaluations of Charles Lyell's geological ideas were those of the philosophers of science, John F.W. Herschel and William Whewell. In this paper I shall argue that the great difference between these evaluations--whereas Herschel was fundamentally sympathetic to Lyell's geologizing, Whewell was fundamentally opposed--is a function of the fact that Herschel was an empiricist and Whewell a rationalist" (Author).

BL39 Wilson, Leonard G. *Charles Lyell: The Years to 1841: The Revolution in Geology*. New Haven: Yale University Press, 1972. xiii + 553 pp., plates, illus., maps, bibl., index.

First volume of the definitive biography of the great English geologist.

BM. Subjects: M

Macelwane, J.B.

BM1 Byerly, Perry, and William V. Stauder. "James B. Macelwane, S.J. September 28, 1883-February 15, 1956." *Biographical Memoirs of the National Academy of Sciences* 31 (1958): 254-281.

"More than any other [he] contributed to establishing [seismology] on the firm, well-established academic footing it enjoys in this country today."

Mahlmann, C.H.W.

BM2 Kassner, C. "Carl Heinrich Wilhelm Mahlmann." *Meteorologische Zeitschrift* 29 (1912): 309-318.

Biographical sketch of Mahlmann (1812-1848), a brilliant protégé of Alexander von Humboldt, who was appointed as first director of the Prussian Meteorological Institute and who published the first world chart of isotherms in 1844.

Maillet, B. de

BM3 Carozzi, Albert. "Maillet, Benoit de" (b. St. Mihiel, France, 12 April 1656; d. Marseilles, France, 30 January 1738). *Dictionary of Scientific Biography* 9 (1974): 26-27.

Maillet proposed a Cartesian cosmology in which heavenly bodies go through alternate luminous (stellar) and dark (planetary) phases. "At present the earth was in a dark phase, and the level of the universal ocean was being lowered by evaporation into outer space, at the rate of three inches per century, until total depletion would occur. This general regression had taken place for at least the past two billion years.... Maillet used all the aspects of his oceanographical knowledge to explain, by the action of the sea, all the ... features of the geological record.... His ultraneptunian theory of the earth, in which everything was explained by the action of a retreating sea, makes him a marine geologist of the eighteenth century."

Makarov, S.O.

BM4 Fedoseyev, I.A. "Makarov, Stepan Osipovich" (b. Nikolayev, Russia, 8 March 1849; d. aboard the battleship *Petro-*

pavlosk, Port Arthur, Russia, 13 April 1904). *Dictionary of Scientific Biography* 9 (1974): 42-43.

"Makarov began his oceanographic studies in 1881-1882 on currents in the Bosporus. [He] proved the existence of a deep current running counter to the surface current.... Makarov organized systematic observations of the water density and temperature at various depths, and of the velocity of the current through the strait." In a later work, a report on a voyage around the world, he "compiled the first water temperature tables for the North Pacific Ocean. He also considered the origin of the deep waters of the North Pacific the reason for the homogeneous temperature and density of the water at every depth of the English channel ... and the general pattern of ocean currents, with an indication of the primary significance of the action of the coriolis force on sea currents."

Mallet, R.

BM5 Cox, Ronald C., ed. *Robert Mallet 1810-1881. Centenary Seminar Papers*. Dublin: The Institution of Engineers of Ireland, 1982. 146 pp.

Includes: R.C. Cox, "Robert Mallet: Engineer and Scientist," pp. 1-33; G.L. Herries Davies, "Robert Mallet: Earth-Scientist," pp. 35-52. Mallet is best known as a founder of seismology (he introduced this and related terms); he also wrote on glacial flowage, geological dynamics, and vulcanology.

BM6 Fischer, Walther. "Mallet, Robert" (b. Dublin, Ireland, 3 June 1810; d. London, England, 5 November 1881). *Dictionary of Scientific Biography* 9 (1974): 60-61.

In 1846, Mallet differentiated four kinds of earthquakes; he considered local elevations of portions of the earth's solid crust to be the cause of earthquakes. He set off explosions in different locations to determine the rate of travel of seismic waves in various substances. He introduced the terms "seismology," "angle of emergence," etc., and prepared an extensive catalog of earthquakes. He speculated on the origin of volcanic energy and eventually attributed it to crushing of the earth's crust caused by secular cooling and contraction.

BM7 Smith, Peter J. "Robert Mallet." *Open Earth* 14 (1981): 12-13.

Manley, G.

BM8 Lamb, Hubert H. "The Life and Work of Professor Gordon Manley (1902-1980)." *Weather* 36 (1981): 220-231.

A biographical sketch with portrait of the British geographer and climatologist, with an account of his career at British universities and an appraisal of his contributions, 35 of which are cited. Most notable was his reconstruction of monthly temperature values for Central England, starting in 1659--the largest instrumental observation series in the world. The essay contains anecdotal notes by Michael Tooley, Jean M. Grove, and Frank Oldfield of associations with Manley. A complete list of his publications was published by M.J. Tooley and J.M. Kenworthy in *Transactions of the Institute of British Geographers*, N.S. 5(4): 514-517.

* Margerie, Emmanuel de. *Critique et géologie: Contribution à l'histoire des sciences de la terre (1882-1942).* Cited below as item H32.

Margules, M. @

BM9 Exner, Felix M. "Max Margules." *Meteorologische Zeitschrift* 37 (1920): 322-324.

An obituary note for the Austrian atmospheric dynamicist Margules (1856-1920).

BM10 Reuter, Heinz. "Max Margules (1856-1920)." *Wetter und Leben* 22 (1970): 221-228.

An appraisal of the fundamental contributions of this Austrian theoretical meteorologist. His investigations on barometric fluctuations and continuity of flow led to the development of the tendency equation which became one of the pillars of numerical weather prediction. His analysis of conversions of potential to kinetic energy also became basic for further progress in atmospheric energetics. A list of his most important papers is given. Some biographical information is included about his tragic end in poverty and starvation.

Mariotte, E.

BM11 Mahoney, Michael S. "Mariotte, Edme" (d. Paris, France, 12 May 1684). *Dictionary of Scientific Biography* 9 (1974): 114-122.

Mariotte discussed the relation between barometric pressure and winds and weather, and gave a complete account of the rainbow. He argued that natural springs and rivers derive their water exclusively from rainfall (an extension of the theory of P. Perrault). He collected detailed information on winds from around the world.

Markham, C.R.

BM12 Glick, Thomas F. "Markham, Clements Robert" (b. Stillingfleet, Yorkshire, England, 20 July 1830; d. London, England, 30 January 1916). *Dictionary of Scientific Biography* 9 (1974): 123-124.

He organized (as president of the Royal Geographical Society) the geophysical exploration of Antarctica; R.F. Scott's expedition (1901-4) "was the crowning achievement of Markham's exploration program."

Maskelyne, N.

BM13 Forbes, E.G. "Maskelyne, Nevil" (b. London, England, 6 October 1732; d. Greenwich, England, 9 February 1811). *Dictionary of Scientific Biography* 9 (1974): 162-164.

"In a famous experiment of 1774 Maskelyne attempted to determine the earth's density from measurements of the deviation of a plumb line produced by the gravitational attraction of Mt. Shiehallion, in Scotland ... [he] concluded the mean density of the earth to be between 4.867 and 4.559 times that of water, a result that compares quite well with the presently accepted value of 5.52."

Mason, C.

BM14 Mason, A. Hughlett. "Mason, Charles" (b. Wherr, Gloucestershire, England, baptized 1 May 1728; d. Philadelphia, Pennsylvania, 25 October 1786). *Dictionary of Scientific Biography* 9 (1974): 164-165.

Mason's name is best known to Americans as one of the surveyors of the Mason-Dixon line in the 1760s. Later he selected the mountain in Scotland used by N. Maskelyne to estimate the earth's density.

Matuyama, M.

BM15 Kumagai, Naoti. "Matuyama (Matsuama), Motonori" (b. Uyeda [now Usa], Japan, 25 October 1884; d. Yamaguchi, Japan, 27 January 1958). *Dictionary of Scientific Biography* 9 (1974): 180-182.

Matuyama "was the first to suggest that the determination of microfeatures of the gravity field of the earth could reveal geological substructure." In 1929 he proposed that the earth's magnetic field had reversed itself in the Miocene and Quaternary Epochs.

Maupertuis, P.L.M. de

BM16 Glass, Bentley. "Maupertuis, Pierre Louis Moreau de" (b. St. Malo, France, 28 September 1698; d. Basel, Switzerland, 27 July 1759). *Dictionary of Scientific Biography* 9 (1974): 186-189.

Maupertuis was one of the leaders of the expedition to Lapland to measure the length of a degree to confirm Newton's theory that the earth is flattened at the poles.

Maurer, J.M.

BM17 Schmeidler, F. "Maurer, Julius Maximillian" (b. Freiburg im Breisgau, Germany, 14 July 1857; d. Zurich, Switzerland, 21 January 1938). *Dictionary of Scientific Biography* 9 (1974): 189.

Maurer wrote on climatology and the flow of radiation through the atmosphere, and constructed meteorological instruments.

Maury, M.F. @

BM18 Brown, Ralph Minthorne. "Bibliography of Commander Matthew Fontaine Maury including A Biographical Sketch." *Bulletin of the Virginia Polytechnic Institute* 24, no. 2 (1930). 61 pp.

The second edition (next item) contains additional references.

BM19 Brown, Ralph Minthorne. "Bibliography of Commander Matthew Fontaine Maury including A Biographical Sketch." (2nd edition). *Bulletin of the Virginia Polytechnic Institute* 37, no. 12 (1944). 46 pp.

Lists many books, pamphlets, and newspaper and periodical articles about Maury's life and work, as well as Maury's own publications and unpublished addresses, maps, and charts.

BM20 Canfield, N.L. "Weather and the Navigation of the Sea." *Weekly Weather and Crop Bulletin* 43, no. 2 (1956): 6-8.

A brief history of the achievements of Matthew Fontaine Maury (1806-1873) in marine meteorology as a tribute on the sesquicentennial of his birth with a portrait and a reproduction of one of his wind and current charts. His efforts for international meteorological cooperation and the basis he laid for hydrographic (later oceanographic) studies of the Navy and the marine weather services of the Weather Bureau are set forth. No documentation or references are given.

BM21 Charlier, Patricia S., and Roger H. Charlier. "Matthew Fontaine Maury, Pioneer of International Cooperation in Marine Sciences." *Actes, XII*^{e} *Congres International d'Histoire des Sciences, Paris 1968* (pub. 1971) 7: 67-71.

BM22 Corbin, Diana Fontaine Maury. *A Life of Matthew Fontaine Maury*. London: Low, Marston, Searle & Rivington, 1888. vi + 326 pp.

Biography by Maury's daughter.

BM23 Cotter, Charles H. "Matthew Fontaine Maury (1806-1873): 'Pathfinder of the Seas.'" *Journal of Navigation* 32 (1979): 75-83.

Biographical sketch based on the work by R.M. Brown (1944) and the book by Maury's daughter, D.F.M. Corbin (1888).

BM24 DuVal, Miles P., Jr. "Matthew Fontaine Maury: Benefactor of Mankind." *Explorer's Journal* 43 (1965): 203-217.

A biography of Maury, renowned oceanographer and ocean climatologist, who organized the first international conference on maritime meteorology (1853).

BM25 Hawthorne, Hildegarde. *Matthew Fontaine Maury--Trailmaker of the Seas*. New York: Longmans, Green & Co., 1943. 266 pp.

A valuable biography of Maury. Clearly established his merit for organizing the first international meteorological conference in 1853 in Brussels, his work as the first head of the Naval Observatory, and his tragic eclipse because he had, as a Virginian, espoused the Confederate cause during the Civil War.

BM26 Landsberg, H.E. "Matthew Fontaine Maury, Advocate of Weather and Crop Reporting." *Weekly Weather and Crop Bulletin* 52, no. 1 (1965): 7-8.

Maury, better known for his work on oceanic meteorology (*The Physical Geography of the Sea*, London: Sampson Low, 1855), after the Civil War became an advocate of an agricultural weather service and made a number of speeches about it, which were favorably received by farmers; but a feud with Joseph Henry of the Smithsonian Insitution and Alexander Bache of the Coast and Geodetic Survey as well as political forces squelched the idea at the time.

* Lewis, C.L. "Maury--First Meteorologist." Cited below as item M37.

BM27 Williams, Frances Leigh. *Matthew Fontaine Maury, Scientist of the Sea*. New Brunswick, N.J.: Rutgers University Press, 1963. xx + 720 pp.

A comprehensive biography, with numerous notes and a bibliography of primary and secondary sources.

McAdie, A.

BM28 McAdie, Mary R.B. *Alexander McAdie: Scientist and Writer*. Charlottesville, Va.: Mary R.B. McAdie, 1949. 421 pp.

This is a reprint of 54 selected meteorological articles by McAdie (1863-1943) who was the founder and long-time director of the Blue Hill Observatory near Boston. It gives a short biography, a list of his publications and the instruments he designed, and tributes to him.

Medici, L. de

BM29 Middleton, William E. Knowles. "A Royal Meteorologist of the 17th Century." *Weather* (London) 28 (1973): 435-438.

Meigs, J.

BM30 Carter, Horace S. "Josiah Meigs, Pioneer Weatherman." *Weatherwise* 13 (1960): 166-167, 181.

Biographical note on the initiator of weather observations in Athens, Georgia, Josiah Meigs (1757-1822), who was the first President of the University of Georgia and who later started observations by the U.S. General Land Office.

BM31 Meigs, W.M. *Life of Josiah Meigs*. Philadelphia, 1887. 132 pp.

A biography of an early advocate of meteorological observation by the government of the United States, who himself made observations while in Bermuda. Later as President of the University of Georgia he started observations at Athens and as Commissioner of the U.S. General Land Office made an effort in 1817 to have the Registers of that office throughout the country make observations.

Meinardus, W.

BM32 Flohn, Hermann. "Wilhelm Meinardus." *Meteorologische Rundschau* 5 (1952): 184.

Obituary of Meinardus (1867-1952), long-time professor of geography at Göttingen, who made major contributions to the meteorology of the oceans and the Antarctic.

BM33 Heyer, Ernst. "Nachruf auf Professor Dr. Wilhelm Meinardus." *Zeitschrift für Meteorologie* 6 (1952): 289-290.

Obituary of Meinardus (1867-1952) giving an appraisal of his contributions to meteorology and climatology.

Meinzer, O.E.

BM34 Stringfield, V.T. "Meinzer, Oscar Edward" (b. near Davis, Illinois, 28 November 1876; d. Washington, D.C., 14 June 1948). *Dictionary of Scientific Biography* 9 (1974): 257-258.

Meinzer "initiated the development of the science of ground-water hydrology."

Mendel, G. @

BM35 Orel, Vítezslav, ed. *Folia Mendeliana Musei Moraviae*. Brno: Moravian Museu, 1967. 48 pp.

Collection of essays on the life and work of Gregor Mendel, including one by Franz Weiling and Vítězslav Orel about his work in geography and meteorology.

Mendeleev, D.I. @

BM36 Mendeleev, Dmitri Ivanovich. *Meteorologicheskie Raboty* [Meteorological Works]. Volume 22 of *Sochinenica* [Collected Works]. Leningrad, 1950. 866 pp.

The famed inventor of the periodic system of elements also produced a number of meteorological papers dealing with air density, thermometry, barometry, and atmospheric physics.

BM37 Karol, B.P. *D.I. Mendeleev i meteorologiia* [D.I. Mendelee and Meteorology]. Leningrad: Gidro Meteorologiches Roe Izdatel'stvo, 1950. 35 pp.

Mendeleev (1834-1907), better known as a chemist, had among his extensive bibliography 32 meteorological publications, dealing with, among other topics, atmospheric composition, upper air temperature, and eclipse meteorology.

Michell, J. @

BM38 Geikie, Archibald. *Memoir of John Michell*. London: Cambridge University Press, 1918. 108 pp.

Includes an account of Michell's researches on earthquakes and on the density of the earth.

BM39 Hardin, Clyde. "The Scientific Work of the Reverend John Michell." *Annals of Science* 22 (1966): 27-47.

Includes a summary of his theory of earthquakes.

Mieghem, J. van

BM40 Newton, Chester W. "Jacques van Mieghem, 1905-1980." *Bulletin of the American Meteorological Society* 62 (1981): 350.

An obituary of the great Belgian dynamic meteorologist, who was director of the Royal Belgian Meteorological Institute from 1962 to 1970. His main scientific contributions were to synoptic analysis of storms, numerical meteorological models, hydrodynamic instability, and energy conversions in the atmosphere. He was author of several books and editor of the International Geophysics series. He also played a major role in the World Meteorological Organization and other international bodies.

BM41 Pearce, R.P. "Jacques van Mieghem (1905-1980)." *Quarterly Journal of the Royal Meteorological Society* 107 (1981): 745.

Milham, W.I. @

BM42 "Willis Isbister Milham." *Bulletin of the American Meteorological Society* 38 (1957): 301.

An obituary with portrait of Milham (1874-1957) who was professor of astronomy at Williams College for 47 years, and who in 1912 wrote a classical textbook on meteorology which dominated in America for two decades.

Milne, J. @

BM43 Herbert-Gustar, L.K., and P.A. Nott. *John Milne, Father of Modern Seismology*. Norbury, Tenterden, Kent, 1980. xvi + 200 pp.

BM44 Wartnaby, John. "The Early Scientific Work of John Milne." *Japanese Studies in History of Science* 8 (1969): 77-124.

On Milne's work between 1874 and 1886 in glaciology, volcanology, and other areas of earth science.

Mitra, S.K.

BM45 Ghar, J.N. "S.K. Mitra (1890-1963)." *Astronautica Acta* 9 (1963): 208-209.

Obituary of a pioneer explorer of the high atmosphere, whose book *The Upper Atmosphere* (1947) was the first comprehensive treatise on the topic.

See also note: "Prof. S.K. Mitra," *Journal of Scientific and Industrial Research* 22 (1963): 427-428.

Mohn, H.

BM46 Pedersen, Olaf. "Mohn, Henrik" (b. Bergen, Norway, 15 May 1853; d. Christiania [now Oslo], Norway, 12 September 1916). *Dictionary of Scientific Biography* 9 (1972): 442-443.

Mohn wrote extensively on thunderstorms, fog signals, and other topics in meteorology and oceanography. "His studies, with C.M. Guldberg, on the motions of the atmosphere (1876-1880) utilized the Coriolis law and also took into account friction between the atmosphere and the earth."

Mohorovičić, A.

BM47 Boswell, S.J. "Discoverer of the Moho." *Open Earth* 13 (1981): 14.

Brief biographical note.

BM48 Bullen, K.E. "Mohorovičić, Andrija" (b. Volosko, Istria, Croatia, 23 January 1857; d. Zagreb, Yugoslavia, 18 December 1936). *Dictionary of Scientific Biography* 9 (1974): 443-445.

"Mohorovičić's fame rests nearly entirely on the results of his very thorough investigation of the destructive earthquake of 8 October 1909 [near Zagreb]. [He] correctly interpreted the observations as showing that the focus of the earthquake was inside a distinct outer layer of the earth." Other seismologists confirmed this discovery (though reducing his estimate that it is 30-35 miles thick) and showed that the layer is worldwide. The boundary between this layer (the "crust") and the mantle is now called the Mohorovičić discontinuity, or (by the journalists) "the Moho."

"As a meteorologist Mohorovičić was noted for his great organizational ability, his insistence on high standards of precision ... and his success in circumventing bureaucracy. His research papers dealt with such subjects as cloud movements, the variation of atmospheric temperature with height, rainfall in Zagreb, and a tornado."

Molchanov, P.A.

BM49 Selezneva, Eugeniya, and E.A. Tudorovskaya. *P.A. Molchanov--vydayushchiisya sovetskii aerolog* [P.A.

Molchanov--An Outstanding Soviet Aerologist]. Leningrad: Gidrometeorologicheskoe Izdatstelsvo, 1958. 101 pp.

Molchanov (1893-1941), an explorer of the upper air, won international fame by construction of the first successful radiosonde, tested in 1930, which revolutionized the analysis of the third dimension in the atmosphere.

Molga, M.

BM50 Mikulski, Zdzislaw, and Stanislaw Zych. "Marian Molga 1906-1980." *Przeglad Geofizyczny* 25 (1980): 209-218.

Obituary of the world-famous Polish agricultural meteorologist, with portrait and a classified list of his 130 publications.

See also: Eulogy for Molga by the Meteorological Committee of the Polish Academy of Science, *Przeglad Geofizyczny* 25 (1980): 219-226.

Montanari, G.

BM51 Tabarroni, Giorgio. "Montanari, Geminiano" (b. Modena, Italy, 1 June 1633; d. Padua, Italy, 13 October 1687). *Dictionary of Scientific Biography* 9 (1972): 484-487.

Montanari was the first to use the term "atmospheric precipitation." He used the barometer as an altimeter and proposed its use in forecasting weather.

Moro, A.L.

BM52 Thomasian, Rose. "Moro, Antonio-Lazzaro" (b. San Vito del Friuli, Italy, 16 March 1687; d. San Vito del Friuli, 12 April 1764). *Dictionary of Scientific Biography* 9 (1974): 531-534.

Moro rejected Neptunist theories and proposed that mountains and islands are volcanic in origin. Moro's importance as a founder of plutonism was pointed out by C. Lyell.

Mushketov, I.V.

BM53 Batiushkova, Irina V. "Mushketov, Ivan Vasilievich" (b. Alekseevskaya, Voronezh, Russia, 21 January 1850;

d. St. Petersburg, Russia, 23 January 1902). *Dictionary of Scientific Biography* 9 (1974): 592-594.

Mushketov collected information on earthquakes and wrote a synthetic work on physical geology. Starting from the nebular hypothesis and the theory of secular cooling and contraction of the earth, he saw the solidification of the earth proceeding both from the center and the surface. He attributed earthquakes and tectonic processes to contraction of the crust.

BN. Subjects: N

Nagaoka, H. @

BN1 Kimura, Tosaku. "Nagaoka's Geophysical Studies and Their Role in His Physical Researches." *Japanese Studies in the History of Science* 11 (1972): 91-98.

Hantaro Nagaoka (1865-1950) participated in gravity measurements in Japan, and published theoretical papers on seismic wave propagation. He also wrote on terrestrial magnetism and radio wave transmission in the ionosphere.

Nakaya, U.

BN2 Magono. Choji. "Ukichiro Nakaya." *Journal of the Meteorological Society of Japan*, 2nd Ser., 41 (1963): 1-4.

Obituary of Nakaya (1900-1962) who made fundamental contributions to the knowledge of snow and ice crystals. Includes portrait and list of 67 papers by Nakaya in English.

BN3 Rigby, Malcolm. "Ukichiro Nakaya, 1900-1962." *Bulletin of the American Meteorological Society* 43 (1962): 580.

An obituary of the pioneer researcher on formation and nature of snow crystals.

See also obituary: James A. Bender, *Arctic* 15 (1962): 242.

Naumann, K.F.

BN4 Burke, John G. "Naumann, Karl Friedrich" (b. Dresden, Germany, 30 May 1797; d. Dresden, 26 November 1873). *Dictionary of Scientific Biography* 9 (1974): 620.

Naumann's *Lehrbuch der Geognosie* contained all the information known about earthquakes at that time. He "held that certain earthquakes occurred independently of any volcanic activity and might therefore be termed 'plutonic.' This view was in opposition to that of Humboldt, who believed earthquakes and volcanoes to be merely different manifestations of the same causes."

Neumayer, G.B.

BN5 Loewe, Fritz. "First Australian 'Government Meteorologist.'" *Australian Meteorological Magazine* no. 48 (1965): 46-48.

Georg Balthasar Neumayer (1826-1909) established the first government observatory near Melbourne and was director from 1857 to 1862, before returning to his native Germany.

BN6 Rodewald, M. "Der Wissenschaftler Dr. Georg Neumayer." *Mitteilungen der Pollichia* (Pollichia Museum, Bad Dürkheim) 15 (1968): 5-12.

Appraisal of the work of Neumayer who started observations in Australia and later founded the Deutsche Seewarte, the central German institute for marine meteorology.

Nikitin, S.N.

BN7 Batyushkova, Irina V. "Nikitin, Sergey Nikolaevich" (b. Moscow, Russia, 4 February 1851; d. St. Petersburg, Russia, 18 November 1909). *Dictionary of Scientific Biography* 10 (1974): 128.

"Nikitin laid the foundation for systematic hydrogeological and hydrological research in Russia."

Nordenskiöld, A.E. @

BN8 Kish, G. "Adolf Erik Nordenskiöld (1832-1901): A Scandinavian Pioneer of the Earth Sciences." *Actes du XIIIe Congres International d'Histoire des Sciences* 1971 (pub. 1974), 8: 53-58.

As a result of his discovery of mineral particles on the Greenland icecap, Nordenskiöld proposed that the earth was formed by slow accumulation of cosmic dust. He wrote extensively on the history of cartography.

Norman, R.

BN9 Kelly, Suzanne. "Norman, Robert" (fl. England, late sixteenth century). *Dictionary of Scientific Biography* 10 (1974): 149-150.

"In making magnetic compasses, Norman noticed that the needle did not remain parallel to the earth's surface but that the north-seeking pole dipped toward the earth.... He measured this deviation to be 71° 50' at London and was interested in finding its value at other points on the earth's surface."

Numerov, B.V.

BN10 Kulikovsky, P.G. "Numerov, Boris Vasilievich" (b. Novgorod, Russia, 17 January 1891; d. 19 March 1943). *Dictionary of Scientific Biography* 10 (1974): 158-160.

"Numerov introduced into practice the pendulum gravimeter and the variograph for studying the upper layers of the earth's crust in geological prospecting.... Numerov's plan for a general gravimetrical survey of the Soviet Union produced extremely valuable results."

BO. Subjects: O

Obruchev, V.A.

BO1 Naumov, G.V. "Obruchev, Vladimir Afanasievich" (b. Klepenino, Rzhev district, Tver [now Kalinin] guberniya, Russia, 19 October 1863; d. Moscow, U.S.S.R., 19 June 1956). *Dictionary of Scientific Biography* 10 (1974): 166-170.

Obruchev studied the tectonics of Siberia.

Ogg, A.

BO2 Van Wijk, A.M. "Alexander Ogg, 1870-1948." *Terrestrial Magnetism and Atmospheric Electricity* 53 (1948): 153-154.

Obituary of the South African scientist, with a list of his geomagnetic publications. (Portrait of Ogg, facing p. 109.)

Oldham, R.D.

BO3 Bullen, K.E. "Oldham, Richard Dixon" (b. Dublin, Ireland, 31 July 1858; d. Llandrindod Wells, Wales, 15 July 1936). *Dictionary of Scientific Biography* 10 (1974): 203.

"Oldham became famous for his report on the great Assam earthquake on June 1897.... The most far-reaching result was the first clear identification on seismograms of the onsets of the primary (P), secondary (S), and tertiary (surface) waves.... This ... showed that the earth could be treated as perfectly elastic to good approximation in studying seismic waves.... Oldham also supplied the first clear evidence that the earth has a central core (1906)."

Omori, F.

BO4 Bullen, K.E. "Omori, Fusakichi" (b. Fukui, Japan, 30 October 1868; d. Tokyo, Japan, 8 November 1923). *Dictionary of Scientific Biography* 10 (1974): 210-211.

"Under [John] Milne's encouragement Omori made the first precise studies of earthquake aftershocks.... Omori is probably most noted today for his work in designing seismological instruments.... Omori's contributions touched on practically all aspects of seismology.... He was also interested in the mechanism of volcanoes and used seismic methods in studying them."

Orlov, A.Y.

BO5 Kulikovsky, P.G. "Orlov, Aleksandr Yakovlevich" (b. Smolensk, Russia, 6 April 1880; d. Kiev, Ukrainian S.S.R., 28 January 1954). *Dictionary of Scientific Biography* 10 (1974): 231-233.

"Orlov's scientific work touched several areas: (1) the motion of the poles and variations in latitude; (2) tidal deformations of the earth; (3) seismology; and (4) geodesy and geophysics."

BP. Subjects: P

Palissy, B.

BP1 Biswas, Margaret R., and Asit K. Biswas. "Palissy, Bernard" (b. La Capelle Biron, France, ca. 1510; d.

Paris, France, ca. 1590). *Dictionary of Scientific Biography* 10 (1974): 280-281.

"He was one of the few men of his century to have a correct notion of the origins of rivers and streams. An early advocate of the infiltration theory, he refuted ... the old theories that streams came from seawater or from air that had condensed into water."

Pallas, P.S.

BP2 Esakov, Vasiliy A. "Pallas, Pyotr Simon" (b. Berlin, Germany, 3 October 1741; d. Berlin, 20 September 1811). *Dictionary of Scientific Biography* 10 (1974): 283-285.

"He formulated the first general hypothesis of the origin of mountains. In his opinion, granite constituted the skeleton of the earth and its nucleus.... The raising of the mountains and the receding of the seas occurred, in Pallas' opinion, as the result of volcanic processes."

Palmen, E.H.

BP3 T[aba], H. "The Bulletin Interviews; Professor E.H. Palmen." *World Meteorological Organization Bulletin* 30 (1981): 92-100.

A brief biographical memoir is followed by an interview between the distinguished Finnish meteorologist and the editor of the *Bulletin*. Recalls the career of Palmen and his collaboration with other meteorologists, among them Jacob Bjerknes (1897-1975), C.-G. Rossby (1889-1957), Herbert Riehl, and Chester Newton.

Parry, C.H.

PB4 Botely, Cecily M. "Parry of the Parry Arc." *Weather* 10 (1955): 443-445.

Biographical sketches of Caleb Hillier Parry (1755-1822) and his son William Edward Parry (1790-1855), who made low-level kite ascents in the 1820s, measured the speed of sound in air, and first observed in 1820 the halo phenomenon named for him.

Patterson, J.

BP5 Thompson, Andrew. "Obituary Notice--Dr. J. Patterson, O.B.E., M.A., LL.D., F.R.S.C." *Quarterly Journal of the Royal Meteorological Society* 82 (1956): 381-382.

Patterson (1872-1956) was director of the Meteorological Service of Canada and contributed notably to the development of meteorological instruments.

Pavlov, A.P.

BP6 Batiushkova, Irina V. "Pavlov, Aleksei Petrovich" (b. Moscow, Russia, 13 November 1854; d. Bad Tolz, Germany, 9 September 1929). *Dictionary of Scientific Biography* 10 (1974): 428-431.

Pavlov conducted research in glacial geology and tectonics. He emphasized the comparison with lunar topography for reconstructing the early history of the earth's surface.

Peirce, B. @

BP7 Lenzen, V.F. *Benjamin Peirce and the U.S. Coast Survey.* San Francisco: San Francisco Press, 1968. vii + 54 pp.

On the American scientist Benjamin Peirce (1809-1880), Superintendent of the Coast Survey from 1867 to 1874; he initiated a geodetic connection between the surveys on the Atlantic and Pacific Coasts.

Peirce, C.S.

BP8 Eisele, Carolyn. "Peirce, Charles Sanders" (b. Cambridge, Massachusetts, 10 September 1838; d. Milford, Pennsylvania, 19 April 1914). *Dictionary of Scientific Biography* 10 (1974): 482-488.

As an employee of the U.S. Coast Survey from 1861 to 1891, Peirce invented a pendulum for geodetic work, and conducted a mathematical study of the relations between the variation of gravity and the figure of the earth.

Penck, A.

BP9 Beckinsale, Robert P. "Penck, Albrecht" (b. Reuditz [near Leipzig], Germany, 25 September 1858; d. Prague, Czechoslovakia, 7 March 1945). *Dictionary of Scientific Biography* 10 (1974): 501-506.

Penck collaborated with E. Brückner on extensive field studies in the Alpine Valleys, undertaken with a view to perfecting a chronology of ice sheet advances and retreats, leading to a classic work, *Die Alpen im Eiszeitalter* (1901-9). Having earlier agreed with Suess on the principle "that secular variations in the relative altitude of land and sea were due to world-wide fluctuations of sea level (eustasism)," by 1900 he "had accepted independent crustal movement (regional or local) as a concomitant factor in elevating or depressing coastlines." He administered an institute and museum in oceanography and organized an expedition (1925) to make sounding traverses in the South Atlantic.

Penck, W.

BP10 Chorley, Richard. "Penck, Walther" (b. Vienna, Austria, 30 August 1888; d. Stuttgart, Germany, 29 September 1923). *Dictionary of Scientific Biography* 10 (1974): 506-509.

Walther Penck (son of Albrecht Penck) studied tectonic movements in Argentina and Turkey; later, he "developed his most influential ideas on the interpretation of landforms through analysis of the relationships between endogenetic (diastrophic) and exogenetic (erosional) processes.

Perrault, P.

BP11 Biswas, Margaret R., and Asit K. Biswas. "Perrault, Pierre" (b. France, 1611; d. France, 1680). *Dictionary of Scientific Biography* 10 (1974): 521-522.

In his book *De l'origine des fontaines* (1674), Perrault showed that "rainfall alone is sufficient to sustain the flow of springs and rivers throughout the year.... His experimental work on the rainfall and runoff of the upper Seine is a milestone in the history of hydrology."

Perrier, G.

BP12 Gougenheim, A. "Perrier, Georges" (b. Montpellier, France, 28 October 1872; d. Paris, France, 16 February 1946). *Dictionary of Scientific Biography* 10 (1974): 523-524.

"Perrier was the son of François Perrier, who revived French geodesy ... he played a major role in preparing and executing a scientific mission sent to Peru and Ecuador to measure an arc of meridian at low latitudes ... from 1919 until his death Perrier served as secretary-general of the organization created in 1919 and several years later named the Association Internationale de Geodesia."

Petterssen, S.

BP13 Bundgaard, Robert C. "Sverre Petterssen, Weather Forecaster." *Bulletin of the American Meteorological Society* 60 (1979): 182-195.

A biography of Petterssen (1898-1974), one of the foremost synoptic meteorologists of the 20th century, trained in the Norwegian school. He taught at MIT and the University of Chicago, wrote influential textbooks, and developed new methods for weather prediction.

Pettersson, H.

BP14 Deacon, G.E.R. "Pettersson, Hans" (b. Kalhuvudet, Marstrand Bohuslan, Sweden, 26 August 1888; d. Goteborg, Sweden, 25 January 1966). *Dictionary of Scientific Biography* 10 (1974): 563-564.

"Like his father, Pettersson became one of the most outstanding oceanographers of his period.... Like his father, he was particularly interested in the differences of flow in stratified water, waves in internal boundary surfaces, and improved methods of measuring water density and flow. Throughout this work he had a compelling interest in changes of sea level brought about by meteorological factors and in the effect of oceans on climate."

Picard, J.

BP15 Taton, Juliette, and Rene Taton. "Picard, Jean" (b. La Fleche, France, 21 July 1620; d. Paris, France, 12 October 1682). *Dictionary of Scientific Biography* 10 (1974): 595-597.

Picard directed the measurement of the arc of meridian from Sourdon to Malvoisine in order to obtain a more accurate figure for the earth's radius. The results, published in 1671, "made possible a great advance in the determination of geographical coordinates and in cartography, and enabled Newton in 1684 to arrive at a striking confirmation of the accuracy of his principle of gravitation."

Porro, I.

BP16 Boley, Bruno A. "Porro, Ignazio" (b. Pinerolo, Italy, 25 November 1801; d. Milan, Italy, 8 October 1875). *Dictionary of Scientific Biography* 11 (1975): 95.

Porro "conceived and constructed several optical surveying instruments that revolutionized topographical and geodetic practice, and brought it essentially to its modern status."

Prandtl, L. @

BP17 Dryden, Hugh L. "Ludwig Prandtl, 1875-1953." *Journal of the Aeronautical Sciences* 20 (1953): 779, 800.

An obituary of the pioneer aerodynamicist who contributed fundamentally to meteorological boundary layer studies.

Proudman, J.

BP18 Cartwright, D.E., and F. Ursell. "Joseph Proudman, 30 Dec. 1888-26 June 1975." *Biographical Memoirs of the Fellows of the Royal Society* 22 (1976): 319-333.

Proudman worked in tidal theory and other areas of physical oceanography.

Przhevalsky, N.M.

BP19 Esakov, Vasiliy A. "Przhevalsky, Nikolay Mikhaylovich" (b. Kimbarovo, Smolensk guberniya, Russia, 12 April 1839; d. Karakol [now Przhevalsk], Russia, 1 November 1888). *Dictionary of Scientific Biography* 11 (1975): 180-182.

"His meteorological observations provided the first basis for a climatology of central Asia."

BR. Subjects: R

Rames, J.B.

BR1 Stockmans, F. "Rames, Jean Baptiste" (b. Aurillac, France, 26 December 1832; d. Aurillac, 22 August 1894). *Dictionary of Scientific Biography* 11 (1975): 269-270.

"It was in the study of volcanoes of the Cantal that he made his outstanding contributions. Following the work of Nicolas Desmarest, he showed that they were not caused by elevation (the *soulevement* Buch) but followed by an injection of material which fell back around the crater."

Ramsay, A.C.

BR2 Beckinsale, Robert P. "Ramsay, Andrew Crombie" (b. Glasgow, Scotland, 31 January 1814; d. Beaumaris, Wales, 9 December 1891). *Dictionary of Scientific Biography* 11 (1975): 276-277.

"... his ideas on glacial action, which caused tremendous contemporary disagreement, have for the most part been vindicated."

Rankine, A.O.

BR3 Knudsen, Ole. "Rankine, Alexander Oliver" (b. Guildford, England, 1881; d. Hampton, Middlesex, England, 20 January 1956). *Dictionary of Scientific Biography* 11 (1975): 290-291.

"Rankine ... improved the Eötvös gravimeter; this work led him to construct a magnetometer of great sensitivity."

Raspe, R.E.

BR4 Carozzi, Albert V. "The Geological Contribution of Rudolf Erich Raspe (1737-1794)." *Archives des Sciences* 22 (1969): 624-644.

"Raspe's *Specimen* (1763) is not a simple illustration of Hooke's system, but an original contribution to some aspects of structural geology ... he develops the dynamic concept of Hooke which is based on the uplifting forces of 'earthquakes,' this last term meaning all vertical movements of the earth's crust. Raspe subdivides this process into four distinct mechanisms which in his opinion are capable of explaining the origin of islands, continents, and mountains within continents and at the bottom of the ocean. Raspe's concept of uniformitarianism is worthy of Hutton and Lyell...."

BR5 Carozzi, Albert V. "Rudolf Erich Raspe (1737-1794) or the Impossible Scientific Dream." *Stechert-Hafner Book News* 24(1) (1969): 1-3.

The dream was to write a complete natural history of the earth based on Robert Hooke's ideas about earthquakes and volcanoes.

BR6 Carozzi, Albert V. "Raspe, Rudolf Erich" (b. Hanover, Germany, 1737; d. Muckross, Ireland, 1794). *Dictionary of Scientific Biography* 11 (1975): 302-305.

Raspe developed a theory of the earth based on Hooke's ideas about earthquakes and subterraneous eruptions. Later he developed his own ideas on volcanoes, and considered them to be responsible "not only for the formation of islands and continents but also for the deposition of fossiliferous shales and limestones ... even for the salty character of seawater."

The Rayleighs

BR7 Howard, J.N. "The Scientific Papers of the Lords Rayleigh." *Actes, XIe Congres International d'Histoire des Sciences, 1965* (pub. 1968) 4: 315-318.

BR8 Howard, J.N. *The Rayleigh Archives Dedication*. Bedford, Mass.: Air Force Cambridge Research Laboratories, 1967. vi + 70 + 5 + 11 pp.

On 30 March 1966 a ceremony was held at the Research Library of the Air Force Cambridge Research Laboratories, Bedford, Massachusetts, to dedicate the Rayleigh Archives: a collection of the notebooks, manuscripts, working papers, and correspondence of the Lords Rayleigh (John William Strutt, the 3rd Baron Rayleigh, and Robert John Strutt, the 4th Baron Rayleigh). Includes comments by S. Chapman on the influence of the 4th Baron Rayleigh on air glow and auroral research.

Rayleigh, R.J.S. @

BR9 Howard, John, ed. *Robert John Strutt, 4th Baron Rayleigh: Unpublished Manuscripts and Reviews of His Work*. Bedford, Mass.: Air Force Cambridge Research Laboratories, 1971. viii + 75 pp.

BR10 Strutt, G.R. *The Airglow Rayleigh: Robert John Strutt, 4th Baron Rayleigh, a Memoir*. Edited by J.N. Howard. Bedford, Mass.: Air Force Cambridge Research Laboratories, 1969. vi + 34 pp.

Reck, H.

BR11 Tobien, Heinz. "Reck, Hans" (b. Wurzburg, Germany, 24 January 1886; d. Lourenco Marques, Mozambique, 4 August 1937). *Dictionary of Scientific Biography* 11 (1975): 335-337.

In his studies of volcanism he treated "Linear eruptions craters of elevation, volcanic horsts, collapse calders, volcanic bombs, and the relationships between volcanism and tectonics."

Reclus, E. @

BR12 "Resurrecting Reclus." *Open Earth* no. 5 (October 1979): 9.

Note on Elisee Reclus, based on the article by J.O. Berkland (below, item BR13).

BR13 Berkland, James O. "Elisee Reclus--Neglected Geologic Pioneer and First (?) Continental Drift Advocate." *Geology* 7 (1979): 189-192.

In *The Earth* (American edition, 1872) Reclus described noncatastrophic separation and movement of the continents. "Reclus also showed great foresight regarding the antiquity of the Earth, convection-subduction, the 'Ring of Fire' around the Pacific, glacial drift, the nature of earthquakes, animal reactions as precursors for earthquakes," etc.

Redfield, W.

BR14 Burstyn, Harold L. "Redfield, William C." (b. Middletown, Connecticut, 26 March 1789; d. New York, N.Y., 12 February 1857). *Dictionary of Scientific Biography* 11 (1975): 340-341.

Redfield proposed that a storm is a "progressive whirlwind," and continued for 25 years to develop his ideas on the rotary nature of storms.

Reichelderfer, F.W.

BR15 Hughes, Patrick. "Francis W. Reichelderfer, Part I: Aerologists and Air-devils." *Weatherwise* 34 (1981): 52-59.

An illustrated biographical article giving a sketch of a pioneer weather officer of the U.S. Navy, his efforts to introduce modern forecasting techniques, developed in Norway, into American meteorology in the post-World War I era, and his exploits as a lighter-than-air pilot and weather officer.

BR16 Taba, H. "The Bulletin Interviews: Dr. F.W. Reichelderfer." *World Meteorological Organization Bulletin* 31 (1982): 171-184.

Following a biographical sketch of the Chief of the U.S. Weather Bureau from 1938 to 1963 after a career as aerologist in the U.S. Navy, Dr. Reichelderfer (b. 1895) discusses many milestones in the recent history of meteorology. Included are the developments leading to the convention creating the World Meteorological Organization as a specialized agency of the United Nations. Reichelderfer was the WMO's first president (1951-1955).

Reid, H.F.

BR17 Byerly, Perry. "Reid, Harry Fielding" (b. Baltimore, Maryland, 18 May 1859; d. Baltimore, 18 June 1944). *Dictionary of Scientific Biography* 11 (1975): 361-362.

"Reid ... may well be said to have been the first geophysicist in the United States.... In his early manhood glaciology was his principal interest.... Reid's greatest contribution to geophysics was undoubtedly his masterful exposition of the 'elastic rebound theory' of ... earthquakes."

Rennell, J.

BR18 Eyles, Joan M. "Rennell, James" (b. Upcott, near Chudleigh, Devon, England, 3 December 1742; d. London, England, 29 March 1830). *Dictionary of Scientific Biography* 11 (1975): 376.

"His work on ocean currents ... was used by many subsequent writers."

Ricco, A.

BR19 Abetti, Giorgio. "Ricco, Annibale" (b. Modena, Italy, 15 September 1844; d. Rome, Italy, 23 September 1919). *Dictionary of Scientific Biography* 11 (1975): 412.

Ricco investigated the relation between magnetic storms and sunspots; "he determined the gravitational anomalies and the terrestrial magnetic constants for Sicily, especially in relation to seismic activity...."

Richardson, L.F.

BR20 Ashford, O.M. "Lewis F. Richardson, D. Sc., F.R.S." *Weather* 4 (1949): 9-10.

An obituary with portrait of this pioneer mathematical modeler of the atmosphere. Richardson's (1881-1949) most famous work was *Weather Prediction by Numerical Process* (Cambridge University Press, 1922; 236 pp.). He abandoned meteorology because it was a science applicable to warfare, which he as a Quaker abhorred.

BR21 Ashford, Oliver M. "The Dream and the Fantasy." *Weather* 36 (1981): 323-325.

Just in advance of the 60th anniversary of the publication of Lewis F. Richardson's book *Weather Prediction by Numerical Process* (1922) this retrospective biographical note brings out many details of Richardson's life and career not contained in the earlier obituaries.

BR22 Gleiser, Molly. "The First Man to Compute the Weather." *Datamation* 26, no. 6 (1980): 181-184.

An appraisal of the life and contribution of Lewis Fry Richardson to numerical weather prediction.

BR23 Gold, Ernest. "Lewis Fry Richardson, 1881-1953." *Obituary Notices of Fellows of the Royal Society* 9 (1954): 217-235.

An elaborate obituary of Richardson, who was the pioneer of numerical weather prediction.

BR24 Richardson, L.F. "Meteorological Publications by L.F. Richardson As They Appear to Him in October 1948." *Weather* 4 (1949): 6-9.

An annotated list of Richardson's meteorological publications compiled by him shortly before his death. Contains 27 entries.

BR25 Richardson, Stephen A. "Lewis Fry Richardson (1881-1953). A Personal Biography." *Journal of Conflict Resolution* 1 (1957): 300-304.

Biographical note on Richardson written by his son, emphasizing his pacifist activities.

Richter, C.F.

BR26 Spall, Henry. "Charles F. Richter--An Interview." *Earthquake Information Bulletin (USGS)* 21, no. 1 (1980): 4-8.

Mostly a discussion of his scale for earthquake magnitudes.

Robitzsch, M.

BR27 Grotewahl, Max. "Prof. Dr. Max Robitzsch." *Polarforschung*, Serie 3, 22 (1952): 145.

Obituary of the explorer of polar meteorology and developer of new meteorological measuring techniques, Robitzsch (1887-1952).

BR28 Hesse, Walter. "Max Robitzsch zum 65. Geburtstag." *Beiträge zur Geophysik* 62 (1952): 162-163.

A biographical sketch of Robitzsch (1887-1952), who was principally a developer of meteorological instruments.

BR29 König, W. "Nachruf auf Professor Dr. Max Robitzsch." *Zeitschrift für Meteorologie* 6 (1952): 195-197.

Obituary of Robitzsch (1887-1952) who made notable contributions to meteorological instrument development and became well known for his aerological ascents with captured balloons launched from a boat on Lake Constance.

Roche, É.A.

BR30 Levy, Jacques R. "Roche, Édouard Albert" (b. Montpellier, France, 17 October 1820; d. Montpellier, 18 April 1883). *Dictionary of Scientific Biography* 11 (1975): 498.

Roche's limit, for the distance at which a satellite becomes unstable under tidal forces, is an essential criterion in cosmogony. He developed hypotheses for the internal density and structure of the earth, and showed that temporary diminutions in solar radiation may be caused by local atmospheric phenomena.

Rodes, L.

BR31 Macelwane, J.B. "Padre Luis Rodes, S.J., 1881-1939." *Terrestrial Magnetism and Atmospheric Electricity* 45 (1940): 87-91.

Obituary of Rodes, known for his work on solar activity and its influence on the earth's magnetic and electrical fields. List of publications; portrait (facing p. 1).

Rogers, H.D. and W.B.

BR32 Rodgers, John. "Rogers, Henry Darwin" (b. Philadelphia, Pennsylvania, 1 August 1808; d. Shawlands, near Glasgow, Scotland, 29 May 1866) and "Rogers, William Barton" (b. Philadelphia, 7 December 1804; d. Boston, Massachusetts, 30 May 1882). *Dictionary of Scientific Biography* 11 (1975): 504-506.

The Rogers brothers proposed tectonic theories of the Appalachian mountains, emphasizing tangential rather than vertical forces.

Romé De L'Isle, J.-B.L.

BR33 Hooykaas, R. "Rome De L'Isle (or Delisle), Jean-Baptiste Louis" (b. Gray, France, 29 August 1736; d. Paris, France, 7 March 1790). *Dictionary of Scientific Biography* 11 (1975): 520-524.

Rome rejected the theory of central heat advocated by Buffon and others, claiming instead that all terrestrial heat derived from the sun. He was best known for his work in crystallography and mineralogy.

Ross, J.C.

BR34 Laurie, P.S. "Ross, James Clark" (b. London, England, 15 April 1800; d. Aylesbury, England, 3 April 1862). *Dictionary of Scientific Biography* 11 (1975): 554-555.

In 1831 Ross discovered the north magnetic pole. He made numerous observations of terrestrial magnetism during his voyages.

Rossby, C.-G. @

BR35 Byers, Horace R. "Carl-Gustav Rossby, the Organizer." *The Atmosphere and the Sea in Motion*. Edited by Bert Bolin. New York: Rockefeller Institute Press, 1959, pp. 56-59.

Rossby (1898-1957), besides making fundamental contributions to dynamic meteorology, was a charismatic organizer of three university departments of meteorology, the Weather Bureau research program, and the training program of weather officers in World War II.

The same volume, pp. 60-64, lists "Publications by Carl-Gustav Rossby."

BR36 Nyberg, Alf. "Professor C.-G. Rossby." *Bulletin of the World Meteorological Organization* 6 (1957): 157-158.

Obituary of Rossby (1898-1957) with portrait.

See other obituaries: B.J. Mason, *Weather* 12 (1957): 351; B. Bolin, *Tellus* 9 (1957): 257-258; U.K. Bose, *Indian Journal of Meteorology and Geophysics* 8 (1957): 465; H. Flohn, *Meteorologische Rundschau* 11 (1958): 1-2; *Science* 126 (1957): 444; R.C. Sutcliffe, *Quarterly Journal of the Royal Meteorological Society* 84 (1958): 88; Hurd C. Willet, *Science* 127 (1958): 686-687; Jacques van Mieghem, *Ciel et Terre* 74 (1958): 143-144; Horace R. Byers, *Bulletin of the American Meteorological Society* 39 (1958): 98-100; Raoul Bilancini, *Rivista di Meteorologia Aeronautica* 18 (1958): 72-73; Frigyes Desi, *Idöjaras* 61 (1957): 461-462; W. Hansen, *Naturwissenschaften* 45/8 (1958): 15; *Meteorologiya i Gidrologiya* No. 10 (1958): 56-57; Horace R. Byers, *National Academy of Sciences Biographical Memoirs* 34 (1960): 249-270.

Rouille, A.-L.

BR37 Allard, Michel. "Antoine-Louis Rouille, secretaire d'État à la Marine (1749-1754): Progrès scientifique et marine." *Revue d'Histoire des Sciences* 30 (1977): 97-103.

Rubey, W.W.

BR38 Ernst, W.G. "William Walden Rubey, December 19, 1898-April 12, 1974." *Biographical Memoirs, National Academy of Sciences* 49 (1978): 205-233.

His scientific contributions included the systematics of stream hydrology, origin of the atmosphere, seawater, and chemical differentiation of the earth, mechanisms of overthrust faulting and mountain building, and factors governing the release of seismic energy. His most frequently cited work was a 1951 presidential address to the Geological Society of America (published 1955) which "demonstrated that the atmosphere and hydrosphere have accumulated gradually near the earth's surface over the course of geological time as a consequence of outgassing of the deep interior."

BS. Subjects: S

Sabine, E.

BS1 Reingold, Nathan. "Sabine, Edward" (b. Dublin, Ireland, 14 October 1788; d. Richmond, Surrey, England, 26 June 1883). *Dictionary of Scientific Biography* 12 (1975): 49-53.

Sabine organized worldwide efforts to gather terrestrial magnetism observations. He "believed in the existence of two magnetic poles in each hemisphere and that terrestrial magnetism was essentially the same as, or closely related to, atmospheric phenomena." The article describes Sabine's work in relation to that of other 19th-century scientists (Gauss, John Herschel) and gives an intriguing glimpse into Victorian science-politics.

Saussure, H.B. de @

BS2 Barry, R.G. "H.B. de Saussure: The First Mountain Meteorologist." *Bulletin of the American Meteorological Society* 59 (1978): 702-705.

De Saussure (1740-1799) of Geneva was one of the most original contributors to meteorology in the 18th century. He constructed the first reliable hygrometer and recognized the dependence of atmospheric water vapor on temperature. He measured temperatures on ascents above the slopes of the Alps and calculated vertical lapse rates. He made measurements all the way up to the top of Mont Blanc, where he also showed the presence of CO_2 in the atmosphere. His cyanometer was the first device to determine sky color, which he found to become deeper blue with increasing elevation.

BS3 Freshfield, D.W. *The Life of Horace Benedict de Saussure.* London: E. Arnold, 1920. xii + 479 pp.

A biography of Saussure (1740-1799) who was the first to place atmospheric humidity measurements on a firm footing. Includes H.R. Mill, "A Note on the Meteorological Work and Observations on Deep Temperatures of H.B. de Saussure," pp. 457-465.

Schardt, H.

BS4 Wegmann, Eugene. "Schardt, Hans" (b. Basel, Switzerland, 18 June 1858; d. Zurich, Switzerland, 3 February, 1931). *Dictionary of Scientific Biography* 12 (1975): 139-141.

"Schardt's research encompassed tectonics, hydrology, stratigraphy, and engineering geology. His most important work, however, lay in his discovery of the older, rootless exotic complexes of the Alps, which, floating on younger series, led him to the hypothesis concerning the great alpine mass displacements that became known as the nappe theory."

Scherhag, R.

BS5 Flohn, H. "Am Grabe von Richard Scherhag (29.9.1907-31.8.1970)." *Meteorologische Rundschau* 24 (1971): 65-67.

A eulogy presented at the funeral of Richard Scherhag giving biographical details and an appraisal of the important contributions of the deceased to meteorology. His work on synoptic analysis and the aerology of the high atmosphere, especially his discovery of sudden stratospheric warmings in the late winter, gave impetus to much research.

BS6 Hugo, Heinz, and Leopold Klauser. "In Memoriam Richard Scherhag." *Beilage zur Berliner Wetterkarte* (Institut für Meteorologie, Freie Universität Berlin) (Unnumbered issue) (1970). 3pp.

An obituary of Professor Richard Scherhag (1907-1970), with portrait, outlining his scientific career, his major contributions to synoptic meteorology, and his organization of a meteorological department at the University in the western zone of Berlin in 1951. By 1970 this had become the largest German meteorological department combining research, teaching, and weather service for West Berlin, employing 130 persons.

BS7 "Veröffentlichungen von Richard Scherhag." *Beilage zur Berliner Wetterkarte* (Institut für Meteorologie, Freie Universität Berlin) 1/71, SO 1/71 (1971). 11 pp.

A list of the 223 publications of the late Professor Scherhag (1907-1970), a principal contributor to the development of synoptic meteorology in the 20th century.

Schimper, K.F.

BS8 Tobien, Heinz. "Schimper, Karl Friedrich" (b. Mannheim, Germany, 15 February 1803; d. Schwetzingen, near Heidelberg, Germany, 21 December 1867). *Dictionary of Scientific Biography* 12 (1975): 167-168.

His study of traces of glaciers led him to the concept of the "Ice Age" which he announced in 1837; a priority dispute with Louis Agassiz was not decided in his favor until much later. He postulated an alternation of warm and cold periods in the earth's history. In hydrology he treated topics such as the refraction and rejection of light in inland waters, and the formation of ice on rivers.

Schmauss, A.

BS9 Bilancini, Raoul. "Augusto Schmauss." *Rivista di Meteorologia Aeronautica* 15 (1955): 64.

Obituary of Schmauss (1877-1954) with an appraisal of his principal books.

BS10 Geiger, Rudolf. "Das Leben von August Schmauss (26.11. 1877-10.10.1954)." *Annalen der Meteorologie* 7 (1955/ 56): 161-172.

A biographical memoir of Schmauss's career and contributions.

BS11 Hesse, W. "August Schmauss." *Beiträge zur Geophysik* 64 (1955): 231-233.

Obituary note.

BS12 König, Willi. "Nachruf auf Professor Dr. August Schmauss." *Zeitschrift für Meteorologie* 8 (1954): 321-322.

Obituary of Schmauss who was head of the Bavarian meteorological service and a prolific author of research papers.

BS13 Weickmann, Ludwig. "Gedächtnisrede für Geheimrat Prof. Dr. August Schmauss Gehalten am 16. November 1954 in München." *Meteorologische Rundschau* 8 (1955): 1-6.

A eulogy for Schmauss with portrait.

Schmidt, A.

BS14 Bartels, J. "Adolf Schmidt, 1860-1944." *Terrestrial Magnetism and Atmospheric Electricity* 51 (1946): 439-447.

Includes a list of 52 of Schmidt's publications on geomagnetism.

Schmidt, C.A. Von

BS15 Wattenberg, Diedrich. "Schmidt, Carl August Von" (b. Diefenbach, Wurttemberg, Germany, 1 January 1840; d. Stuttgart, Germany, 21 March 1929). *Dictionary of Scientific Biography* 12 (1975): 186-187.

Schmidt demonstrated that seismic waves move in curved paths inside the earth and introduced the time-distance curve into seismology. "As early as 1894 he pointed out the separation of seismic waves into longitudinal and magnetism and the shape of the earth ... contributed to meteorology through works on the application of thermodynamics and the kinetic theory of gases to the study of the atmosphere ... introduced the concept of barometric tendency into weather forecasting and investigated the mechanism of thunderstorms."

Schneider-Carius, K.

BS16 Faust, H. "Karl Schneider-Carius." *Mitteilungen des Verbands Deutscher Meteorologen* 11 (1960): 13-14.

Obituary of Schneider-Carius (1896-1959) who worked on problems of boundary-layer meteorology and published a notable history of meteorology since the 17th century (*Wetterkunde-Wetterforschung*, München, 1955).

See also obituary: W. Boer, *Zeitschrift für Meteorologie* 14 (1960): 161-162; and bibliography: ibid., 162-165.

Schott, C.A. @

BS17 Abbe, C. "Biographical Memoir of Charles Anthony Schott." *National Academy of Sciences, Biographical Memoirs* 8 (1915): 87-133.

Biographical sketch of Schott (1826-1901), an employee of the U.S. Coast and Geodetic Survey since immigration from Germany in 1848; he was persuaded by Joseph Henry to prepare statistical analyses of U.S. hydrometeors for the Smithsonian Institution.

Schumacher, H.C. @

BA18 Andersen, Einar. *Heinrich Christian Schumacher: Et mindeskrift.* Geodaetisk Instituts historiske boger 3. Copenhagen: Deodaetisk Institut, 1975. 207 pp.

Schumacher (1780-1850) directed geodetic surveys which provided accurate data for determining the shape of the earth and differences in longitude between various places.

Schuster, A. @

BS19 Chapman, S. "Arthur Schuster, 1851-1934." *Terrestrial Magnetism and Atmospheric Electricity* 39 (1934): 341-345.

Summarizes his work on geomagnetism; followed by a note by J. Bartels, "Arthur Schuster's Work on Periodicities," *ibid.*, 345-346. (Portrait, facing p. 265.)

Scrope, G.J.P.

BS20 Page, L.E. "Scrope, George Julius Poulett" (b. London, England, 10 March 1797; d. Fairlawn [near Cobham], Surrey, England, 19 January 1876). *Dictionary of Scientific Biography* 12 (1975): 261-264.

Scrope wrote "the earliest systematic treatise in vulcanology" (1825); though his views were catastrophist, he influenced Lyell's formulation of uniformitarian geology and helped to undermine the doctrine of the Flood. He continued to speculate on the causes of volcanism and the geological effects of internal heat into the 1870s.

See, T.J.J. @

BS21 Ashbrook, Joseph. "Astronomical Scrapbook: The Saga of Mare Island." *Sky and Telescope* 24 (1962): 193-202.

Biographical note on Thomas Jefferson Jackson See (1866-1962).

BS22 Webb, William Larkin. *Brief Biography and Popular Account of the Unparalled Discoveries of Thomas J.J. See.* Lynn, Mass.: Nichols, 1913.

Thomas Jefferson Jackson See (1866-1962) was an eccentric American astronomer who published some interesting speculations on the origin of the earth and moon, causes of earthquakes, etc. According to Ashbrook (item BS21) this biography was probably commissioned and written mostly by See himself.

Sezawa, K.

BS23 Bullen, K.E. "Sezawa, Katsutada" (Yamaguchi, Japan, 21 August 1895; d. Tokyo, Japan, 23 April 1944). *Dictionary of Scientific Biography* 12 (1975): 341-342.

Through mathematical analysis of seismic surface waves, "Sezawa derived useful estimates of layering in the earth's crust and produced evidence indicating that the Pacific crust is thinner than the Eurasian."

Sharonov, V.V.

BS24 Kulikovsky, P.G. "Sharonov, Vsevolod Vasilievich" (b. St. Petersburg, Russia, 10 March 1901; d. Leningrad, U.S.S.R., 26 November 1964). *Dictionary of Scientific Biography* 12 (1975): 352-354.

Sharonov developed instruments for astronomical and geophysical observations. He measured the color of the clear sky and determined the solar light constant.

Shaw, W.N.

BS25 Gold, E. "William Napier Shaw (1854-1945)." *Obituary Notices of Fellows of the Royal Society of London* 5 (1945): 203-230.

BS26 Kutzbach, Gisela. "Shaw, William Napier" (b. Birmingham, England, 4 March 1854; d. London, England, 23 March 1945). *Dictionary of Scientific Biography* 12 (1975): 366-367.

Shaw was Director of the British Meteorological Office from 1905 to 1920. His *Life-History of Surface Air-Currents* (1906), written with R.G.K. Lempfert, "pointed the way toward air-mass analysis and the concept of fronts.... Shaw introduced the principle of isentropic analysis ... and devised a thermodynamic diagram (the tephigram) that is widely used in meteorology.... [His] *Manual of Meteorology* [is] a unique account of the historical roots and the physical and mathematical basis of the subject."

BS27 *William Napier Shaw, M.A., Sc. D., F.R.S.* London: The Biographical Press Agency, 1904. 20 pp.

Shokalsky, Y.M.

BS28 Plakhotnik, A.F. "Shokalsky, Yuly Mikhaylovich" (b. St. Petersburg, Russia, 17 October 1856; d. Leningrad, U.S.S.R., 26 March 1940). *Dictionary of Scientific Biography* 12 (1975): 411-413.

Shokalsky "made a major contribution to cartography, geomorphology, terrestrial hydrology, glaciology, and geodesy. In *Okeanografia* (1917), his most important work, he postulated the mutual dependence of all marine phenomena."

Shtokman, V.B.

BS29 Plakhotnik, A.F. "Shtokman, Vladimir Borisovich" (b. Moscow, Russia, 10 March 1909; d. Moscow, 14 June 1968). *Dictionary of Scientific Biography* 12 (1975): 414-416.

"By applying probability theory and random functions to the study of ocean turbulence, Shtokman was a pioneer ... he also developed direct techniques of measuring turbulent pulsation and current velocity at a series of points.... By the mid-40s Shtokman had become widely known as an eminent Soviet specialist in the theory of ocean currents.... Shtokman was the first to show clearly the important role in the dynamics of ocean currents of the transverse irregularities of tangential stress exerted by wind on water."

Simpson, G.

BS30 Gold, Ernest. "Sir George Simpson." *Nature* 205 (1965): 1156-1157.

Simpson (1878-1964) made his principal scientific contributions to atmospheric electricity, especially the electrification of cloud droplets, and a widely discussed hypothesis on radiative factors as causes of ice ages.

Somerville, M. @

BS31 Sanderson, Marie. "Mary Somerville--Her Work in Physical Geography." *The Geographical Review* 54 (1974): 410-420.

A brief biography and appraisal of the work of Somerville (1780-1872), an early woman scientist with broad interests who already in the 1830s clearly recognized the importance of the heat balance of incoming and outgoing radiation for air temperature and climate in her book *On the Connexion of the Physical Sciences* (1834), which went through eight editions. She also was quite specific on air-sea interactions in her *Physical Geography* (1848).

Sorby, H.C.

BS32 Smith, Cyril Stanley. "Sorby, Henry Clifton" (b. Woodbourne, near Sheffield, England, 10 May 1826; d. Sheffield, 10 March 1908). *Dictionary of Scientific Biography* 12 (1975): 542-546.

By laboratory experiments Sorby deduced the temperature and pressure at which various rocks had been formed, and recognized "the great role played in rock formation by water-bearing magma at high temperature and pressure."

Sprung, A.F.W.

BS33 Grunow, J. "Sprung, Adolf Friedrich Wichard" (b. Kleinow, near Perleberg, Germany, 5 June 1848; d. Potsdam, Germany, 16 January 1909). *Dictionary of Scientific Biography* 12 (1975): 594-596.

"In his daily concern with atmospheric conditions, Sprung became the first to apply the theorems of mathematical physics to the interpretation of meteorological processes. He thereby laid the foundations for the theory of the dynamics of the atmosphere...."

Stagg, J.M.

BS34 Sutcliffe, J.M. "James Martin Stagg, C.B., O.B.E., D. Sc." *Quarterly Journal of the Royal Meteorological Society* 102 (1976): 273-274.

Obituary notice for Stagg (1900-1975) whose principal claim to meteorological fame was his performance as coordinator and meteorological briefing officer for General Eisenhower before the Normandy invasion in 1944 (see his book *Forecast for Overlord* [1971]).

Sternberg, P.K.

BS35 Kulikovsky, P.G. "Sternberg, Pavel Karlovich" (b. Orel, Russia, 2 April 1865; d. Moscow, U.S.S.R., 31 January 1920). *Dictionary of Scientific Biography* 13 (1976): 45-46.

Sternberg promoted cooperative research on the motion of the earth's pole leading to the establishment of the International Latitude Service.

Stewart, B.

BS36 Siegel, Daniel M. "Stewart, Balfour" (b. Edinburgh, Scotland, 1 November 1828; d. Drogheda, Ireland, 19 December 1887). *Dictionary of Scientific Biography* 13 (1976): 51-53.

As director of the Kew Observatory, Stewart was responsible for observations of terrestrial magnetism, meteorology, and solar physics; he proposed hypotheses on electric currents in the upper atmosphere to explain variations of the geomagnetic field.

Stille, W.H.

BS37 Kay, Marshall, "Stille, Wilhelm Hans" (b. Hannover, Germany, 8 October 1876; d. Hannover, 26 December 1966). *Dictionary of Scientific Biography* 13 (1976): 63-65.

"Stille's synthesis of global tectonics has been considered a worthy successor to that of Edward Suess.... His magmatic or volcanic geosynclinal belts of the earth (eugeosynclinal belts) gained wide application."

Störmer, F.C.M. @

BS38 Chapman, Sidney. "Prof. Carl Störmer, For. Mem. R.S." *Nature* 180 (1957): 633-634.

Fredrik Carl Mülertz Störmer (1874-1957) was a Norwegian pioneer in the study of the aurora and the upper atmosphere.

BS39 Chapman, Sydney. "Fredrik Carl Mülertz Störmer, 1874-1957." *Biographical Memoirs of the Royal Society of London* 4 (1958): 257-279, port.

Includes an outline of Störmer's main results on the trajectories of electric changes in the geomagnetic field, and comments on his auroral studies. This memoir is of considerable historical value because of Chapman's extended personal and professional interactions with Störmer. The bibliography is selective. (There is a shorter notice by Chapman in *Nature* 180 (1957): 633-634.)

Stommel, H.

* Warren, Bruce A., and Carl Wunsch, eds. *Evolution of Physical Oceanography. Scientific Surveys in Honor of Henry Stommel*. Cited below as item L56.

Stoneley, R.

BS40 Jeffreys, Harold. "Robert Stoneley, 14 May 1894-2 Feb. 1976." *Biographical Memoirs of the Fellows of the Royal Society* 22 (1976): 555-564.

Stoneley contributed to the theory of seismic wave propagation; this article sets his work in the context of research on the internal structure of the earth.

Strabo

BS41 Warmington, E.H. "Strabo" (b. Amasia, Asia Minor, 64/63 B.C.; d. Amasia, ca. A.D. 25). *Dictionary of Scientific Biography* 13 (1976): 83-86.

In his *Geographica*, "Strabo followed Eratosthenes in showing the known world as a single ocean-girt landmass (*oikoumene*, that is, inhabited). [He] devoted much discussion to the forces that had formed the *oikoumene*." He attributed the formation of islands to volcanic action or earthquakes, and concluded that most of the *oikoumene* had once been submerged.

Süring, R.

BS42 Conrad, Victor. "Reinhard Süring, 1866-1950." *Bulletin of the American Meteorological Society* 33 (1952): 52, 55.

Obituary of the important meteorological researcher, administrator, and co-author (with J. v. Hann, 1839-1921) of an influential textbook.

BS43 Kleinschmidt, E. "Reinhard Süring: Ein Nachruf." *Annalen der Meteorologie* 3 (1950): 368-370.

BS44 König, W. "Nachruf auf Professor Dr. Reinhard Süring." *Zeitschrift für Meteorologie* 5 (1951): 33-34.

An obituary of Süring (1866-1950), who was director of the central meteorological observatory in Potsdam. He participated in a number of free balloon flights for the exploration of the atmosphere (1881-1900) and made major contributions to radiation measurements and cloud physics. He was editor of the *Meteorologische Zeitschrift* for many years and co-author of the influential textbook *Lehrbuch der Meteorologie*, together with J. v. Hann (1839-1921), which went through many editions and introduced many generations of meteorologists to the subject. Its thorough annotation and reference material make it a very important document.

BS45 Steinhauser, Ferdinand. "Reinhard Süring." *Archiv für Meteorologie, Geophysik und Bioklimatologie* A3 (1951): 339-340.

BS46 [Süring, R]. "Veröffentlichungen von Reinhard Süring." *Zeitschrift für Meteorologie* 5 (1951): 129-133.

A list of Süring's publications, 154 in all.

See also the obituary by H. Flohn in *Petermanns Geographische Mitteilungen* 95 (1951): 263.

Suess, E.

BS47 Wegmann, E. "Suess, Eduard" (b. London, England, 20 August 1831; d. Marz, Burgenland, Austria, 26 April 1914). *Dictionary of Scientific Biography* 13 (1976): 143-149.

Suess is known for his treatise on global tectonics, *Das Antlitz der Erde* (1883-1909), which linked together many branches of the earth sciences. He concluded that earthquakes are manifestations of mountain chains in motion and are produced along great faults. "From his conclusion that the ancient massifs have remained stable, Suess deduced that it was the level of the sea that had varied; as a result he proposed the principle of eustatic levels."

Sutcliffe, R.C.

BS48 T[aba], H. "The Bulletin Interviews: Professor R.C. Sutcliffe." *World Meteorological Organization Bulletin* 30 (1981): 169-181.

The editor of the organ of the World Meteorological Organization obtains a "living history" account of the career, experiences, and contributions of a senior British meteorologist with important comments on 20th-century developments in meteorology and leading personalities in the field. Considerable emphasis is placed on his role in the WMO.

Sverdrup, H.U. @

BS49 Wordic, J.M. "Prof. H.U. Sverdrup." *Nature* 180 (1957): 1023.

Obituary of Sverdrup (1888-1957).

See other obituaries: C.F. Brooks, *Bulletin of the American Meteorological Society* 38 (1957): 486; *Bulletin of the World Meteorological Organization* 6 (1957): 158; J.N. Carruthers, *Quarterly Journal of the Royal Meteorological Society* 84 (1958): 89-90; L.P. Kirwan, *Geographical Journal* 123 (1957): 579-580; Olav Listöl, *Journal of Glaciology* 3 (1958): 276; Hans W:son Ahlmann, *Geographical Review* 48 (1958): 284-285; Ludwig Weickmann, *Meteorologische Umschau* 11 (1958): 73; Hans W:son Ahlmann, *Akademia NAUK SSSR Izvestiya Setiya Geograficheska*, No. 2 (1958): 130-132; Raoul Bilancini, *Rivista di Meteorologia Aeronautica* 18 (1958): 67.

Swinden, J.H. Van

BS50 Hackmann, Willem D. "Swinden, Jan Henrik Van" (b. The Hague, Netherlands, 8 June 1746; d. Amsterdam, Netherlands, 9 March 1823). *Dictionary of Scientific Biography* 13 (1976): 183-184.

"Van Swinden was best known outside the Netherlands for his extraordinarily accurate meteorological observations." He also with his students made hourly observations of terrestrial magnetism over a 10-year period.

Swoboda, G.J.H.

BS51 Ashford, Oliver M. "Dr. Gustav Swoboda." *Quarterly Journal of the Royal Meteorological Society* 83 (1957): 414.

Obituary of Swoboda (1893-1956).

BS52 Gordon, A.H. "Obituary--Gustav J.H. Swoboda, Ph.D." *Meteorological Magazine* 85 (1956): 349-350.

Swoboda (1883-1956) made contributions to forecasting as head of the Prague weather office. He became Secretary of the International Meteorological Organization in 1938 and when this became the World Meteorological Organization in 1951 was its first Secretary-General.

BT. Subjects: T

Tannehill, I.R.

BT1 Tenenbaum, Oscar. "Ivan R. Tannehill." *Bulletin of the American Meteorological Society* 40 (1959): 431.

This obituary of Tannehill (1890-1959) recalls his contributions to the knowledge of hurricanes and droughts.

Taylor, F.B.

BT2 Aldrich, Michele L. "Taylor, Frank Bursley" (b. Fort Wayne, Indiana, 23 November 1860; d. Fort Wayne, 12 June 1938). *Dictionary of Scientific Biography* 13 (1976): 269-271.

Taylor proposed that the planets were originally comets captured by the sun and that the moon's capture by the earth pulled the continents away from the poles. He published a detailed theory of continental drift in 1910. He distinguished his theory from Wegener's, relying on sliding of continents along a narrow shear zone rather than isostatic adjustments.

Taylor, G.I.

BT3 Sheppard, P.A. "Sir Geoffrey Taylor, O.M., F.R.S." *Quarterly Journal of the Royal Meteorological Society* 102 (1976): 271-273.

Taylor's (1886-1975) major contributions were on the dynamics of fluid flow. His analyses of turbulence and diffusion have become basic principles in meteorology. He also contributed during World War II to the work on dispersal of fog from runways.

Thompson, A.

BT4 Thomas, M.K. "Andrew Thompson: A Profile." *Atmosphere* 11 (1975): 127-133.

A sympathetic account of Dr. Thompson's work as Deputy and later Head of the Candian meteorological service from 1934 to 1959. This was an extraordinary task during World War II, which required a ten-fold expansion. He had a notable combination of administrative skill and scientific appreciation and also played a major role in international meteorological relations. A bibliography of his writings is included.

Thomson, Sir C.W.

BT5 Thomas, Phillip Drennon. "Thomson, Sir Charles Wyville" (b. Bonsyde, Linlithgow, Scotland, 5 March 1830; d. Bonsyde, 10 March 1882). *Dictionary of Scientific Biography* 13 (1976): 360-361.

Thomson was one of the organizers of the *Challenger* expedition (1872-76), a milestone in the history of oceanography though it was relatively weak on the physical side (Thomson's main interests were in natural history).

Thornthwaite, C.E.

BT6 Flohn, Hermann. "Charles Warren Thornthwaite." *Petermanns Geographische Mitteilungen* 107 (1963): 267-268.

Obituary of a leading American climatologist.

BT7 Landsberg, Helmut E. "C.W. Thornthwaite." *Bulletin of the World Meteorological Organization* 12 (1963): 232-233.

Obituary of Thornthwaite (1899-1963), who devised a widely acclaimed classification of climate and carried out fundamental studies on rainfall distribution and evaporation.

See also: B.L. Dzerdzeevskii, "Charles W. Thornthwaite (1899-1963)," *Meteorologiya i Gidrologiya* 12 (1963): 232-233.

Tilton, J.

BT8 Hagarty, J.H. "Dr. James Tilton: 1745-1822." *Weatherwise* 15 (1962): 124-125.

Biographical note on the Surgeon General of the U.S. Army who in 1814 ordered all post surgeons to make daily meteorological observations. This laid the groundwork for an admirably uniform set of observations, which were the first governmentally sponsored meteorological data collection.

Toulmin, G.H.

BT9 Eyles, V.A. "Toulmin, George Hoggart" (b. Southwark, Surrey, England, September 1754; d. Wolverhampton, England, July 1817). *Dictionary of Scientific Biography* 13 (1976): 441-442.

In his *Antiquity and Duration of the World* (1780), Toulmin rejected attempts to establish a chronology of the earth's history and accepted the Aristotelian belief in the eternity of the world. Though some of his views are similar to those of J. Hutton, they had no influence on the development of geology.

Truman, H.S.

BT10 Beebe, Robert G. "Reminiscences of Harry S. Truman." *Bulletin of the American Meteorological Society* 54 (1973): 44.

Truman was an Associate Member of the American Meteorological Society. While he was President, a weather map had to be on his desk every morning at 8:00 A.M. so he could make his own forecasts, "because the Weather Bureau's were no damn good." Some other anecdotes are reported.

Turner, H.H.

BT11 Bullen, K.E. "Turner, Herbert Hall" (b. Leeds, England, 13 August 1861; d. Stockholm, Sweden, 20 August 1930). *Dictionary of Scientific Biography* 13 (1976): 500-501.

Turner edited the *International Seismological Summary* (1918-27) and revised the travel-time tables of Zoppritz.

BV. Subjects: V

Van der Stok, J.P.

BV1 Van Everdingen, E. "Johannes Paulus van der Stok." *Terrestial Magnetism and Atmospheric Electricity* 39 (1934): 239-240.

Van der Stok (1851-1934) "succeeded in completely disentangling the very complicated tidal phenomena in the East Indian Archipelago and thanks to his work the tidal movements in the Dutch East Indies are better known than in any other like region of the Earth." He also contributed to geomagnetism and climatology. (Portrait, facing p. 173.)

Vangengeim, A.F.

BV2 Suvorov, N.P., and S.P. Khromov. *A.F. Vangengeim, Organizator gidrometeorolicheskoi sluzhby SSSR* [A.F. Vangengeim (Wangenheim), Organizer of the Hydrometeorological Service of the U.S.S.R.]. *Meteorologiya i Gidrologiya* 6 (1965): 44-45.

Vangengeim (1881-1942), who made contributions to agricultural meteorology and long-range forecasting, also wrote a meteorological textbook. In 1929 he reorganized the meteorological service of the Soviet Union.

Vangengeim, G.I.

BV3 Girs, A.A., and L.A. Dydina. "Life and Scientific Activity of G. la. Vangengeim." *Contributions to Long-Range Weather Forecasting in the Arctic*. Jerusalem: Israel Program for Scientific Translations, 1966, pp. 2-12. (A translation from the Russian *Trudy Articheskii i Antartichesskii*, Nauchno-Isledovatelskii Institut No. 255, 1963.)

Vangengeim (1896-1961) established a system of long-range forecasting for the Arctic, based on large-scale atmospheric flow patterns and weather types.

Venetz, I.

BV4 Beer, Gavin de. "Venetz, Ignatz" (b. Visperterminen, Valais, Switzerland, 21 March 1788; d. Saxon-les-Bains, Valais, Switzerland, 20 April 1859). *Dictionary of Scientific Biography* 13 (1976): 604-605.

Confirming and extending the observations of J.P. Perraudin, Venetz proposed that glaciers had extended more widely in earlier ages; he "thus provided the link between the observations of a peasant and the great discovery of a scientist--the Ice Age."

Vening Meinesz, F.A.

BV5 Nieuwenkamp, W. "Vening Meinesz, Felix Andries" (b. Scheveningen, Netherlands, 30 July 1887; d. Amersfoort, Netherlands, 10 August 1966). *Dictionary of Scientific Biography* 13 (1976): 605-611.

The author gives a detailed account of Vening Meinesz's efforts to improve the accuracy of pendulum measurements made without stable support, leading to his successful determinations of gravity anomalies near island arcs. Although he rejected continental drift, he argued that convection currents must flow in the earth's mantle, and his gravity results "challenged every inventive mind to discover their significance."

Vernadsky, V.I.

BV6 Fedoseyev, I.A. "Vernadsky, Vladimir Ivanovich" (b. St. Petersburg, Russia, 12 March 1863; d. Moscow, U.S.S.R., 6 January 1945). *Dictionary of Scientific Biography* 13 (1976): 616-620.

Vernadsky studied the formation of minerals under various physical conditions inside the earth and the distribution of elements in the crust. He "examined the sources of energy under the influence of which tectonic motions and transfers of substance in the earth's crust take place" and "was one of the first to recognize radioactivity as a powerful source of energy."

Vines, B.

BV7 DeAngelis, Richard M. "The Hurricane Priest." *ESSA World* 3(4) (1966): 16-17.

Fr. Benito Vines, director of Belen Observatory near Havana, Cuba, was the 19th century's outstanding authority on West Indian tropical storms. The observational network he established in the 1870s is still the skeleton of the present system. He made the first successful hurricane track predictions. His 1877 book on West Indian hurricanes was a best seller.

Vize, V.Y. @

BV8 "Obituary--Vladimir Yulevich Vize." *Polar Record* 7 (1955): 431-432.

Vize (1888-1954) was a specialist in polar meteorology and pioneered long-range weather predictions and forecasts of sea-ice movements.

Voeikov, A.I. @

BV9 Voeikov, Aleksandr Ivanovich. *Izbrannye sochinennia* [Selected Works]. Moscow: Akademiia Nauk, 1948. 750 pp.

Voeikov (1842-1916), Russia's foremost climatologist, published many important books and papers which are listed here and partially reproduced. An appraisal of his work by A.A. Grigoriev and a biographical sketch by G.D. Rikhter are included.

BV10 Davitaya, F.F. "Aleksandr Ivanovich Voeikov (k sorakaletiiu so dnia smerti)" [Fortieth Anniversary of the Death of Aleksandr Ivanovich Voeikov]. *Meteorologiya i Gidrologiya*, No. 3 (1956): 7-11.

Appraises the achievements of Voeikov in meteorology and climatology.

BV11 Nezdiurov, D.F. "Vospominaniia ob Aleksandro Ivanovich Voeikovo" [Recollections of A.I. Voeikov]. *A.I. Voeikov i sovremennye problemy klimatologii* [A.I. Voeikov and Contemporary Problems in Climatology]. Edited by M.I. Budyko. Leningrad: Gidrometeorolichesko Izdatelstvo, 1956, pp. 22-28.

Personal reminiscences of his teacher Voeikov at St. Petersburg.

BV12 Voeikov, I.D. "Aleksandr Ivanovich Voeikov (Lichny vospominaniya)" [Aleksandr Ivanovich Voeikov (Personal Recollections)]. *Meteorologiya i Gidrologiya* No. 12 (1958): 15-22.

A surviving nephew tells about family, education, and characteristics of the famous Russian climatologist Voeikov.

BW. Subjects: W

Walker, G.T.

BW1 Simpson, George C. "Obituary of Sir Gilbert T. Walker." *Weather* 14 (1959): 67-68.

Renowned for his application of statistics to forecast problems, Walker (1868-1958) as director of the Indian meteorological service attempted long-range forecasts of monsoon intensity. He also discovered the Pacific long-term pressure changes, known as the "southern oscillation."

Other obituaries: Sir Charles Normand, *Nature* 182 (1958): 1706; S. Basu, *Meteorological Magazine* 88 (1959): 90-91.

Walker, J. @

BW2 Walker, John [1731-1803]. *Lectures on Geology, Including Hydrography, Mineralogy, and Meteorology, with an Introduction to Biology*. Edited with notes and introduction by Harold W. Scott. Chicago: University of Chicago Press, 1966. xliv + 280 pp., illus.

John Walker (1731-1803) was Professor of Natural History and Keeper of the Museum, University of Edinburgh, from 1779 to 1803. This edition is based on manuscript notes taken in or before 1788.

Washington, G.

BW3 Hodge, William. "George Washington as a Farmer and Weather Observer." *Weekly Weather and Crop Bulletin* 65, no. 8 (1978): 9.

Points out that descriptions of the weather are found in all his writings. They were fairly systematic in his farming years at Mt. Vernon. An entry on the day of onset of his last illness, 13 December 1799, notes temperature, wind direction, snowfall, and sky condition.

Wegener, A. @

BW4 Benndorf, H. "Alfred Wegener." *Gerlands' Beitrage zur Geophysik* 31 (1931): 337-377.

Biography, list of 170 publications.

BW5 Dornsiepen, V., and V. Haak, eds. *Internationales Alfred Wegener Symposium, 1980: Kurzfassungen der Beitrage*. Berliner Geowissenschaftliche Abhandlungen: Reihe A, Bd. 19. Berlin: Reimer, 1980. 263 pp.

BW6 Georgi, Johannes. *Im Eis Vergraben; Erlebrisse auf Station "Eismitte" der letzten Groeland-Espedition Alfred Wegeners*. München: Verlag des Blodigschen Alpen Kalendars, P. Muller, 1933. 224 pp. English translation: *Mid-Ice, the Story of the Wegener Expedition to Greenland*. Translated by F.H. Lyon, revised and supplemented by the author. London: Kegan, Paul, Trench, Trübner & Co., 1934. xiii + 247 pp.

BW7 Georgi, Johannes. "Memorial to Alfred Wegener on His 80th Anniversary." *Polarforschung*, Beiheft No. 2. Kiel, 1960. 104 pp.

Monograph on Wegener's Greenland expedition.

BW8 Jacobshagen, Volker, ed. *Alfred Wegener 1880-1930, Leben und Werk: Ausstellung anlässlich der 100. Wiederkehr seines Geburtsjahres (katalog)*. Berlin: Dietrich Reimer Verlag, 1980. 60 pp.

A profusely illustrated catalog of the exhibit held in honor of the 100th anniversary of Alfred Wegener's birth in West Berlin. It contains a large number of biographical items, pictures from his expeditions, illustrations from his works, list of his principal contributions, and biographies written by others. Although best known for his theory of continental drift, he also made very substantial contributions to atmospheric science, especially thermodynamics, and to paleoclimatology, the latter jointly with his father-in-law, Wladimir Koeppen (1846-1940).

BW9 Körber, H.-G. "Alfred Wegener (1880-1930): Zum 100. Geburtstag und 50. Todestag des Gelehrten." *Zeitschrift für Meteorologie* 31 (1981): 327-341.

1980 was the 100th anniversary of Alfred Wegener's birth and the 50th of his death. This biographical sketch appraises his contributions to geophysics and meteorology and outlines the career of the originator of the theory of continental drift.

BW10 Schmauss, August. "Alfred Wegeners Leben und Wirken las Meteorologie." *Annalen der Meteorologie* 4 (1951): 1-13.

A memorial to Wegener with a summary of his principal meteorological publications.

BW11 Schwarzbach, Martin. *Alfred Wegener und die Drift der Kontinente*. Stuttgart: Wissenschaftliche Verlagsgesellschaft, 1980. 159 pp.

BW12 Wegener, Else. *Alfred Wegener: Tagebücher, Briefe, Erinnerungen*. Wiesbaden: Brockhaus, 1960. 260 pp.

Alfred Lothar Wegener, who won fame for his theory of continental drift but made important contributions to atmospheric thermodynamics, left behind many diaries and notes which are here excerpted with other biographical information by his widow.

BW13 Wegener, Else, ed. *Alfred Wegeners letzte Groenlandfahrt; Die Erlebnisse der Deutschen Groenlandexpedition 1930/1931 geschildert von seine Reisegelfahrten und nach Tagekuchern des Forchers*. Leipzig: Broekhaus, 1932. 303 pp. English translation: *Greenland Journey: The Story of Wegener's German Expedition to Greenland in 1930-31 as Told by Members of the Expedition and the Leader's Diary*. Translated by W.M. Deans. London: Blackie, 1939. xx + 239 pp.

Weickmann, L.

BW14 Schwerdtfeger, Werner. "Ludwig Weickmann." *Meteorologische Rundschau* 15 (1962): 1-3.

Weickmann (1881-1961) contributed particularly to knowledge of waves in the atmosphere and to analysis of periodicities. He reorganized the West German Weather Service after World War II.

See also obituaries: Raoul Bilancini, *Revista di Meteorologia Aeronautica* 22 (1962): 62-63; Hermann Flohn, *Beiträge zur Physik der Atmosphäre* 35 (1962): 141-144.

Wells, W.C.

BW15 Dock, William. "Wells, William Charles" (b. Charleston, South Carolina, 24 May 1757; d. London, England, 18 September 1817). *Dictionary of Scientific Biography* 14 (1976): 253-255.

"Wells' most important contribution was his meticulous study of the formation of dew...."

Werner, A.G. @

BW16 Guntau, Martin. "Der Aktualismus bei A.G. Werner. Mit einen Fragment zum Aktualismus aus dem handschriftlichen Nachlass von A.G. Werner)." *Bergakademie, Wissenschaftliche Zeitschrift für den Berghan* 19 (1967): 294-297.

Werner, J.

BW17 Folkerts, Menso. "Werner, Johann(es)" (b. Nuremberg, Germany, 14 February 1468; d. Nuremberg, May [?] 1522). *Dictionary of Scientific Biography* 14 (1976): 272-277.

"... he can be regarded as a pioneer of modern meteorology and weather forecasting."

Wexler, H.

BW18 Chapman, Sidney. "Harry Wexler 1911-1962, Excellent and Devoted Student of Our Wonderful Atmosphere." *Bulletin of the American Meteorological Society* 46 (1965): 226-240.

Wexler memorial lecture: after an appraisal of Wexler's contributions to meteorology, discusses author's work on the high atmosphere.

BW19 Neiburger, Morris. "Harry Wexler, 1911-1962." *Bulletin of the American Meteorological Society* 43 (1962): 579-580.

Obituary of Wexler who directed the research of the U.S. Weather Bureau after World War II and was Chief Scientist of the U.S. Antarctic Program during the International Geophysical Year.

See also obituaries: Sir Oliver Graham Sutton, *Nature* 196 (1962): 318-319; Frank J. Malina, *Astronautica Acta* 8 (1962): 401; G. de Q. Robin, *Journal of Glaciology* 4 (1963): 496-497, with portrait.

Whitehurst, J.

BW20 Challinor, John. "Whitehurst, John" (b. Congleton, Chesire, England, 10 April 1713; d. London, England, 18 February 1788). *Dictionary of Scientific Biography* 14 (1976): 311-312.

Challinor mentions Whitehurst's book *An Inquiry into the Original State and Formation of the Earth* (1778) as being "well known" but dismisses it with the comment that "the first part is entirely speculative."

Wiechert, E.

BW21 Bullen, K.E. "Wiechert, Emil" (b. Tilsit, Germany, 26 December 1861; d. Göttingen, Germany, 19 March 1928). *Dictionary of Scientific Biography* 14 (1976): 327-328.

At Göttingen, Wiechert founded "one of the world's most famous schools of geophysics," whose pupils included B. Gutenberg, K. Zöppritz, L. Geiger, and G. Angenheister. He was "one of the first to suggest the presence of a dense core in the earth and ... produced the first theoretical model of the planet that allowed for it." He organized research in seismology, invented the inverted-pendulum seismograph, and developed the mathematical theory of seismic wave propagation.

Wild, H.

BW22 Biswas, Asit K., and Margaret R. Biswas. "Wild, Heinrich" (b. Uster, Zurich canton, Switzerland, 17 December 1833; d. Zurich, Switzerland, 5 September 1902). *Dictionary of Scientific Biography* 14 (1976): 356-357.

"Wild was an active meteorologist who played a major part in the development of the science." He improved several instruments and extended the network of observation stations.

Willis, B.

BW23 Waters, Aaron C. "Willis, Bailey" (b. Idlewild-on-Hudson, New York, 31 March 1857; d. Palo Alto, California, 19 February 1949). *Dictionary of Scientific Biography* 14 (1976): 402-403.

Willis designed scaled-down experiments in a "pressure box" to stimulate mountain formation. He became well known for his research on earthquakes in California but could not persuade politicians to pass a building code requiring better construction of structures susceptible to earthquake damage.

Wilson, J.T.

BW24 Wilson, J. Tuzo. "Early Days in University Geophysics." *Annual Review of Earth and Planetary Sciences* 10 (1982): 1-14.

The author presents autobiographical reflections and comments on some of his colleagues.

Woltman, R.

BW25 Frazier, Arthur H. "Woltman, Reinhard" (b. Axstedt, Germany, December 1757; d. Hamburg, Germany, 20 April 1837). *Dictionary of Scientific Biography* 14 (1976): 494-495.

Woltman invented one of the first water current meters suitable for measuring the velocity of water flowing in rivers.

Woodward, J.

BW26 Eyles, V.A. "Woodward, John" (b. Derbyshire, England, May 1665; d. London, England, 25 April 1728). *Dictionary of Scientific Biography* 14 (1976): 500-503.

In his *Essay Toward a Natural History of the Earth* (1695), Woodward "assumed that the earth formerly had been submerged beneath a universal deluge, the waters of which had originated in a central abyss...."

Woodward, R.S.

BW27 Reingold, Nathan. "Woodward, Robert Simpson" (b. Rochester, Michigan, 21 July 1849; d. Washington, D.C., 29 June 1924). *Dictionary of Scientific Biography* 14 (1976): 503-504.

Woodward "calculated the effects on shore lines of the removal of superficial masses" and criticized Kelvin's estimates of the cooling time of the earth. As President of the Carnegie Institution of Washington he was responsible for its research-support policies.

Wright, G.F.

BW28 Morison, William J. "Wright, George Frederick" (b. Whitehall, New York, 22 January 1838; d. Oberlin, Ohio, 20 April 1921). *Dictionary of Scientific Biography* 14 (1976): 516-518.

"Wright advocated the relatively late end of the Ice Age, approximately ten thousand years ago. Further, he contended that there had been only one Ice Age."

BZ. Subjects: Z

Zenkevich, L.A.

BZ1 *Okeanologiia* 20, no. 5 (September-October 1980).

Issue commemorating Zenkevich (1889-1970); includes Z.A. Filatova and N.G. Vinogradova, "Academician L.A. Zenkevich and His Role in the Development of the Ocean Science"; A.P. Lisitzin, "Works of L.A. Zenkevich and the Development of Geology of the Seas and Oceans."

Zubov, N.N.

BZ2 Plakhotnik, A.F. "Zubov, Nikolay Nikolaevich" (b. Izmail, Russia [now Ukrainian S.S.R.], 23 May 1885; d. Moscow, U.S.S.R., 11 November 1960). *Dictionary of Scientific Biography* 14 (1976): 634-636.

Zubov's research concentrated on three major problems: the vertical mixing of seawater, ocean currents, and sea ice. Rejecting the term "oceanography" as implying mere description, he considered himself an "oceanologist" who "attempted to penetrate the very essence of the processes that he studied."

C. SOCIAL AND INSTITUTIONAL HISTORY; INTERNATIONAL PROJECTS

C1 Atwood, Wallace Walter. *The International Geophysical Year in Retrospect*. Washington: Public Services Division, Bureau of Public Affairs, 1959. 8 pp.

"Reprinted from the Dept. of State Bulletin of May 11, 1959."

C2 Biermann, Kurt R. "Alexander von Humboldt als Initiator und Organisator internationaler Zusammenarbeit auf geophysikalischem Gebiet." *Human Implications of Scientific Advance. Proceedings of the XVth International Congress of the History of Science, Edinburgh 10-15 August 1977*. Edited by E.G. Forbes. Edinburgh: Edinburgh University Press, 1978, pp. 126-138.

C3 Chapman, Sydney. *IGY: Year of Discovery: The Story of the International Geophysical Year*. Ann Arbor: University of Michigan Press, 1959. 111 pp.

"This book offers a popular account of some scientific aspects of the earth and sun--in special connection with the 1957-58 enterprise...." Numerous illustrations, no bibliography.

C4 Corby, G.A. "The First International Polar Year (1882/83)." *World Meteorological Organization Bulletin* 31 (1982): 197-214.

Inspired by Karl Weyprecht, an Austrian naval officer, 12 Arctic stations for meteorological and geophysical studies were established inside the Arctic Circle by 11 cooperating nations by special expeditions. Aside from the first regular weather observations from the Arctic, they provided indisputable proof of the link between auroral and geomagnetic phenomena.

* Dettwiller, Jacques. *Chronologie de quelques événements météorologiques*.... Cited below as item M9.

C5 Dettwiller, J. "La révolution de 1789 et la météorologie." *Bulletin d'Information, Direction de la Météorologie, Ministère des Transports* 40 (1978): 24-33.

Summarizes the weather conditions that contributed to the French revolution. Contains contemporary observations and quotations on weather events.

* Ficker, H. *Die Zentralanstalt für Meteorologie und Geodynamik, 1851-1951*. Cited below as item NB17.

C6 Fleming, John A. "Origin and Development of the American Geophysical Union." *Transactions, American Geophysical Union* 35 (1954): 1-46.

Founded under the aegis of the National Academy of Sciences in 1919 to represent the United States in the International Union of Geodesy and Geophysics, this scientific society has become the principal forum for geophysics in the country.

C7 Fraser, Ronald George Juta. *Regard the Earth: The Story of the International Geophysical Year, 1957-1958*. Foreword by James Wordie. London: Science Club, 1957. 24 pp.

C8 Gerson, N.C. "From Polar Years to IGY." *Advances in Geophysics* 5 (1958): 1-52.

Karl Weyprecht was primarily responsible for fostering the First Polar Year (1882-83). Johannes Georgi proposed the second one, held in 1932-33. Lloyd Berkner proposed a Third Polar Year but since (following a suggestion of S. Chapman) tropical observations were also included, the name was changed to International Geophysical Year. This article describes the participation of various countries in these programs and summarizes their achievements. While they contributed to the advances of geophysics, they "accepted as a measure of accomplishment the standards of the production factory" (number of men employed, cost, etc.) and did not, up to the time of writing, produce any great breakthrough. "Follow-up in terms of research on the accumulated data has been noticeably small" (p. 54).

* Hall, D.H. *History of the Earth Sciences During the Scientific and Industrial Revolutions, with Special Emphasis on the Physical Geosciences*. Cited above as item A48.

C9 Ingersoll, Robert C. "The Effect of the Coast Survey on 19th Century American Science." *Synthesis* (Cambridge, Mass.) 2, no. 4 (Winter 1975): 22-32.

"This essay explores one theme, seen throughout the history of the Coast Survey, of how the Survey was a leader in establishing new, higher scientific standards and encouraging original scientific research in many areas.... These trends began with the Coast Survey's first superintendent, Ferdinand Rudolph Hassler; they were expanded and further established by Hassler's successor, Alexander Dallas Bache, and they continue today" (Author's summary).

C10 Koenig, M. "Die Deutsche Geophysikalische Gesellschaft, 1922-1974." *Zur Geschichte der Geophysik* (item A13), pp. 3-13.

C11 Korotkevich, E.S. "The Polar Regions in World Geophysical Programmes." *World Meteorological Organization Bulletin* 31 (1982): 231-234.

An appraisal of the results of the International Polar and the International Geophysical Years.

C12 Laursen, V. "The Second International Polar Year (1932/33)." *World Meteorological Organization Bulletin* 31 (1982): 214-221.

19 nations participated in the Second International Polar Year. 27 scientific teams in the Arctic were occupied, many of them at the sites of the First International Polar Year, many of them on the coast of Greenland. Equipment and program were more sophisticated than 50 years earlier and aircraft reconnaissance became part of the effort. Although World War II interrupted the final evaluation of the data, these are preserved for further study in a Danish archive.

C13 Marshack, Alexander. *The World in Space: The Story of the International Geophysical Year*. New York: T. Nelson, 1958. 176 pp.

A popular account of the historical background of the various problems to be studied during the IGY, with a description of some of the instruments and techniques to be used.

* Neumann, J., and D.A. Metaxas. "The Battle Between the Athenian and Peloponnesian Fleets, 429 B.C., and Thucydides' *Wind from the Gulf* (of Corinth)." Cited below as item S16.

C14 Nicolet, Marcel. "The International Geophysical Year (1957/58)." *World Meteorological Organization Bulletin* 31 (1982): 222-231.

This beautifully illustrated article treats the first global geophysical observation program, including both arctic and antarctic observations, with 67 participating nations. Aside from many important scientific results, the effort led to the establishment of cooperative world data centers and initiation of the space age.

C15 Perrier, Georges. *La coopération internationale en géodésie et en géophysique: Troisième assemblée générale de l'Union Géodésique Internationale à Prague, Septembre 1927*. Paris: Gauthier Villars, 1927. 120 pp.

C16 Reingold, Nathan. *Preliminary Inventory of the Records of the Coast and Geodetic Survey*. National Archives Publication, 59:3. Washington, D.C.: National Archives and Records Service, 1958. v + 83 pp.

"Described in this inventory are primary source materials for the history of the scientific fields of interest to the Coast and Geodetic Survey [including geomagnetism, seismology, hydrography, tides, and currents as well as geodesy]; for the biographies of the many notable scientists who worked for the agency [e.g., F.R. Hassler, A.D. Bache, B. Peirce]; and for information on the development of particular instruments and techniques." The records pertain mainly to the 19th century.

C17 Roberts, Elliott B. "Coast and Geodetic Survey. Highlight of 150 Years." *U.S. Naval Institute Proceedings* (February 1957): 188-211.

A brief history with no references but many illustrations.

C18 Roberts, E.B. "The Service of the United States Coast and Geodetic Survey." *Boletín Bibliográfico de Geofísica y Oceanografía Americanas*, 1, parte Geofísica (1958): 141-146.

Brief history with list of selected general sources.

C19 Roberts, Elliot B. "The IGY in Retrospect." *Smithsonian Institution Annual Report, 1959*. Washington, D.C.: Smithsonian Institution, 1960, pp. 263-284.

Brief survey of the major results (not historically presented).

C20 Robin, G. de Q. "The Scott Polar Research Institute." *World Meteorological Organization Bulletin* 31 (1982): 234-237.

The Scott Polar Research Institute was established in 1920 at Cambridge, England. It specializes in the conduct, documentation, and publication of polar research, mostly of a geophysical nature.

C21 Ross, Frank [Xavier]. *Partners in Science: The Story of the International Geophysical Year*. New York: Lothrop, Lee & Shepard, 1961. 192 pp.

A popular account of the major scientific problems investigated in the IGY, with brief sections on their historical background. Numerous photographs show how observations were made.

C22 Shrock, Robert Rakes. *Geology at M.I.T., 1865-1965: A History of the First Hundred Years of Geology at Massachusetts Institute of Technology*. Vol. 1, The Faculty and Supporting Staff. Cambridge, Mass.: M.I.T. Press, 1977. 1032 pp., ports., bibl.

Includes chapters on W.B. Rogers (1804-1882), T.S. Hunt (1826-1892), R.A. Daly (1871-1957), L.B. Slichter (1891-1969), G.J.F. MacDonald, R.R. Doell, W.S. Von Arx, and F. Press.

C23 Spasskii, Boris I., Leonid V. Levshin, and V.A. Krasil' nikov. "Physics and Astronomy at Moscow University: On the University's 225th Anniversary." *Soviet Physics Uspekhi* 23 (1980): 78-93. Translated from *Uspekhi Fizicheskikh Nauk* 130 (1980): 149-175.

Monitors briefly the research on the Kursk Magnetic Anomaly and contributions to physical oceanography of V.V. Shuleikin, and gravimetric measurements by A.A. Miklhailov and L.V. Sorokin, in the 1930s. In the 1960s, research on the earth's radiation belts was pursued by S.N. Vernov, A.E. Chudakov, and B.A. Tverskii. Current geophysical research is noted at the end of the article.

C24 Special Committee for the International Geophysical Year. *Annals of the International Geophysical Year, 1957-1958*. Vol. 1: *The Histories of the International Polar Years and the Inception and Development of the International Geophysical Year*. New York: Pergamon Press, 1959. xi + 446 pp.

Contents: Introduction by S. Chapman; "The First International Polar Year" by N.H. de V. Heathcote and A. Armitage; "The Second International Polar Year" by J. Bartels, V. Laursen, C.E.P. Brooks, J. Paton, W.J.G. Beynon, E.H. Vestine, T. Nagata; "The Inception and Development of the International Geophysical Year" by H.S. Jones. Topics include meteorology, magnetism, atmospheric electricity, aurora and ionosphere observations. Each article is followed by a French translation.

C25 Stanley, Albert A. "Sesquicentennial of Coastal Charting. 150 Years Service of the United States Coast and Geodetic Survey." *The Military Engineer* 327 (January-February 1957): 1-5.

A brief history.

C26 Sullivan, Walter. *Assault on the Unknown: The International Geophysical Year*. New York: McGraw-Hill, 1961. 460 pp.

A comprehensive account with some historical material and adequate references.

C27 Taba, H. "An Interview with Dr. F.W.G. Baker, Executive Secretary of ICSU." *World Meteorological Organization Bulletin* 31 (1982): 187-196.

The International Council of Scientific Unions will commemorate in 1982/83 the centenary of the first International Polar Year 1882/83, the second Polar Year 1932/33, and the International Geophysical Year 1957/58. The Secretary of ICSU reviews the historical events leading to these cooperative international undertakings, with some historical photographs.

C28 Toperczer, Max. *Die Geschichte der Geophysik an der Zentralanstalt für Meteorologie und Geodynamik*. Wien: Zentralanstalt für Meteorologie und Geodynamik (Arbeiten, Heft 17; publikation nr. 208), 1975. 24 pp.

For additional information on this topic see works cited in the biography sections on the following scientists:

Davidson, G. Woodward, R.S.

D. ORIGIN AND DEVELOPMENT OF THE EARTH; PLANETARY COSMOGONY

D1 Allan, D.W. "Kelvin on the Origin of the Earth." *Nature* 177 (1956): 801-802.

Notes that T. Gold's "pore" theory is similar to one proposed by Lord Kelvin in 1864. Gold replies briefly (p. 802).

D2 Bargrave-Weaver, D. "The Cosmogony of Anaxagoras." *Phronesis* 4(2) (1959): 77-91.

An interpretation of some disputed phrases, leading to the conclusion that the separation and regrouping of the elements was conceived as a set of simultaneous rather than sequential processes.

D3 Becker, George F. "Kant as a Natural Philosopher." *American Journal of Science* 4th series, 5 (1898): 97-113.

A survey of the writings of the German philosopher Immanuel Kant (1724-1804) on the origin of the solar system and related topics in planetary physics.

D4 Berlage, H.P. *The Origin of the Solar System*. Oxford: Pergamon Press, 1968. vii + 130 pp.

A survey and critique of all the major theories including his own.

D5 Bhattacharyya, Narendra Nath. *History of Indian Cosmogonical Ideas*. New Delhi: Munshiram Menoharlal, 1971. xvi + 138 pp.

Contents: Rigvedic cosmogony, cosmogonies of the later Samhitas and the Brahmanas, of the principal Upanishads, the Sankhya; philosophical cosmogonies; Buddhist and Jain, theistic, materialistic, and atheistic cosmogonies; brief comparison with other countries.

D6 Both, Ernst E. *A History of Lunar Studies*. Buffalo: Buffalo Museum of Science, 1961. 34 pp. Published in part in

Science on the March (Buffalo Museum of Science) 41 (1960-61).

Includes views of Marshall von Bieberstein, Moll, and Gruithuisen on the origin of the moon.

D7 Brooke, J.H. "Nebular Contraction and the Expansion of Naturalism." *British Journal for the History of Science* 12 (1979): 200-211.

Essay review of item D55, with remarks on works by Merleau-Ponty, Jaki, Schweber, P. Lawrence, P. Baxter, etc.

D8 Brush, Stephen G. "The Origin of the Planetesimal Theory." *Origins of Life* 8 (1977): 3-6.

On the theory of T.C. Chamberlin.

D9 Brush, Stephen G. "A Geologist Among Astronomers: The Rise and Fall of the Chamberlin-Moulton Cosmogony." *Journal for the History of Astronomy* 9 (1978): 1-41, 77-104.

At the end of the 19th century the Laplacian nebular hypothesis was still generally accepted by astronomers, though known to be inadequate in many respects. The American geologist Thomas Chrowder Chamberlin (1843-1928) was led by his research on the ice age to doubt the assumption that the earth was initially a hot liquid ball, and he adopted instead the hypothesis that it had been formed by aggregation of cold solid particles which he called "planetesimals." A young astronomer, Forest Ray Moulton (1872-1952), helped in working out the technical details of Chamberlin's theory and in putting together in a systematic and convincing way the objections to the nebular hypothesis. Chamberlin and Moulton subsequently concluded that the solar system had formed as a result of the interaction of the sun with a passing star. This assumption was also adopted in the "tidal" theory later proposed by Harold Jeffreys and James Jeans in England.

Astronomers generally accepted some version of the Chamberlin-Moulton or Jeans-Jeffreys theory of the origin of the solar system until the mid-1930s, when H.N. Russell and others published effective criticisms. For several years there was no satisfactory alternative. This case history provides a counter-example to the contention of philosopher Imre Lakatos that scientists will not abandon a theory until a better one is available; but the Chamberlin-Moulton theory can be interpreted as an "interfield theory" of the type recently discussed by other philosophers of science such as Lindley Darden and Nancy Maull.

D10 Brush, Stephen G. "From Bump to Clump: Theories of the Origin of the Solar System 1900-1960." *Space Science Comes of Age: Perspectives in the History of the Space Sciences*. Edited by Paul A. Hanle and Von Del Chamberlain. Washington, D.C.: National Air and Space Museum/Smithsonian Institution Press, 1981, pp. 78-100.

The article concentrates on the work of T.C. Chamberlin (summarizing the more detailed account in the preceding item), H.N. Russell, and H.C. Urey. After Russell's refutation of the encounter theory (Chamberlin-Moulton and Jeans-Jeffreys) in 1935, the major developments were the revival of the planetesimal hypothesis, the concept of "magnetic braking" of the sun's rotation, suggestions that the sun had encountered an interstellar cloud and captured from it the material that later formed planets, claims for discovery of extrasolar planetary systems, and research on cosmic abundances of the elements. The nebular hypothesis was revived in a new form by C.F. von Weizsäcker in 1944, and later modified by Gerard Kuiper. Urey started from Kuiper's theory and emphasized the formation of the earth from a cold dust cloud. In 1951 and in several later papers he stressed the significance of the moon's present surface as a record of conditions in an earlier stage of the history of the solar system. He proposed that the moon was one of the primary objects of the solar system, perhaps older than the earth. His theory could make specific predictions to be tested by lunar exploration, and thus influenced the early space program.

D11 Brush, Stephen G. "Nickel for Your Thoughts: Urey and the Origin of the Moon." *Science* 217 (1982): 891-898.

"The theories of Harold C. Urey (1893-1981) on the origin of the moon are discussed in relation to earlier ideas, especially George Howard Darwin's fission hypothesis. Urey's espousal of the idea that the moon had been captured by the earth and has preserved information about the earliest history of the solar system led him to advocate a manned lunar landing. Results from the Apollo missions, in particular the deficiency of siderophile elements in the lunar crust, led him to abandon the capture selenogony and tentatively adopt the fission hypothesis" (Author's summary).

D12 Chambers, Robert. *Vestiges of the Natural History of Creation*. London, 1844. Reprint, with Introduction by Gavin de Beer, Leicester: Leicester University Press; New York: Humanities Press, 1969. 38 + vi + 390 pp.

This book, published anonymously, stimulated much public discussion of evolution and established a connection between views of the origin and early development of the earth (based on the nebular hypothesis) and the concept of biological evolution.

D13 Chilmi, G.F. *200 Jahre wissenschaftlicher Kosmogonie.* Moscow: Verlag "Das Wissen," 1955. 32 pp.

For Russian edition see Khil'mi (item D40).

D14 Clerke, Agnes. *Modern Cosmogonies.* London: Black, 1905. 287 pp.

A lively survey of the development and criticisms of the nebular hypothesis, and some of the alternatives proposed in the late 19th century. Highly recommended.

* Cloud, Preston, ed. *Adventures in Earth History.* Cited above as item A21.

D15 Cohen, I.B. "Galileo, Newton, and the Divine Order of the Solar System." *Galileo, Man of Science.* Edited by Ernan McMullin. New York: Basic Books, 1968, pp. 207-231.

On the "Galileo-Plato problem" of creating the solar system by dropping each planet from a definite place, allowing it to accelerate to its present speed, then changing its linear motion to circular.

D16 Collier, Katharine Brownell. *Cosmogonies of Our Fathers. Some Theories of the Seventeenth and Eighteenth Centuries.* New York: Columbia University Press, 1934. Reprint, New York: Octagon Books, 1968. 500 pp.

Includes sections on Descartes, Kircher, S. Bochart, E. Stillingfleet, T. Burnet, J. LeClerc, W. Whiston, J. Woodward, J. Witty, W. Derham, E. Swedenborg, Buffon, Thomas Wright of Durham, J.A. De Luc.

D17 Cowen, Richard, and Jere H. Lipps, eds. *Controversies in the Earth Sciences, A Reader.* St. Paul, Minn.: West Publishing Co., 1975. x + 439 pp.

Anthology of research papers, mostly from the early 1970s, including theories of the origin of earth and moon by A.G.W. Cameron (1970), J.A. O'Keefe (1970), S.F. Singer (1970), K.K. Turekian and S.P. Clark (1969), D.L. Anderson and T.C. Hanks (1972), W.M. Kaula and A.W. Harris (1973); papers on plate tectonics and hot spots by A. Hammond (1971), W. Jason Morgan (1972), K. Burke, W.S.F. Kidd, and J.T. Wilson (1973), R.B. Hargraves and R.A. Duncan (1973), P. Molnar and T. Atwater (1973).

D18 Davies, Gordon L. "Robert Hooke and His Conception of Earth History." *Proceedings of the Geological Association* 75, no. 4 (1964): 493-498.

D19 Downing, A.M.W. "Swedenborg as Cosmologist." *Journal of the British Astronomical Association* 22 (1912): 196-197.

Author awards priority for the nebular hypothesis to Emanuel Swedenborg (1688-1722), whose *Principia Rerum Naturalium* appeared in 1734.

D20 Drake, Stillman. "Galileo's 'Platonic' Cosmogony and Kepler's *Prodromus*." *Journal for the History of Astronomy* 4 (1973): 174-191.

"In the opening pages of the *Dialogue*, Galileo did offer a speculation about the creation of the planets at some place from which they were dropped in uniform acceleration to reach their present speeds and distances from the sun. Ancestral in a way to later 'capture' theories, Galileo's cosmogony differed from those in that it required a special act of the Creator to turn each planet into circular orbit at the moment it reached the speed at which it was destined forever to revolve" (p. 174). "Galileo's cosmogony originated in a mechanical theory involving arithmetical calculations from astronomical data.... It was not, as some have thought, a mere armchair speculation by a credulous Platonist ..." (p. 190).

D21 Eberhard, Gustav. *Die Cosmogonie von Kant*. Wien: Frick, 1893. 34 pp.

Presents a comparison of the Kant and Laplace theories, and a mathematical development of Kant's propositions.

D22 Eksinger, D. "Kosmogonicka hipoteza Thomas Jeffersona See-a." *Zemlja i Svemir* 2 (1959): 37-40.

On the cosmogonic theory of the American astronomer T.J.J. See (1866-1962).

* Engelhardt, Wolf von. "Krafte der Tiefe--Zum Wandel Erdgeschichtlicher Theorien." Cited below as item F14.

D23 Eyles, V.A. "Bibliography and the History of Science." *Journal of the Society for the Bibliography of Natural History* 3 (1955): 63-71.

An account of his study of early theories of the earth (T. Burnet, J. Woodward, J. Hutton).

D24 Fedossev, Ivan A. "Development of Conceptions on the Origin and Accumulation of Water on the Earth." *Actes, XIIe Congrès International d'Histoire des Sciences, Paris 1968* 7 (pub. 1971): 31-34.

A brief survey of ideas from Plato to W. Rubey (1951) and A. Vinogradov (1959). For further details see the author's book (item K11).

D25 Felber, H.-J. "Kants Beitrag zur Frage der Verzögerung der Erdrotation." *Die Sterne* 50 (1974): 82-90.

D26 [Fesenkoff, V.G.] Struve, O. "Review of *Cosmogony of the Solar System* (in Russian) by V.G. Fesenkoff, Moscow, 1944." *Astrophysical Journal* 102 (1945): 264-266.

Includes the reviewer's recollections of a colloquium 30 years earlier at which Fesenkoff reported on Poincaré's *Leçons*.

D27 Günther, S. "Die Entstehung der Lehre von der meteoritischen Bildung des Erdkörpers." *Sitzungsberichte der mathematisch-physikalischen Klasse der Königlichen Bayerischen Akademie der Wissenschaften zu München* 38 (1908): 21-39.

On the theories of Karl Ehrenbert v. Moll.

D28 Haar, D. ter, and A.G.W. Cameron. "Historical Review of Theories of the Origin of the Solar System." *Origin of the Solar System*. Edited by R. Jastrow and A.G.W. Cameron. New York: Academic Press, 1963, pp. 1-37.

Includes a discussion of Descartes' vortex theory, the Kant-Laplace nebular hypothesis, several theories of H.P. Berlage, Jr., ideas about the encounter of two stars, role of the solar magnetic field as studied by H. Alfvén, turbulence, F.L. Whipple's dust-cloud hypothesis, variations of C. von Weizsäcker's theory, and the Russian school of cosmogony. This review is more useful for its technical criticism of the theories from the viewpoint of two modern researchers than for its insight into historical development.

D29 Herczeg, Tibor. "Planetary Cosmogonies." *Vistas in Astronomy* 10 (1968): 175-206.

A review of theories developed since the mid-1930s.

D30 Hobbs, William. *The Earth Generated and Anatomized. An Early Eighteenth Century Theory of the Earth.* Edited and with an introduction by Roy Porter. Ithaca, N.Y.: Cornell University Press, 1981. 158 pp.

Based on a manuscript acquired by the British Museum (Natural History) in 1973, previously unknown. In addition to a theory of the origin and development of the earth, Hobbs presents a discussion of the origin and cause of tides (which he denies are due to lunar influence), and the raising of mountains by the internal heat of the earth. His general vision of nature allowed no sharp distinctions between animate and inanimate.

See also Roy Porter, "William Hobbs of Weymouth and His *The Earth Generated and Anatomized* (?1715)." *Journal of the Society for the Bibliography of Natural History* 7 (1976): 333-341.

D31 Hoppe, H. "Die Kosmogonie Emanuel Swedenborgs und die Kant'sche und Laplace'sche Theorie." *Archiv für Geschichte der Philosophie* 25 (1911): 53-68.

D32 Jaki, Stanley L. "The Five Forms of Laplace's Cosmogony." *American Journal of Physics* 44 (1976): 4-11.

"Between 1796 and 1824, Laplace published five forms of his theory of the origin of the solar system--a fact ignored in the literature on Laplace and cosmogony. Yet, those five forms not only reveal Laplace's undue fondness for his theory, but also give a glimpse of his biased handling of recent discoveries in astronomy and of his reluctance to test in a quantitative way some grave dynamical problems implied in his theory, later called the nebular hypothesis" (Author's summary).

D33 Jaki, Stanley L. *Planets and Planetarians: A History of Theories of the Origin of Planetary Systems.* New York: Wiley, 1978. vi + 266 pp.

This is the only comprehensive account of theories of the origin of the solar system from Plato to Hoyle, including speculations about life on other planets. It is marred by the author's hypercritical attitude toward almost all writings on the subject.

D34 Jones, Kenneth Glyn. "The Observational Basis for Kant's *Cosmogony*: A Critical Analysis." *Journal for the History of Astronomy* 2 (1971): 29-34.

D35 Kant, Immanuel. *Kant's Cosmogony, as in His Essay on the Retardation of the Rotation of the Earth and His Natural History and Theory of the Heavens*. Translated by W. Hastie. Revised and edited with an introduction and appendix by Willey Ley. New York: Greenwood, 1968. xx + 183 pp.

D36 Kant, Immanuel. *Universal Natural History and Theory of the Heavens*. New Introduction by Milton K. Munitz. Ann Arbor: University of Michigan Press, 1969. xxii + 180 pp.

Reprint of Hastie's (incomplete) translation, including Kant's 1754 essay "Whether the Earth Has Undergone an Alteration of Its Axial Rotation."

D37 Kant, Immanuel. *Kant's Cosmogony*. Translated by W. Hastie, with a New Introduction by G.J. Whitrow. New York: Johnson Reprint, 1970. xl + 205 pp.

D38 Kant, Immanuel. *Allgemeine Naturgeschichte und Theorie des Himmels*. Mit einem wissenschaftshistorischen Nachwort, herausgegeben von Fritz Krafft. München: Kindler Verlag GmbH, 1971. 211 pp.

A reprint of Kant's 1755 book on cosmogony, with historical notes and bibliography by a modern scholar.

D39 Kant, Immanuel. "An English Translation of the Third Part of Kant's Universal Natural History and Theory of the Heavens, by S.L. Jaki." *Cosmology, History, and Theology*. Edited by W. Yourgrau and A.D. Beck. New York: Plenum, 1977, pp. 387-403.

The third part was omitted from Hastie's translation and modern reprints thereof.

D40 Khil'mi, G[enrikh] F[rantsevich]. *Dvesti Let Nauchnoi Kosmogonii*. Vsesoiuznoe Obshestvo po Rasprostraneniu Politicheskikh i Nauchnykh Znanii, Seriia III, No. 25. Moscow, 1955. 31 pp.

History of cosmogony from Kant to O.Iu. Shmidt, concentrating on the nebular hypothesis with some attention to its relations to the structure of the earth. For German translation see item D13.

D41 Lambrecht, H. "Die Kosmogonie Immanuel Kants." *Die Sterne* 50 (1974): 74-81; 51 (1975): 29-39.

D42 Lambrecht, Hermann. "Kant und die moderne Kosmologie." *Wissenschaftliche Zeitschrift der Friedrich-Schiller Universität, Jena, Ges.-Sprach. R.* 24 (1975): 167-173.

"For Kant there is no difference between cosmogony and cosmology. Kant's highly topical teaching of the evolution of the cosmos, of the historicity of the universe, is a valuable heritage. Going beyond the Copernican principle, Kant's natural laws are not only spatially but also temporally universal. Kant replaced the Newtonian ordering God by materialistic proofs for natural changes" (Author's summary).

D43 Lambrecht, H. "Ergänzende Bemerkungen zum Beitrag 'Die Kosmogonie Immanuel Kants." *Die Sterne* 52 (1976): 51-53.

D44 Lawrence, Philip. "Heaven and Earth--The Relation of the Nebular Hypothesis to Geology." *Cosmology, History, and Theology*. Edited by Wolfgang Yourgrau and Allen D. Breck. New York: Plenum, 1977, pp. 253-281.

While the nebular hypothesis gained acceptance in the decade 1810-20 there was disagreement on the tenability of its geological implications. Work in historical geology "presented increasingly persuasive evidence for a directional Earth history in which diminishing igneous forces played a crucial role." J.B.J. Fourier's theory of heat conduction was directly applicable to geothermal observations and "established the groundwork for a comprehensive theory of geological dynamics predicated on the existence of a 'central heat.'" A theoretical synthesis was developed by Leonce Elie de Beaumont, attributing orogenies to successive collapses of the earth's crust onto a shrinking core. His theory was a powerful rival to that of C. Lyell, who rejected progressive cooling. The "directionalist synthesis," based on the nebular hypothesis, continued to influence geological thought throughout the 19th century.

D45 Lebedinsky, A. "Recent Soviet Theories on the Origin of the Solar System." *Journal of the British Astronomical Association* 63 (1953): 274-277.

Gives special emphasis to the theory of Otto Schmidt.

D46 Levin, B.Iu. "Kosmogonicheskaia gipotoza Laplasa (Istoriia sozdaniia i publikatsii)." *Voprosy Istorii Estestvoznaniia i Tekhniki* 54, no. 1 (1976): 18-30.

On the history of the creation and publication of the cosmogonic hypothesis of Laplace; discussion of its relation to I. Kant's theory.

D47 Levin, B.Iu. "K Istorii Termina 'Gipoteza Kanta-Laplasa.'" *Voprosy Istorii Estestvoznaniia i Tekhniki* 64-66 (1979): 8-15.

On the history of the term "Kant-Laplace hypothesis."

D48 Maillet, Benoit de. *Telliamed or Conversations Between an Indian Philosopher and a French Missionary on the Diminution of the Sea.* Translated, with introduction and notes, by Albert V. Carozzi. Urbana: University of Illinois Press, 1968. xiv + 465 pp.

Telliamed, first published in French in 1748, starts with Descartes' cosmogony and constructs a billion-year history of the earth governed by the eruption and recession of oceans.

D49 Merleau-Ponty, J. "Situation et rôle de l'hypothèse cosmogonique dans la pensée cosmologique de Laplace." *Revue d'Histoire des Sciences et de Leurs Applications* 29 (1976): 21-49; 30 (1977): 71-72.

"Arriving in Paris, in 1770, at the age of 21, Laplace had undertaken to prove, by solving some great pending problems in celestial mechanics, that the Newtonian theory of gravitation was able to explain, without any residue, all the known observations on the solar system, and all the terrestrial phenomena from gravitational origin. When the French Revolution began he thought he had reached that goal, and, by the way, that he had also proved that the solar system, as a whole, and in all its parts was stable; moreover he had exactly measured, with new techniques of probability calculus, the improbability of some structural features of that system--precisely those that fulfilled the analytical conditions of stability. That, clearly, pointed out a 'regular cause.' But which one? In 1796, at last, he tries to find out, in the past history of the solar system, the explanation of its present state; cautiously, first. Then in the successive editions of the *Exposition du système du monde*, he develops his theory, more and more emphatically condemning any recourse to final causes, mainly after 1812, when the new theory of nebulae of William Herschel had enhanced his belief in the nebular hypothesis" (Author's summary).

D50 Merleau-Ponty, J. "Laplace as a Cosmologist." *Cosmology, History, and Theology*. Edited by Wolfgang Yourgrau and Allen D. Beck. New York: Plenum, 1977, pp. 283-291.

"The naturalist-philosophers of the nineteenth century misunderstood Laplace's cosmogony on an important point: They construed, and I think misconstrued, it in a transformist outlook; but Laplace's view is not transformist; he does not imagine the Universe to be engaged in an endless process of more or less creative evolution..." (p. 289).

D51 Numbers, Ronald L. "The Nebular Hypothesis in American Thought." Ph.D. dissertation, University of California, Berkeley, 1969. 262 pp. University Microfilms order no. 70-6182.

For summary see *Dissertation Abstracts International* 30 (1970): 4380-A. A revised version was published as *Creation by Natural Law* (cited below as item D55).

D52 Numbers, Ronald L. "Daniel Kirkwood's Analogy: An American Confirmation of the Nebular Hypothesis." *Proceedings of the XIII International Congress of the History of Science, Moscow, 1971* (pub. 1974) 6: 352-357.

Kirkwood (1814-1895) found in 1848 a simple approximate relation between the number of rotations performed by a planet in its orbit around the sun, and the width of the region it dominates gravitationally. This regularity was regarded at the time as evidence in favor of Laplace's cosmogony.

D53 Numbers, Ronald L. "The Nebular Hypothesis of Isaac Orr." *Journal for the History of Astronomy* 3 (1972): 49-51.

Orr's cosmogony was similar to Laplace's; his publication (in *American Journal of Science*, 1823) indicates that the nebular hypothesis was not known to most American scientists at that time.

D54 Numbers, Ronald L. "The American Kepler: Daniel Kirkwood and His Analogy." *Journal for the History of Astronomy* 4 (1973): 13-21.

Gives further details on Kirkwood's analogy (see above, item D52) and its reception by other scientists.

D55 Numbers, Ronald L. *Creation by Natural Law: Laplace's Nebular Hypothesis in American Thought.* Seattle: University of Washington Press, 1977. xi + 184 pp.

"Since the nebular hypothesis involved a naturalistic account of the origin of the Earth in place of *Genesis*, one might have expected it to provoke the same religious objections as Darwin's theory, and popular acceptance of the Laplace-Herschel theory might in some sense have paved the way for the acceptance of Darwinism. That is the challenging thesis which Ronald Numbers attempts to document.... He argues that scientists and religious leaders first confronted evolutionary naturalism in the astronomical context; having swallowed it there, they acquired a taste for it and were thus willing to contemplate naturalistic explanations in biology with less resistance than might have been expected.... Numbers has made an excellent contribution to intellectual history by presenting the evidence for his thesis, even if he doesn't convince all readers ..." (from the review by S.G. Brush in *Journal for the History of Astronomy* 9 [1978]: 70-71).

D56 Nyrén, Magnus. "Swedenborg and the Nebular Hypothesis." *Vierteljahrschrift der Astronomischen Gesellschaft* (Leipzig) (1879): 80-91.

"Ueber die von Emanuel Swedenborg augestellte Kosmogonie, als Beitrag zur Geschichte der s.g. Kant-Laplace'schen Nebular-Hypothese; nebst einen Résumé von Thomas Wrights 'New Hypothesis of the Universe'" (Author's summary).

D57 Ogilvie, Marilyn Bailey. "Robert Chambers and the Nebular Hypothesis." *British Journal for the History of Science* 8 (1975): 214-232.

Vestiges of the Natural History of Creation, published anonymously by Robert Chambers (1802-1871) in 1844, popularized and stimulated criticism of evolutionary ideas. By associating the nebular hypothesis with evolution, Chambers caused some of the disapproval of the latter to be attached to the former theory. Changes in successive editions of the *Vestiges*, and comments of reviewers, illuminate popular ideas of the origin of the earth in mid-19th-century Britain.

D58 Page, Thornton, and Lou Williams Page, eds. *The Origin of the Solar System: Genesis of the Sun and Planets, and Life on Other Worlds*. New York: Macmillan, 1966. xvi + 336 pp.

Extracts from popular articles published in *The Sky* by H.N. Russell (1937), W.H. Barton, Jr. (1939), L. Spitzer, Jr. (1941), and others, and in *Sky and Telescope* by O. Struve (1945, 1949, 1956) and others.

D59 Porter, Roy. *The Making of Geology*. New York: Cambridge University Press, 1977. 288 pp.

Points out that "specialization in geology was perhaps aided ... by strictures against geo-physical cosmogony, which permitted study of the Earth independently of the exact and physico-chemical sciences" (p. 140; see also pp. 203-206 and chapter 8).

D60 Porter, Roy. "Creation and Credence: The Career of Theories of the Earth in Britain, 1660-1820." *Natural Order: Historical Studies of Scientific Culture*. Edited by B. Barnes and S. Shapin. Beverly Hills, Calif.: Sage, 1979, pp. 97-123.

"Investigates the rise and fall of the genre of scientific cosmogonies ... by locating their roots in Christian Scriptural accounts of creation ... then analyses the ideological functions of the leading ideas of such theories, as a prelude to the main discussion that is, why the scientific genre had such a chequered career." Concludes that "naturalistic cosmogonies were a form of knowledge arising out of the need to provide new foundations for a conception of Man in the cosmos at a time of social, political, and intellectual change when traditional religious dogmas had ceased to command unanimity." After the French Revolution natural science and human values were separated, and the discipline was "ignominiously superseded by 'geology.'"

* Porter, Roy. "William Hobbs of Weymouth and His *The Earth Generated and Anatomized* (?1715)." Cited above under item D30.

D61 Regnell, Gerhard. "The Primeval Earth as Conceived by Swedish 18th Century Naturalists." *History of Concepts in Precambrian Geology*. Edited by W.O. Kupsch and W.A.S. Sarjeant. Toronto: Geological Association of Canada, Special Paper 19, 1979, pp. 171-180.

"Speculations on the early state of the Earth by the Swedish 18th century naturalists U. Hiärne, C. Polhem, E. Swedenborg, C. Linnaeus, D. Tilas, J.G. Wallerius, and T. Bergman are briefly commented upon. If viewed in their contemporary European context, these authors may be characterized largely as exponents of the prevailing ideas of their times in this particular field. Bergman alone stands out as an independent thinker who exercised a certain influence upon the development of geology" (Author's abstract). See also item A30 above.

D62 Roger, Jacques. "La théorie de la terre au XVIIe siècle." *Revue d'Histoire des Sciences et de Leurs Applications* 26 (1973): 23-48.

"The 'Theory of the Earth,' which was the intellectual frame of the sciences of the earth from the XVIIth to the beginning of the XIXth century, is not a preliminary sketch to modern geology. Made possible by the Copernican Revolution, it is a peculiar way of thinking of the earth, which appeared, as a merely theoretical explanation of the nature of things, in Descartes' *Principia*, and became an historical account of the past of the earth with Burnet's *Telluris Theoria Sacra*. Born from the conjunction of Cartesian physics and Sacred History, it gave fossils their true significance, and paved the way to an entirely historical interpretation of Nature" (Author's summary).

* Runcorn, S.K., ed. *Earth Sciences*. Cited above as item A86.

D63 Sambursky, S. "Galileo's Attempt at a Cosmogony." *Isis* 53 (1962): 460-464.

The hypothesis that all the planets were once assembled at one place far from the sun, and then had their linear accelerated motion converted into circular motion when they had fallen into their present orbits, was invented by Galileo himself but attributed by him to Plato on the basis of a rather liberal interpretation of passages in the *Timaeus*.

D64 Smart, W.M. *The Origin of the Earth*. Cambridge: At the University Press, 1951; second edition, 1953. vi + 239 pp.

Although primarily expository, this book includes some historical chapters on theories in the 19th and early 20th centuries.

D65 Sticker, Bernhard. "Leibniz' Beitrag zur Theorie der Erde." *Sudhoffs Archiv* 51 (1967): 244-259.

On Leibniz's *Protogaea*.

D66 Sticker, Bernhard. "Leibniz et Bourguet: Quelques lettres inconnues sur la théorie de la terre." *Actes du XIIe Congrès International d'Histoire des Sciences 1968* 3b (pub. 1971): 143-147.

Correspondence with Louis Bourguet on Leibniz's theory of the formation of earth at a high temperature.

D67 Struve, Otto. "Henri Poincaré and His Cosmogonical Studies." *Sky and Telescope* 7 (1958): 226-228.

D68 Sullivan, John W.N. "Evolution of Worlds: The Work of Sir George Howard Darwin." *Scientific American* 108 (15 February 1913): 164-165.

On Darwin's theory of the origin of the moon.

D69 Treder, H.J. "Kants Kosmologie und der physische Teil des naturwissenschaftlichen Weltbildes." *Die Sterne* 50 (1974): 65-73.

D70 Von Braumüller. "Geschichtliche Darstellung der haupsächlichsten Theorien über die Entstehung des Sonnensystem." *Himmel und Erde* 10 (1898): 289-300, 357-374.

D71 Whiston, William. *A New Theory of the Earth, from its Original, to the Consummation of All Things. Wherein the Creation of the World in Six Days, the Universal Deluge, and the General Conflagration, As Laid Down in the Holy Scriptures, Are Shown to Be Perfectly Agreeable to Reason and Philosophy. With a Large Introductory Discourse Concerning the Genuine Nature, Stile, and Extent of the Mosaick History of the Creation.* London: B. Tooke, 1696. 388 pp.; reprint, New York, Arno Press, 1978.

D72 Whitrow, G.J. "The Nebular Hypothesis of Kant and Laplace." *Actes, XIIe Congrès International d'Histoire des Sciences, Paris, 1968* (pub. 1971) IIIB: 175-180.

Kant's theory was both historically prior to Laplace's and more like modern views. Since the principle of conservation of angular momentum was not formulated in full generality until 1775 it is not surprising that Kant failed to use it.

D73 Williams, I.P. *The Origin of the Planets*. New York: Crane Russak, 1975. 108 pp.

Semi-historical review and critique of the major theori

D74 Woolfson, M.M. "The Evolution of the Solar System." *Reports on Progress in Physics* 32 (1969): 135-185.

A review of the major theories proposed since 1955.

D75 Wylie, Charles Clayton. "The Nebular Hypothesis in the Nineteenth Century." *Scientific Monthly* 27 (1928): 260-263.

For additional information on this topic see works cited in the biography sections on the following scientists:

Boullanger, N.
Bourguet, L.
Buffon, G.
Burnet, T.
Chamberlin, T.C.
Faye, H.
Jeans, J.H.
Jeffreys, H.
Laplace, P.
Maillet, B. de
Mushketov, I.V.
Nordenskiöld, A.E.
Roche, E.A.
Rubey, W.W.
See, T.J.J.
Taylor, F.B.
Whitehurst, J.
Woodward, J.

E. GEOCHRONOLOGY

E1 Albritton, Claude C. *The Abyss of Time: Changing Conceptions of the Earth's Antiquity after the Sixteenth Century*. San Francisco, Calif.: Freeman, Cooper, 1980. 251 pp.

"During the seventeenth century the age of the earth was reckoned in the thousands of years. Estimates in the millions and even thousands of millions followed in the eighteenth century. During the late nineteenth century there was a general shrinkage to limits that fell as low as a few tens of millions. Now we think in terms of thousands of millions. These fluctuations have not been capricious, but have been prompted by increase in knowledge concerning the nature of matter in general and of that particular matter available for inspection in the outermost parts of the solid earth.... To trace successive changes in perceptions of the earth's antiquity is the purpose of this book" (from the author's introduction).

E2 Badash, Lawrence. "Rutherford, Boltwood, and the Age of the Earth: The Origin of Radioactive Dating Techniques." *Proceedings of the American Philosophical Society* 112 (1968): 157-169.

Following a brief review of the 19th-century controversy on the age of the earth, the author discusses the contributions of E. Rutherford, B.B. Boltwood, R.J. Strutt, and A. Holmes to the development of radiometric dating of rocks and minerals. Whereas earlier the geologists wanted a longer time scale than the physicists would allow, in the first decades of the 20th century such geologists as J. Joly resisted the much longer time scale introduced by radioactivity. Geologists accepted radiometric dating after 1925, partly as a result of Holmes' efforts.

E3 Berry, William B.N. *Growth of a Prehistoric Time Scale Based on Organic Isolation*. San Francisco, Calif.: Freeman, 1968. 158 pp.

Survey of aspects of the history of geology and paleontology primarily in the 18th and early 19th centuries.

E4 Burchfield, Joe D. "Presuppositions and Results: The Age of the Earth in Late Victorian England." *Actes du XIIIe Congrès International d'Histoire des Sciences, 1971* (pub. 1974) 8: 91-96.

A preliminary account of the controversy, discussed in the author's book cited below (item E6).

E5 Burchfield, Joe D. "Darwin and the Dilemma of Geological Time." *Isis* 65 (1974): 301-321.

Discusses the effect on evolutionary theory of the limits on the age of the earth proposed by physicists, treated more fully in the author's book cited below (item E6).

E6 Burchfield, Joe D. *Lord Kelvin and the Age of the Earth.* New York: Science History Pubs., 1975. xii + 260 pp.

Presents a detailed account of the controversy started by William Thomson (later known as Lord Kelvin) when he estimated that the earth has cooled down from a hot liquid state in less than 100 million years. This estimate, based on Fourier's heat conduction theory and supported by the prestige of physics, conflicted with the geologists' assumption that indefinitely long periods of time were available, and indirectly undermined the credibility of Darwinian evolution. The discovery of radioactivity at the end of the 19th century invalidated one of Kelvin's assumptions (that no heat is generated to replace that lost from the inside of the earth) and provided a new method for estimating the age of rocks, some of which were found to be billions of years old. The author also discusses contributions of A. Geikie, J. Phillips, J. Croll, T.H. Huxley, T.M. Reade, S. Haughton, W.J. McGee, J. Perry, T.C. Chamberlin, J. Joly, W.J. Sollas, P. Curie and A. Laborde, E. Rutherford, F. Soddy, and B.B. Boltwood.

E7 Dean, Dennis R. "The Age of the Earth Controversy: Beginnings to Hutton." *Annals of Science* 38 (1981): 435-456.

"Speculation concerning the age of the earth begins with civilisation itself. The creation myths of ancient Egypt and other early cultures were soon expanded into elaborate cosmologies by Indian, Persian and Greek philosophers. Jewish and, more insistently, Christian scholars long believed that the Bible provided an exact chronology beginning with the Creation (and even foretelling the world's end).

Such truncated apocalyptic chronologies were opposed first by Aristotelian advocates of an eternal earth and then by deistic freethinkers who regarded the earth's age as indefinite but immense. As textual scholarship cast doubt upon the literal reliability of *Genesis*, alternative chronologies arose which depended increasingly upon geological evidence. James Hutton's assertions about the earth's age reflect his awareness of this broader context and define his own important contribution to it" (Author's summary).

E8 Eicher, Don L. *Geologic Time*. Englewood Cliffs, N.J.: Prentice-Hall, 1968. viii + 149 pp. Second edition, 1976. 150 pp.

"Large parts of this book deal with the history of geology from the standpoint of a development of ideas about geologic time--from the earlier days when the earth was believed to have a history of only a few thousand years to the present time when it has almost as many billions of years of history. Nowhere else can one find in such convenient summaries the various estimates of the age of the earth, the various methods for determining geologic time, the history of the founding of the various geologic systems, the history of the various kinds of geological and paleo-geographic maps, the history of the development of the concept of continental drift, and the history of the development of radiometric dating. The bibliographies, in the main, refer to original sources rather than to secondary ones. Includes historical survey of ideas about the age of the earth, and quantitative development of the time scale in the 20th century" (G.W. White).

E9 Engel, A.E.J. "Time and the Earth." *American Scientist* 57 (1969): 458-483.

Survey of estimates of the age of the earth in the 20th century, and their implications for geology.

E10 Fabian, Hans-Georg. "Zur Geschichte der Bleimethoden für die geologische Altersbestimmung." *NTM. Schriftenreihe für Geschichte der Naturwissenschaften, Technik und Medizin* 4, no. 10 (1967): 63-72.

On the use of the radioactive decay of uranium and thori to lead isotopes in geochronology form Rutherford's initia suggestion (1905) to the work of F.G. Houtermans (1951).

E11 Fabian, Hans-Georg. "Zur Geschichte der Kalium-Argon- und der Rubidium-Strontium-Methode für die geologische

Alterbestimmung." *NTM, Schriftenreihe für Geschichte der Naturwissenschaften, Technik und Medizin* 11, no. 2 (1974): 69-81.

A survey of attempts (since 1937) to use the beta-decay of ^{40}K and ^{87}Rb in geochronology.

E12 Faul, Henry. "A History of Geologic Time." *American Scientist* 66 (1978): 159-165.

On the estimates of Lord Kelvin, J. Joly, B.B. Boltwood, Arthur Holmes, and F.G. Houtermans, with remarks on C. Lyell, T.C. Chamberlin, A.C. Lane, G.P. Baxter, and A.O. Nier.

E13 Haber, Francis C. *The Age of the World, Moses to Darwin.* Baltimore: Johns Hopkins Press, 1959. xi + 303 pp.

"The work is intended as a study in the history of ideas, and ... I hope that its total effect will be to shed some light on 'historicism.'" Concentrating on the interactions between religious and scientific views in the 17th and 18th centuries, the author includes descriptions of the works of Buffon, T. Burnet, G. Cuvier, Descartes, E. Hitchcock, Baron von Holbach, R. Hooke, D. Hume, J. Hutton, I. Kant, W. Thomson (Lord Kelvin), R. Kirwan, Lamarck, Laplace, Leibniz, Leonardo da Vinci, C. Lyell, B. de Maillet, I. Newton, M. de Serres, J.P. Smith, Voltaire, W. Whiston, and J. Woodward.

E14 Harper, C.T., ed. *Geochronology: Radiometric Dating of Rocks and Minerals*. Stroudsburg, Pa.: Dowden, Hutchinson & Ross, 1973. xv + 469 pp.

Anthology of research papers and extracts from books by E. Rutherford (1906, 1929), B. Boltwood (1907), R.J. Strutt (1910), A. Holmes (1911, 1927, 1946), V.S. Dubey and A. Holmes (1929), W.D. Urry (1936), D. Goodman and R.D. Evans (1941), F.W. Aston (1929), C.N. Fenner and C.S. Piggot (1929), A.O. Nier, et al. (1941), E.K. Gerling (1942), F.G. Houtermans (1946), C. Patterson (1956), G.R. Tilton and R.H. Steiger (1965), R.D. Russell and R.M. Farquhar (1960), G.W. Wetherill (1955, 1956, 1971), G.R. Tilton (1960), L.T. Silver and S. Deutsch (1963), O. Hahn and E. Walling (1938), L.H. Ahrens (1949), W. Compston and P.M. Jeffery (1959), H.W. Fairbairn, et al. (1961), L.D. Nicolaysen (1961), W. Compston and R.T. Pidgeon (1962), P.M. Hurley, et al. (1963), C.F. von Weizsäcker (1937), L.T. Aldrich and A.O. Nier (1948), F. Smits and W. Gentner (1950), G.J. Wasserburg and R.J. Hayden (1955),

G.W. Wetherill, et al. (1955), S.R. Hart (1961), A. Cox and G.B. Dalrymple (1967), J.F. Evernden and R.K.S. Evernden (1970), A.K. Baksi, et al. (1967), P.E. Damon (1968), L. Merrihue and G. Turner (1966), S. Moorbath (1967), P.M. Hurley and J.R. Rand (1971).

E15 Hattiangadi, J.N. "Alternatives and Incommensurables: The Case of Darwin and Kelvin." *Philosophy of Science* 38 (1971): 502-507.

A philosophical discussion of the dispute about geological time. "The fact that Darwin's *biology* can clash with Kelvin's *physics* shows how commensurable theories may yet be incompatible. But it also shows that they may not be alternatives...."

E16 Hooykaas, R. "James Hutton und die Ewigkeit der Welt." *Gesnerus* 23 (1966): 55-56.

G.H. Toulmin published *The Antiquity and Duration of the World* in 1780 and *The Eternity of the World* in 1785 just before Hutton's *Theory of the Earth* (1788). In these works the geological idea of eternity was associated with atheism and materialism.

E17 [Kant, I.] Reinhardt, O., and D.R. Oldroyd. "Kant's Thoughts on the Ageing of the Earth." *Annals of Science* (1982): 349-369.

A translation of Kant's early paper "Die Frage, ob die Erde veralte, physikalisch erwogen" is presented, and the main features of his position on this question in 1754 are summarized. He believed that the earth was about 6000 years old. The relation of this theory to his 1755 cosmogony is discussed, as well as his ideas on the processes of erosion and the formation of rivers, deltas, and sandbanks.

E18 Libby, W.F. "History of Radiocarbon Dating." *Radioactive Dating and Methods of Low-Level Counting*, Proceedings of a Symposium organized by the International Atomic Energy Agency in Cooperation with the Joint Commission on Applied Radioactivity (ICSU) and held in Monaco, 2-10 March, 1967. Vienna: International Atomic Energy Agency, 1967, pp. 3-25.

* Rudwick, Martin J.S. "Lyell on Etna and the Antiquity of the Earth." Cited below as item H42.

* Rudwick, Martin J.S. "Poulett Scrope on the Volcanoes of Auvergne: Lyellian Time and Political Economy." Cited below as item HB8.

* Runcorn, S.K., ed. *Earth Sciences*. Cited above as item A86.

E19 Tasch, Paul. "Lyell's Geochronological Model: Published Year Values for Geological Time." *Isis* 68 (1977): 440-442.

Contrary to the usual view that Lyell "spoke only in broad generalities about geological time," the author points out that in his lectures before 1859 Lyell speculated about geological periods lasting millions of years. An estimate of 240 million years since the start of the Cambrian period can be inferred from his statements.

E20 Wager, Lawrence Richard. "The History of Attempts to Establish a Quantitative Time-Scale." *Quarterly Journal of the Geological Society of London* 120S (1964): 13-28. Also published as *The Phanerozoic Time Scale*. Edited by W.A. Harland et al. London: Geological Society of London, 1964.

Mostly on the work of Arthur Holmes.

For additional information on this topic see works cited in the biography sections on the following scientists:

Gessner, J.
Holmes, A.
Joly, J.
King, C.R.
Larsen, E.S., Jr.
Lyell, C.
Toulmin, G.H.

F. PHYSICS OF THE EARTH'S INTERIOR

F1 Adams, Frank Dawson. "Origin and Nature of Ore Deposits: An Historical Study." *Bulletin of the Geological Society of America* 45 (1934): 375-424.

A survey of the views of Aristotle (action of solar rays that penetrate the earth's crust), Renaissance alchemists (central fire produced by convergence of solar and stellar rays), J.J. Becher (1653), the "Golden Tree" of Peter Martyr (1577) and others, propagation by seeds, and maturing of base metals into gold. The geophysical aspect of this topic is the origin and function of the earth's central fire. Robert Boyle's development of questionnaires for managers of mines, to check the stories that certain metals mature or are regenerated from refuse, should interest historians of scientific method.

F2 Barus, Carl. "Certain Suggestions by J. Willard Gibbs on Geophysical Research." *American Journal of Science* series 4, 21 (1906): 461-462.

On the law of corresponding states and properties of matter near the critical point.

F3 Batiushkova, Irina V. "Regularities in the Development of Science as Manifested in the Evolution of Notions of the Structure of the Earth." *Actes, XIIe Congrès International d'Histoire des Sciences, Paris 1968* (pub. 1970) 2: 9-12.

A brief sketch, mostly about pre-1800 theories.

F4 Brush, Stephen G. "Nineteenth-Century Debates about the Inside of the Earth: Solid, Liquid or Gas?" *Annals of Science* 36 (1979): 225-254.

"In the first part of the 19th century, geologists explained volcanoes, earthquakes and mountain-formation on the assumption that the earth has a large molten core underneath a very thin (25-50 miles) solid crust. This assumption was attacked on astronomical grounds by William Hopkins, who argued that the crust must be at least 800 miles thick, and on physical grounds by William Thomson, who showed that the earth as a whole behaves like a solid with high rigidity. Other participants in the debate insisted that there is evidence for a fluid or plastic layer not far below the crust. By the end of the century many geologists had incorporated the doctrine of a completely solid earth into their theories. Acceptance of a relatively small liquid core, indicated by seismological research, was delayed for another two decades" (Author's summary).

F5 Brush, Stephen G. "Discovery of the Earth's Core." *American Journal of Physics* 48 (1980): 705-724.

"In 1896 when Emil Wiechert proposed his model of the Earth with an iron core and stony shell, scientists generally believed that the entire earth was a solid as rigid as steel. R.D. Oldham's identification of P and S waves in seismological records allowed him to detect a discontinuity corresponding to a boundary between core and shell (mantle) in 1906, and Beno Gutenberg established the depth of this boundary as 2900 km. But failure to detect propagation of S waves through the core was not sufficient evidence to persuade seismologists that it is fluid (contrary to modern textbook statements). Not until 1926 did Harold Jeffreys refute the arguments for solidity and establish that the core is liquid. In 1936 Inge Lehmann discovered the small inner core. K.E. Bullen argued, on the basis of plausible assumptions about compressibility and density, that the inner core is solid. Attempts to find seismic signals that have passed through the inner core as S waves have so far failed (with one possible exception), but analysis of free oscillations provided fairly convincing evidence for its solidity" (Author's summary).

F6 Brush, Stephen G. "Chemical History of the Earth's Core." *Eos* 47 (1982): 1185-1186, 1188.

"The history of ideas about the earth's core is reviewed, from the 19th century to the present. Following the determination of the outer core boundary by B. Gutenberg and the establishment of the fluidity of the core by H. Jeffreys, the current model for the overall physical structure was completed by Inge Lehmann's discovery of the inner core, and the proposal by F. Birch and K.E. Bullen that the inner core is solid.

"The traditional assumption that the core is primarily iron was challenged by several scientists in the 1940s, especially W.H. Ramsey who proposed that the core boundary marks a change in physical but not chemical state. His hypothesis, that the core is a liquid 'metallized silicate,' was refuted by research on the properties of silicates at high pressures, but it raised the question whether a theory of the present state of the earth's interior should be consistent with some plausible theory of its origin and development. While Western geophysicists tended to ignore this criterion, a group of Russian scientists developed a theory which satisfied it, although it was difficult to maintain the metallized silicate hypothesis. A compromise model, proposed in the 1970s, involves iron and oxygen, in proportions chosen to satisfy density conditions but also derivable by physico-chemical evolution from an initially homogeneous earth" (Author's summary).

F7 Bullard, Edward. "Historical Introduction to Terrestrial Heat Flow." *Terrestrial Heat Flow*. Edited by W.H.K. Lee. Geophysical Monograph Series No. 8. Washington, D.C.: American Geophysical Union, Pub. No. 1288, 1965, pp. 1-6.

Discusses ideas of J.B. Morin (1619), Robert Boyle (1671), and others on high temperatures inside the earth; British Association committees were active 1868-83 and 1935-39.

F8 Bullen, K.E. "Matthew Flinders Lecture 1969. Researches on the Internal Structure of the Earth." *Records of the Australian Academy of Science* 1, no. 4 (November 1969): 39-58.

Historical sketch of 20th-century progress in seismology, free oscillations of the earth, and development of models for internal density variation, including the author's own important contributions.

* Burchfield, Joe D. *Lord Kelvin and the Age of the Earth*. Cited above as item E6.

F9 Burke, John G. "Romé De L'Isle and the Central Fire." *XIIe Congrès International d'Histoire des Sciences, Paris 1968* (pub. 1971) 7: 15-17.

Jean-Baptiste Louis Romé de L'Isle published in 1779 *L'Action de feu central* (revised 1781), which disputed the theory of Buffon and Dortous de Mairan that the earth's central heat affects its surface temperature. Instead he emphasized the influence of the sun mediated by the atmosphere, and "he appears to have had a notion of the atmosphere's greenhouse effect."

F10 Burke, John G. "The Earth's Central Heat: From Fourier to Kelvin." *Actes du XIIIe Congrès International d'Histoire des Sciences, 1971* (pub. 1974) 8: 118-123.

Brief survey of the views of J. Fourier, H. Davy, C. Daubeny, B. Silliman, W. Hopkins, and W. Thompson (Lord Kelvin) on the Thermal state of the earth's interior and the cause of volcanic action.

F11 Byerly, Perry. "Subcontinental Structure in the Light of Seismological Evidence." *Advances in Geophysics* 3 (1956): 105-152.

Includes historical account of research since 1909 by A. Mohorovičić, V. Conrad, H. Jeffreys, B. Gutenberg, P. Byerly, and others.

F12 Byerly, Perry. "The Earth's Mantle in the Early Days." *Union Geogesique et Géophysique Internationale, Comptes Rendus No. 14, Reunie à Berkeley. Août 1963*, pp. 32-42.

Surveys papers on seismology and the structure of the earth by A. Schmidt (1888), O. Fisher (1889), E. Wiechert (1897), R.D. Oldham (1906), H. Benndorf (1905, 1906), G. Herglotz (1907), B. Gutenberg (1912, 1914, 1926), A. Mohorovičić (1910, 1922), etc., with comments on their relation to current ideas.

* Cloud, Preston, ed. *Adventures in Earth History*. Cited above as item A21.

F13 Debus, Allen G. "Edward Jorden and the Fermentation of the Metals: An Iatrochemical Study of Terrestrial Phenomena." *Toward a History of Geology*. (item A89), pp. 100-121.

Jorden (1569-1632) argued against the hypothesis of a central fire in the earth and suggested that the generation of metals could be explained instead by fermentation.

F14 Engelhardt, Wolf von. "Krafte der Tiefe--Zum Wandel Erdgeschichtlicher Theorien." *Janus* 69 (1982): 119-140.

A survey of theories of the structure and development of the earth from the 17th to 20th centuries.

F15 Gerstner, Patsy A. "The Reaction to James Hutton's Use of Heat as a Geological Agent." *British Journal for the History of Science* 5 (1971): 353-362.

A hitherto unpublished account of Hutton's ideas written by John Thomson as a student at Edinburgh illustrates criticisms of the hypothesis that a central fire can continue to burn indefinitely. The problem of the source and distribution of central heat was also raised in reviews of Hutton's theory by J.A. Deluc, R. Kirwan, and J. Murray. The defense by J. Playfair was ineffective.

F16 Günther, Siegmund. "Die Entwickelung der Lehre vom Gasförmigen Zustande des Erdinneren." *Jahresbericht der Geographischen Gesellschaft in München für 1890 und 1891* 14 (1892): 1-20.

On the ideas of B. Franklin (1793), Lichtenberg, Chladni, Muncke, and H. Spencer, as precursors of the theory popularized by Günther and others that the center of the earth is gaseous.

* Kopal, Zdenek. "The Contribution of Boscovich to Astronomy and Geodesy." Cited below as item G41.

F17 Lawrence, Philip. "Charles Lyell versus the Theory of Central Heat: A Reappraisal of Lyell's Place in the History of Geology." *Journal of the History of Biology* 11 (1978): 101-128.

The theory of Leonce Elie de Beaumont, based on the cooling and contraction of the earth's interior, was a powerful competitor to Lyell's uniformitarian (non-directionalist) system.

* Mather, Kirtley F., ed. *Source Book in Geology, 1900-1950*. Cited above as item A65.

F18 Stoneley, R. "The Interior of the Earth." *Occasional Notes of the Royal Astronomical Society* 8 (1940): 103-116.

Semi-historical survey, based mostly on seismological results.

* Thams, J.C., ed. *The Development of Geodesy and Geophysics in Switzerland*. Cited above as item A103.

For additional information on this topic see works cited in the biography sections on the following scientists:

Adams, F.D.
Anderson, E.M.
Barus, C.
Becke, F.J.K.

Becker, G.F.
Berzelius, J.J.
Buffon, G.
Bullen, K.E.
Chamberlin, T.C.
Cordier, P.
Daly, R.A.
Day, A.L.
Doelter, C.A.S.
Elie de Beaumont, J.
Eskola, P.E.
Fenner, C.N.
Gilbert, G.K.
Gutenberg, B.
Hall, J.
Hauy, R.-J.
Helmholtz, H. v.
Holmes, A.
Hopkins, W.
Jeffreys, H.
Leibniz, G.W. v.
Leybenzon, L.S.
Love, A.E.H.
Maskelyne, N.
Michell, J.
Mohorovičić, A.
Mushketov, I.V.
Oldham, R.D.
Orlov, A.Y.
Roche, E.A.
Romé De L'Isle, J.-B.L.
Sorby, H.C.
Sternberg, P.K.
Ştoneley, R.
Vernadsky, V.I.
Wiechert, E.
Woodward, R.S.

G. GEODESY, TERRESTRIAL GRAVITATION, SIZE AND SHAPE OF THE EARTH

G1 Andersen, Einar. *200 ars videnskabelig geodaetisk virksomhed i Danmark, 1757-25. Februar - 1957.* Geodactisk Institut, Meddeleke, no. 32. Kobenhavn, 1957. 63 pp.

With French translation, pp. 23-40; chronology, pp. 41-45; publications, pp. 46-50. Also English translation of 9 leaves: "200 Years of Scientific Geodetic Activity in Denmark."

G2 Arneson, Edwin P. "The Early Art of Terrestrial Measurement and Its Practice in Texas." *Southwestern Historical Quarterly* 24 (1925): 79-97.

G3 Baranov, A.N., and M.K. Kudriavtsev. *50 let sovetskoi geodezii i kartografii, 1917-1967.* Moscow: Nedra, 1967. 446 pp.

Historical-review articles commemorating 50 years of Soviet geodesy and cartography (1917-67).

G4 Bialas, Volker. *Praxis Geometrica. Zur Geschichte der Geodäsie am Beginn der Neuzeit.* Deutsche Geodätische Kommission bei der Bayerischen Akademie der Wissenschaften, Reihe E: Geschichte und Entwicklung der Geodäsie, Heft nr. 11. Munich: Verlag der Bayerischen Akademie der Wissenschaften, 1970. ii + 27 pp.

A sketch of the history of geodesy in the 15th-17th centuries, with extracts from *Eratosthenes Batavus* by Willebrord Snellius (1617), in Latin with German translation.

G5 Bialas, Volker. *Der Streit um die Figur der Erde: zur Begründung der Geodasie im 17. und 18. Jahrhundert.* Deutsche Geodätische Kommission bei der Bayerischen Akademie der Wissenschaften. Series E: Geschichte und Entwicklung der Geodäsie, Issue no. 14. Munich: Verlag der Bayerischen Akademie der Wissenschaften, 1972. 39 pp.

"In the present paper the development of Geodesy, from 1670 until nearly 1750, is expounded on the basis of the main publications. This period, famous for its controversy about the figure of the earth, was very important for the foundation of modern Geodesy. Following their traditional procedures the geometers overrated the precision of the results. On the other side the physicists, above all Newton and Huygens, were developing their own procedures in determining the figure of the earth. The results were contradictory, and seemed not clear enough to prove if the earth is an oblongum or an oblatum. In this situation, also politically rather confused after the culmination of absolutism, two expeditions were started to the northern polar circle and to the equator. The new measurements of the meridians refuted Cassini and verified Newton's theories. Henceforth geometers and physicists have had to work commonly in the exact determination of the figure of the earth. Just this synthesis, to which also Clairant considerably contributed, made possible the formation of modern Geodesy" (Author's summary).

G6 Boccardi, Jean. "Un coup d'oeil sur l'évolution de la géodesie." *Revue Générale des Sciences* 36 (1925): 390-397.

Deals primarily with the past 100 years.

G7 Boss, Valentin. *Newton and Russia: The Early Influence, 1698-1796*. Cambridge, Mass.: Harvard University Press, 1972, xviii + 309 pp.

Chapter 14, "The Shape of the Earth," refers to the views of Jakob Hermann and Leonhard Euler.

G8 Breithaupt, G. "Zur Geschichte der Dosenlibelle." *Zeitschrift für Instrumentenkunde* 51 (1931): 256-259.

On the contributions of H.C.W. Breithaupt and J.C. Breithaupt (see also K. Ludemann's paper on this subject, item G44).

G9 Item deleted.

G10 Brockamp, Bernhard. "Entwicklung der Geophysik im 19. Jahrhundert." *Naturwissenschaft, Technik und Wirtschaft im 19. Jahrhundert*. Edited by W. Treue and K. Mauel. Göttingen: Van den Hoeck & Ruprecht, 1976, pp. 756-761.

Mostly on the figure of the earth. A similar article was published in *Geschichte der Naturwissenschaften und der Technik im 19. Jahrhundert*, by B. Sticker et al., Dusseldorf: VDI-Verlag, 1978, pp. 25-31.

G11 Brown, Harcourt. "From London to Lapland: Maupertuis, Johann Bernoulli I, and La terre applatie, 1728-1738." *Literature and History in the Age of Ideas*. Edited by C.G.S. Williams. Columbus: Ohio State University Press, 1975, pp. 69-96. Reprinted in Brown's *Science and the Human Comedy*, Toronto: University of Toronto Press, 1976, Chapter 8.

The author states that the determination of the oblate shape of the earth by the Maupertuis expedition "is perhaps the event that did most to consolidate the position of Newtonian physics" in the Enlightenment. Johann Bernoulli I was the leader of the Cartesians, who expected the earth to be prolate.

G12 Butterfield, Arthur Dexter. *A History of the Determination of the Figure of the Earth from Arc Measurements*. Worcester, Mass.: The Davis Press, 1906. 168 pp.

Includes Eratosthenes, Posidonius, Al Mamun (814 A.D.), Fernel (1528), Snell (1617), R. Norwood (1633-35), Ricioli and Grimaldi (ca. 1650), Picard (1669), Jacques Cassini (1700-18), Maupertuis and the Lapland expedition (1736), Bouguer and the Peru expedition (1735-39), the Meridian of Paris (1734-42), the Italian arc (1750-51), Lacaille (1752), Liesganig (1762), Beccaria (1762-64), Mason and Dixon, the Greenwich-Paris connection (1784), Svanberg (1802), determination of the meter, Struve and Tenner's work in Russia (1816), von Zach (1809), Lambton and Everest in India (1800-50), other 19th-century measurements. Technical details and results are given in each case.

G13 Cajori, Florian. "Swiss Geodesy and the United States Coast Survey." *Scientific Monthly* (August 1921), pp. 117-129.

On F.R. Hassler's career and the founding of the U.S. Coast Survey.

G14 Cajori, F. "History of Determinations of the Heights of Mountains." *Isis* 12 (1929): 482-514.

Quotes references to heights of mountains in classical Greek authors, with discussion of instruments used by the Greeks; mentions geodetic methods of Edward Wright (1589), W. Snell (1617), J. Picard. The height of the Teneriffe Peak, conspicuous to navigators, was determined by Louis

Feuillee (1724). The barometric method was introduced in the 17th century and refined by Edmund Halley (1686). A thermometric method (based on change in boiling point of water) was used by J.A. de Luc in 1762.

G15 Cajori, Florian. "A Century of American Geodesy." *Isis* 14 (1930): 411-416.

A brief survey of the work of the U.S. Coast Survey from F.R. Hassler to the early 20th century.

G16 Clarke, Alexander Ross, and Frederick Robert Helmert. "Earth, Figure of the." *Encyclopedia Britannica*, 11th Edition, 8 (1910): 801-813.

Includes a long historical section.

G17 Cohen, I.B. "Query No. 125. Proof of the Sphericity of the Earth. What Is the Earliest Appearance of the Proof of the Sphericity of the Earth Based on the Fact That the Masts of a Distant Ship Are Visible Above the Horizon When the Body of the Ship Is No Longer Visible?" *Isis* 41 (1950): 198-199.

In reply to the Query by J.W. Abrams, Cohen quotes Pliny's *Natural History*, the source of a remark by Copernicus in *De Revolutionibus* (preface to Book I). In a later note [*Isis*, 42 (1951): 47] Cohen gives a further discussion of Copernicus' statements on this subject.

G18 Comité national Belge de géodésie et de géophysique. *Aperçu historique de l'activité scientifique de la Belgique en géodésie et en géophysique, publié à l'occasion de la 9e Assemblée générale de l'Union Géodésique et Géophysique Internationale, tenue à Bruxelles du 21 août au 1er septembre 1951*. Bruxelles, 1951. 103 pp.

G19 Cubranic, Nikola. *Geodetski rad Rutera Boskovica; prigodom dvijestopedeset godisnjice rotenja*. Zagreb: Zavod za visu geodeziju Agb Fakulteta, 1961. 148 pp.

Geodetic work of R. Boskovic on the occasion of the 250th anniversary of his birth.

G20 Dadić, Žarko. "Patricius and Dominis on the Shape of the Earth." *XIIe Congrès International d'Histoire des Sciences, Paris 1968, Actes* (pub. 1971) 7: 23-25.

Ferrara F. Patricius (1591), rejecting Aristotelian views, tried to prove that the surface of water cannot be spherically curved and concluded that the earth is an irregularly shaped body delimited by planes rather than a sphere. M.A. Donimis (1624) supported the sphericity of earth and water by Aristotelian arguments about the tendency of elements to seek their natural place.

G21 Delambre, J.B.J. *Rapport historique sur les progrès des sciences mathématiques depuis 1789, et sur leur état actuel*. Paris: Imprimene Imperiale, 1810. vii + 272 pp.

Includes sections on geodesy, density of the earth, and length of the seconds pendulum at different locations.

G22 Delambre, J[ean]-B[aptiste]-J[oseph]. *Grandeur et figure de la terre. Ouvrage augmenté de notes, de cartes, et publié par les soins de G. Bigourdan*. Paris: Gauthier-Villars, 1912. viii + 401 pp.

First publication of an historical work left in manuscript at the author's death in 1822. Contents: Première mesure de la méridienne de France (J. Cassini, Maupertuis); première mesure de l'arc de Laponie (LeMonnier, Outhier); deuxième mesure de la méridienne de France (Cassini de Thury, LeMonnier); première mesure de l'arc du Pérou (Bouguer, LaCondamine); mesures de divers degrés de 1750 à 1780 (Maire, Boscovich, Liesganig, Beccaria, Mason and Dixon); troisième mesure de la méridienne de France, base du système metrique décimal (Méchain and Delambre); mesures de degrés apres 1800 (Cassini, Méchain, Legendre, W. Mudge and I. Dalby, W. Lambton).

G23 Diller, Aubrey. "The Ancient Measurements of the Earth." *Isis* 40 (1949): 6-9.

A survey of estimates of the earth's size found in the writings of Aristotle, Archimedes, Eratosthenes, Posidonius, and Ptolemy.

G24 Item deleted.

G25 Dusek, Val. "Geodesy and the Earth Sciences in the Philosophy of C.S. Peirce." *Two Hundred Years of Geology in America* (item A90), pp. 265-276.

"This is ... an investigation of ways in which the philosophy of the founder of pragmatism and America's greatest philosopher may have been moulded by the particular sorts of professional scientific work in which he engaged."

G26 Fel', S.E. "Kritika nekotorykh istochnikov po istorii russkoi geodezii i kartografii." *Trudy Institut Istorii Estestvoznaniia i Tekhniki* 42 (1962): 172-181.

"A critique of several sources on the history of Russian geodesy and cartography."

G27 Fischer, Irene. "The Figure of the Earth--Changes in Concepts." *Geophysical Surveys* 2 (1975): 3-54.

"The search for the Figure of the Earth is rooted in the age-old wonder about the world we live in. The ancient mythical notions have developed, under the tutelage of Greek systematic thought, into specific questions of the shape and size of the Earth. The history of these questions, now spanning more than two millennia, shows a continuous refinement in formulation, as each bona fide answer prompted refined measurements which, in turn, started a new round of problems. The change in concept of what one was looking for was, and still is, closely connected with the technical capability and its potential for further development.

"This change in concept is traced here from ancient times to the present through some of its milestones: Anaximander's partly invisible celestial sphere and Pythagoras' spherical Earth; Eratosthenes' exemplary combination of astronomy and land surveying for the dual purpose of determining the size of the spherical Earth and making a map of the habitable world; the identical dual purpose followed by the Académie Royale des Sciences in the 17th century, but on a more sophisticated technical level; the challenge to the spherical concept by Newton's gravitational theory and the absorption of that theory into a wider concept of the Figure of the Earth; the acknowledgment of the irregular geoid by Gauss, Bessel, and Helmert, and the changed significance of the ellipsoid; Heiskanen's and Vening Meinesz' 'Basic Hypothesis of Geodesy' concerning the hydrostatic equilibrium of the Earth, and O'Keefe's contradiction on the basis of satellite data; the dream of global surveys come true through satellite geodesy and the accompanying inclusion of the outer gravity field of the Earth into the concept of its Figure; and lastly, the present and projected capability of high precision measurement and the widening of the concept of the Figure of the Earth from 3-dimensional rigidity to 4-dimensional time-dependence" (Author's summary).

G28 Fischer, Irene. "Another Look at Eratosthenes' and Posidonius' Determinations of the Earth's Circumference." *Quarterly Journal of the Royal Astronomical Society* 16 (1975): 152-167.

"Hipparchus' correction to Eratosthenes' result has been characterized by historians as an 'egregious blunder' of the reporting Pliny, but has been shown here to be quite reasonable within its limits. The derivation of the obliquity of the ecliptic as the side of an inscribed 15-gon, as quoted by the ancient authors, has been shown to be a natural approach at the time, as was Eratosthenes' improved determination of the distance between the tropics as 11/83 of the full circle.... The comparison with modern geodetic results has been clarified. Finally, the reasons for rejecting Posidonius' determination of the size of the Earth are explained, and the persistent references to such a determination are shown to be based on confusions, which have been analysed here by geodetic considerations" (Author's summary).

G29 Gore, James H[oward]. "A Bibliography of Geodesy." Appendix 16 to *U.S. Coast and Geodetic Survey, Report for 1887*. Washington, D.C.: Government Printing Office, 1889, pp. 313-512.

"The intention was to include only such works as treated directly of the figure of the earth, or described operations which could be used in determining that figure, the only digression from this plan will be seen in the case of the pendulum, where the theoretical side is also included.... After each book title, and after the full title of each serial publication, there appears in parentheses the name of the owner of the work.... As a rule, those accredited to European libraries could not, at the time of trial, be found in any library in America." Approx. 6000 items, including listings under both author and subject for many of the titles.

G30 Gore, James Howard. *Geodesy*. Boston and New York: Houghton Mifflin and Company, 1891. 218 pp.

Historical survey of ideas about the shape of the earth, with detailed accounts of work by W. Snell, Picard, 18th-century expeditions, and the U.S. Coast Survey.

G31 Gudde, Erwin G. "A Century of Astronomy and Geodesy in California." *A Century of Progress in the Natural Sciences, 1853-1953*. San Francisco: California Academy of Sciences, 1955, pp. 65-74.

On the work of George Davidson.

G32 Günther, S. "Zur Entwicklungsgeschichte der Lehre von der Erdgestalt." *Archiv für die Geschichte der Naturwissenschaften und der Technik* 3 (1912): 451-464.

Mainly on 17th and 18th centuries.

G33 Gupta, S.L. "Hindus on the Shape of the Earth." *National Geographical Journal of India* 10, no. 2 (1964): 130-132.

The Rigveda, which dates back to the 13th century B.C. or earlier, mentions the spherical shape of the earth, long before Philolaos who is usually credited with this idea by Europeans.

* Hall, D.H. *History of the Earth Sciences During the Scientific and Industrial Revolutions, with Special Emphasis on the Physical Geosciences*. Cited above as item A48.

G34 Harradon, H.D. "Some Early Contributions to the History of Geomagnetism--V." *Terrestrial Magnetism and Atmospheric Electricity* 48 (1943): 197-199.

Translation (by J. de Sampaio Ferraz) of a description by Pedro Nunes (1537) of an instrument to observe the sun's altitude and a method of determining the latitude at any hour of the day.

G35 Hartner, Willy. "An Unusual Value for the Length of the Meridian Degree: 66 1/2 Miles, in Ibn Yūnus' Hākimitic Zīj." *Centaurus* 24 (1980): 148-152.

G36 Helbronner, P[aul]. "La genèse de l'opération de la jonction géodésique directe de la Corse à la chaîne méridienne des Alpes." *Revue Générale des Sciences pures et appliquées* 42 (1931): 359-366, 499-507, 676-680, 703-713.

On the work of François Perrier in the late 18th century; extensive quotations from the original reports.

G37 [Hinks, A.R.]. "Maupertuis and the Flattening of the Earth." *Geographical Journal* 98 (1941): 291-293.

Presents a graph showing that Maupertuis' result was very inaccurate compared to those of most of his contemporaries.

G38 [Honeyman, Robert Brodhead]. *The Size and Shape of the Earth. An Exhibition of Books from the Private Collection of Robert Brodhead Honeyman, Lehigh, Class of 1902.* Rare Book Room, The Library, Lehigh University, 1958. 14 pp.

A brief introduction is followed by a list of 23 books with bibliographic descriptions by J. Zeitlin.

G39 Jones, Tom B[ard]. *The Figure of the Earth*. Lawrence, Kans.: Coronado Press, 1967. 181 pp.

An account of the early 18th-century controversy between the Cassinis and the Newtonians about whether the earth is an oblate or prolate spheroid, and the expeditions led by Maupertuis and La Condamine which settled the question. "The tale is admittedly far from new, yet no historian has ever treated it in its entirety. Most of the 'facts' have long ago been set forth by the chroniclers of science, and my only claim to novelty is somewhat feeble and perhaps not substantiated by what follows: it is merely that I have tried to breathe some life into the carcass of a great tragi-comedy nearly trampled to death by the shambling and humorless pachyderms of scholarship" (Author's preface).

G40 Jones, Tom B. "The French Expedition to Lapland, 1736-37." *Terrae Incognitae* 2 (1970): 15-24.

On the expedition led by P.-L.M. de Maupertuis, to determine the length of a degree of latitude.

G41 Kopal, Zdenek. "The Contribution of Boscovich to Astronomy and Geodesy." *Roger Joseph Boscovich*. Edited by L.L. Whyte, New York: Fordham University Press, 1961, pp. 173-182.

In 1750, Boscovich and his Jesuit colleague Father Maire began to measure an arc of the meridian in the Papal States. Boscovich had earlier published a book, *De Telluris Figura*, on this subject. His 1755 report on the arc includes a chapter on the figure of the Earth based on a core-mantle model.

G42 Lenzen, Victor F. "An Unpublished Scientific Monograph by C.S. Peirce." *Transactions of the Charles S. Peirce Society* 5 (1969): 5-24.

Peirce submitted a "Report on Gravity at the Smithsonian, Ann Arbor, Madison, and Cornell" to the superintendent of the U.S. Coast and Geodetic Survey on Nov. 20, 1889. It was never published but was recently found by Albert Whimpey in the Survey's Archives. (An account of the discovery by Mr. Whimpey is published on pp. 3-4.) Lenzen explains the historical and scientific background of the problem and gives a detailed summary, with extracts, of the report, which he calls "the best work of its kind in the nineteenth century."

G43 Lenzen, Victor F. "Charles S. Peirce as Mathematical Geodesist." *Transactions of the Charles S. Peirce Society* 8 (1972): 90-105.

Peirce, as a member of the U.S. Coast and Geodetic Survey, "contributed notably to the practice and theory of research in gravity. Peirce is best known in the field for his measurements of the intensity of gravity by pendulums at stations in America. But he also contributed to the method of utilizing the results of surveys of gravity for the determination of the figure of the earth.... His calculation of the ellipticity of the earth has a place in the history of such determinations. He introduced new terms in the theoretical formula for gravity, and worked at an extension of Clairaut's theorem."

G44 Ludemann, Karl. "Zur Geschichte der Dosenlibelle. Beiträge zur Geschichte des geodätischen und markscheiderischen Messungswesens und der vermessungstechnischen Instrumentenkunde, 12." *Zeitschrift für Instrumentenkunde* 51 (1931): 136-144.

"Die Dosenlibelle war 1777 Johann Tobias Mayer Sohn bekannt. Vermutlich hat er sie erfunden, wahrscheinlich als Erster sie als Bestandteil eines geodätischen Vermessungsinstrumentes benutzt. Sie blieb vielen Feinmechanikern, Geodäten und Markscheidern noch längere Zeit unbekannt. Von 1792 ab hat Johann Gotthelf Studer sie bei seinen Instrumenten als Mayersche Wasserwage zielbewusst verwendet."

G45 Makemson, Maud W. "Old and New Beliefs about the Earth's Figure." *Proceedings of the 10th International Congress of History of Science, Ithaca, 1962* (pub. 1964) 2: 1027-1029.

A brief survey from the 17th century to modern satellite research.

G46 Markovic, Zeljko. *R.J. Boskovic et la théorie de la figure de la terre*. Paris: Université de Paris, Palais de la Découverte, 1961. 45 pp.

G47 Martonne, Edmond de. "La géodésie en Russie soviétique." *Revue Scientifique* 68 (1930): 737-745.

Contents: (I) Ancienne triangulation Russe (1817-1909); (II) La Service Cartographique Militaire Russe; (III) Les Zapisky [published 1837-1917]; (IV) La Commission [K.W.] Scharnhorst (1897-1907); (V) Nouveau Triangulation Russe (1910); (VI) Travaux géodésiques dans l'U.R.S.S. depuis 1920; (VII) Raccord des résultats Scharnhorst avec la nouvelle triangulation.

G48 Marussi, Antonio. "Italian Pioneers in the Physics of the Universe. Third Part: Geodesy." *Cahiers d'Histoire Mondiale* 7 (1963): 471-482.

On G.B. Riccioli, G.D. Cassini, R.G. Boscovich, S.G.B. Beccaria, F. Carlini, P. Pizzetti, C. Somigliana, C. Mineo, E. Fergola, and others.

G49 Mathieson, John. "Geodesy: A Brief Historical Sketch." *Scottish Geographical Magazine* 42 (1926): 328-347.

From antiquity through the 19th century.

G50 Mueller, C. "Zur Geschichte der Roehrenlibelle." *Zeitschrift fuer Vermessungswesen* 35 (1906): 673-678.

Discusses whether the invention should be attributed to Chapatot or Thevenot.

G51 Mueller, C. "Weiteres zur Geschichte der Roehrenlibelle." *Zeitschrift fuer Vermessungswesen* 36 (1907): 254-259.

G52 Mueller, Quodvultdeus. *Geschichte der Breiten-Gradmessungen bis zur Peruanischen Gradmessung*. Rostock, 1871. 52 pp.

History of research on the size and shape of the earth, from Aristotle to the 18th century.

G53 Newton, Robert R. "The Sources of Eratosthenes' Measurement of the Earth." *Quarterly Journal of the Royal Astronomical Society* 21 (1980): 379-387.

G54 Niemela, Lauri J. *Bibliography of the History of Geodesy and Mapping in Finland*. Copenhagen: U.S. Embassy, Foreign Service Despatch no. 252, 1960. Typescript, 10 pp. Copy in Library of National Oceanic and Atmospheric Administration, Rockville, Md.

G55 Perrier, G. "La Figure de la terre. Les grandes opérations géodésiques. L'ancienne et la nouvelle mesure de l'arc meridien de Quito." *Revue de Géographie Annuelle* 2 (1908): 201-508. Also published as separate pamphlet (Paris: Delgrave, 1908).

Part 1, review of the history of geodesy during the last three centuries; Part 2, the Peru expedition of 1735-1744; Part 3, the mission to the equator in 1899-1906; Part 4, description of modern methods; Part 5, current problems.

G56 Perrier, Georges. "Historique sommaire de la géodésie." *Thalès* 2 (1935): 117-129.

G57 Perrier, Georges. "Le développement de la géodésie de ses origines à nos jours." *Rice Institute Pamphlet* 24, no. 3 (July 1937): 168-188.

A brief survey of French contributions (mostly 17th and 18th centuries) followed by remarks on recent work in the U.S. and other countries.

G58 Perrier, Georges. *Petite Histoire de la géodésie, comment l'homme a mesuré et pesé la terre*. Paris: Alcan, Presses universitaires de France, 1939. 187 pp.

Starts with Newton, Huygens, Clairaut; deals mostly with 18th- and 19th-century researches. There is a section on the work of F.A. Vening-Meinesz.

G59 Perrier, Georges. *Wie der Mensch die Erde gemessen und gewogen hat; kurze Geschichte der Geodäsie*. Bamberg: Bamberger Verlagshaus, 1949. 190 pp.

Translation by E. Gigas of *Petite Histoire de la Géodésie*. Published simultaneously as Band 2 of *Veröffentlichungen des Instituts für Erdmessung* under the title "Kurze Geschichte der Geodäsie...." The translator has added a brief biographical note on Perrier, and a "Nachtrag" on international organizations.

G60 Poincaré, Henri. *Science and Method*. 1908. Translated from French by F. Maitland. New York: Dover, 1952. 288 pp.

Chapter II of Part IV is a popular essay on the history of French geodesy.

G61 Randles, W.G.L. *De la terre plate au globe terrestre: Une mutation epistemologique rapide (1480-1520)*. Paris: Armand Colin, 1980. 120 pp.

"This elegant and scholarly disquisition deals with a fascinating problem: the general acceptance of a spherical earth by the world of scholars during the Age of Discovery."--G. Kish, *Isis* 73 (1982): 136-137.

G62 Rawlins, Dennis. "Eratosthenes' Geodesy Unraveled: Was There a High-Accuracy Hellenistic Astronomy?" *Isis* 73 (1982): 259-265.

Contrary to his reputation as "the first great geodesist who attained an accurate size of the Earth," Eratosthenes injected an erroneous observation into a set of accurate data obtained by his predecessors.

G63 Reingold, Nathan. "Research Possibilities in the U.S. Coast and Geodetic Survey Records." *Archives Internationales d'Histoire des Sciences* 11 (1958): 337-346.

With some exceptions the bulk of the records are from the years 1844-1905. They consist mainly of long-time series of observations of the coastline, geomagnetism, tides, currents, gravity, and astronomical observations. There is also a considerable amount of information about scientists such as F.R. Hassler, A.D. Bache, B. Peirce, and C.S. Peirce, and about the relations between pure and applied science in America.

G64 Sarton, George. "Le monument des missions géodésiques, françaises, à Quito." *Isis* 2 (1914): 163-164.

Commemoration of the expeditions of 1735-43 and 1899-1906.

G65 Sarton, George. "La XVIIe Conférence générale de l'Association Géodésique Internationale (Hambourg, 1912)." *Isis* 2 (1914): 191-193.

A review of the history of the Association since its foundation in 1862.

G66 Sawicki, Kazimierz. *Piec' wiekow Geodezji Polskiej; szkice Historyczne od XV do XIX wieku*. Warszawa: Panstwowe Przedsiebiorstwo Wydawnictm Kartograficznych, 1960. 258 pp. Second edition, 1964. 335 pp.

Geodesy in Poland. Includes chapters on Jan Brozek, Joachim Stegmann, Maciej Gtoskowski, Tytus Liwiusz Boratyni, Jozef Narnowicz-Naronski, Stanislaw Solski, Ignacy Zaborowski, Antoni Szahin.

G67 Schmid, Hans. *Vom Anbeginn über Eratosthenes zum Weltnetz*. Antrittsvorksungen der Technischen Hochschule in Wien, 28. Wien: Verlag der Technischen Hochschule, 1973. 20 pp.

G68 Scrutton, Colin T., and Roger G. Hipkin. "Long-Term Changes in the Rotation Rate of the Earth." *Earth-Science Reviews* 9 (1973): 259-274.

A review of research in the last decade; "the most outstanding development is the tentative measurement of the mean rotation rate in the remote geological past, made possible by the study of periodicities in the skeletal growth of fossil organisms."

G69 Shibanov, F.A. "Iz istorii prakticheskoi astronomii, geodezii, i kartografii v Rossii" [The History of Practical Astronomy, Geodesy, and Cartography in Russia]. *Istoriko-Astronomicheskie Issledovaniia* 9 (1966): 235-260.

On the work of P.M. Smyslov, V.F. Lemm, P.I. Kuznetsov, and P.N. Alexandrov in the period 1840-80.

G70 Shibanov, F.A. "I.I. Khodz'ko--pervyi issledovatel' Kavkaza v matematicheskom otnoshenii. Novye materialy k ego biografii." *Istoriko-Astronomicheskie Issledovaniia* 10 (1969): 185-198.

On Iosif Ivanovich Khodzko (1800-1881), "the first investigator of the Caucasus from a mathematical point of view."

G71 Stevens, Wesley M. "The Figure of the Earth in Isidore's *De Natura Rerum*." *Isis* 71 (1980): 268-277.

"The earliest extant diagram of the earth as a globe occurs in manuscripts of the *De rerum natura* by Isidore of Seville, a schoolbook intended to outline the organized knowledge proper for an educated man in seventh-century Visigothic Spain." The author discusses this diagram in the context of earlier Greco-Roman depictions of the globe.

G72 Stone, J.C. "Pioneer Geodesy: The Arc of the 30th Meridian in the Former Northern Rhodesia." *Cartographic Journal* 13 (1976): 122-128.

On the project initiated by David Gill in 1903 and carried out by Sven Tryggve S. Rubin.

G73 Taton, R. "Sur la diffusion des théories Newtoniennes en France. Clairaut et le problème de la terre." *Vistas in Astronomy* 22 (1978): 484-509.

On the contributions of A.C. Clairaut (1713-1765) to geodesy and the mathematical theory of the shape of the earth.

* Thams, J.C., ed. *The Development of Geodesy and Geophysics in Switzerland.* Cited above as item A103.

G74 Turner, A.J. "Hooke's Theory of the Earth's Axial Displacement: Some Contemporary Opinion." *British Journal of Historical Science* 7 (1974): 166-170.

Letters from J. Wallis to E. Halley (1686/7) indicate his views on R. Hooke's hypothesis of displacement of the poles of a non-spherical earth.

G75 Von Hagen, Victor Wolfgang. *South America Called Them: Explorations of the Great Naturalists: La Condamine, Humboldt, Darwin, Spruce.* New York: Knopf, 1945. xii + 311 + ix pp. Second edition, 1945; third edition, 1955 (same pagination). Translated into French (Paris, 1947), Spanish (Mexico, 1947) and German (Vienna, 1951).

Includes an account of the expedition led by C.-M. de la Condamine in the 1730s to determine the length of the degree of latitude near the equator.

For additional information on this topic see works cited in the biography sections on the following scientists:

Bessel, F.W.
Boscovich, R.J.
Bouguer, P.
Bowie, W.
Bugge, T.
Cassini, J.
Cassini de Thury, C.
Daly, R.A.
Davidson, G.
Delambre, J.J.
Dutton, C.E.
Eratosthenes
Faye, H.
Foster, H.
Gauss, C.F.
Gilbert, G.K.
Hassler, F.R.
Hayford, J.F.
Helmert, F.R.
Ibañez e Ibañez de Ibero, C.
Kater, H.
Kavraysky, V.V.
Krasovsky, T.N.
Krayenhoff, C.R.T.
Lacaille, N. de
Laplace, P.
Lawson, A.C.
Maskelyne, N.
Mason, C.
Matuyama, M.
Maupertuis, P.L.M. de
Numerov, B.V.
Orlov, A.Y.
Peirce, B.
Perrier, G.
Picard, J.
Porro, I.
Rankine, A.O.
Schmidt, C.A. Von
Schumacher, H.C.
Shokalsky, Y.M.
Vening Meinesz, F.A.

H. FORMATION OF THE EARTH'S SURFACE FEATURES: TECTONICS, UNIFORMITARIAN-CATASTROPHIST DEBATE, PLUTONIST-NEPTUNIST DEBATE

H1 Aubouin, Jean. *Geosynclines*. Amsterdam: Elsevier, 1965. xv + 335 pp.

Part I (pp. 7-40) is a historical review of the geosynclinal concept from J. Hall (1859) to A.V. Peyre and V.M. Sinitzin (1950).

H2 Bailey, E[dward] B[attersby]. *Tectonic Essays, Mainly Alpine*. Oxford: Clarendon Press, 1935. xii + 200pp.

Includes "Chronology of Tectonic Discovery: 1775-1893" and chapters on the researches of Arnold Escher von der Linth (1807-1872), Bernhard Studer (1794-1887), H. Schardt, Eduard Suess (1831-1914), Marcel Bertrand, and others.

H3 Baker, Victor R., and Stephen Pyne. "G.K. Gilbert and Modern Geomorphology." *American Journal of Science* 278 (1978): 97-123.

A review of Gilbert's contributions to geomorphology (see item BG13 for a detailed exposition) is used to provide several suggestions for future geomorphic research.

H4 Baumgärtel, Hans. "Alexander von Humboldt: Remarks on the Meaning of Hypothesis in His Geological Researches." *Toward a History of Geology* (item A89), pp. 19-35.

Humboldt proposed a law of "loxodromism of the strata" which he thought might represent universal features of the crust.

H5 Beloussov, V.V. *Basic Problems in Geotectonics*. Translation of the Russian edition of 1954. English edition edited by John C. Maxwell. Paul T. Broneer, principal translator. New York: McGraw-Hill, 1962. 809 pp., fig.

Chapter 2 (pp. 19-54) is a history of geotectonics, with emphasis on Russian work. Chapter 37 (pp. 739-765) is a review and critique of the principal hypotheses of the 20th century. The author is an influential (in the U.S.S.R.) opponent of continental drift and plate tectonics.

H6 Cannon, Walter F. "The Uniformitarian-Catastrophist Debate." *Isis* 51 (1960): 38-55.

Charles Lyell, the leader of the Uniformitarian school, was supported by Charles Darwin in the Geological Society of London in the 1830s, although Lyell was later to view his own acceptance of evolution as a retreat from his principles. The catastrophists were led by William Whewell (who introduced the two terms), Roderick Murchison, and William Hopkins. Both uniformitarians and catastrophists relied on the earth's internal heat as a cause of geological change, but Lyell refused to accept arguments based on a continual decrease in this heat. Hopkins, "the effective founder in England of what became known as physical geology," showed that the popular idea of a hot fluid interior under a thin crust is inconsistent with astronomical evidence. Yet, while the cooling of the earth is very slow, it is inevitable in the long run. By the 1850s there seemed to be a basis for compromise between the two schools, although their leaders did not yet recognize it.

H7 Carozzi, Albert V. "Lamarck's Theory of the Earth: *Hydrogéologie*." *Isis* 55 (1964): 293-307.

Analysis of a book published by J.B. Lamarck in 1802, giving his views on the effect of water on geological phenomena including volcanoes, the formation of mountains, and the displacement of ocean basins. Carozzi's translation of this work is cited below, item H29.

H8 Carozzi, Albert V. "De Maillet's *Telliamed* (1748): An Ultra-Neptunian Theory of the Earth." *Toward a History of Geology* (item A89), pp. 80-99.

"De Maillet's cosmogonic system is based on the Cartesian theory of vortices combined with Fontenelle's concept of the plurality of inhabited worlds. It assumes that heavenly bodies go through an endless process of renovation consisting of alternate luminous phases, during which they become suns, and dark phases when they behave as planets. At present the earth is undergoing a dark phase during which the sea, after having covered the entire globe, gradually diminishes by evaporation until there is total depletion. De Maillet considered the sea entirely responsible for all the physiographic, lithological and structural characteristics of the earth's crust. These were generated, according to his own estimate, during billions of years, a concept completely unheard of at the time."

H9 Carozzi, Albert V. "De Maillet's *Telliamed* (1748): The Theory of the Diminution of the Sea." *Endeavor* 29 (1970): 140-143.

"De Maillet's theory of the diminution of the sea, which spans the biological and earth sciences, is a remarkable achievement for a career diplomat and a traveller. His ingenious approach to the history of the Earth, according to which the sea--during millions of years--is entirely responsible for all the physiographic, lithologic, and structural features of the Earth's crust, makes him both a forerunner of generalized transformism and a marine geologist and sedimentologist" (Author's summary).

H10 Carozzi, Albert V. "[An Essay Review of] Gordon L. Davies, *The Earth in Decay: A History of British Geomorphology, 1578-1878*. (New York, 1969)." *Studies in History and Philosophy of Science* 4 (1973): 289-299.

Carozzi disagrees with the author of item H17 on a number of detailed points, e.g., "I feel that Hooke's concept of earthquakes is much more far-reaching than Davies would like to assume. In fact, Hooke, followed by R.E. Raspe, visualized almost an orogenic mechanism," and accuses him of making self-contradictory statements about Hutton. Yet his judgment is generally favorable: "The remarkable skill of the author lies in presenting a well-documented and easily readable narrative in which human actions and events develop within their mutual context. It is a historical work in the modern sense--a model to be followed by those who contemplate writing a new history of geology."

H11 Chorley, Richard J. "Diastrophic Background to Twentieth-Century Geomorphological Thought." *Geological Society of America Bulletin* 74 (1963): 953-970.

"The eustatic theory of Edward Suess provided the major diastrophic influence which conditioned geomorphological thought for the first 3-4 decades of the twentieth century. Assisted by the translation of De Margerie, the work of A. Penck, the results of the Challenger expedition, and the concept of "borderlands," this theory had become a basic article of geological faith by the time Chamberlin's (1909) influential statement on diastrophism and correlation appeared. During the following quarter of a century the theory was consolidated as the result of intensification of interest in North American stratigraphy, studies of glacial eustatism, and the investigation of submarine cannons. The high point of the geomorphological influence of the eustatic theory was marked by the publications of Baulig (1928; 1935), who was partly responsible for its subsequent importance in British geomorphological thought.

"By the late 1930's the increasing awareness of continental instability, largely resulting from ideas bearing on the epeirogenic and isostatic theories, had begun to cast serious doubt on the geomorphological value of Suess' 'great unifying generalization.' Of the variety of patterns of crustal movement which had been proposed, that of W.M. Davis came to be largely accepted both by tectonic geologists and geomorphologists alike, although problems of instability of baselevel and insufficiency of time for peneplanation are proving increasing impediments to its retention. It is ironical, too, that many of the geological advances which first appeared to strengthen the eustatic theory have helped more lately to discredit it, at least in the extreme form proposed by Suess" (Author's summary).

H12 Chorley, Richard J., Antony J. Dunn, and Robert P. Beckinsale. *The History of the Study of Landforms, or the Development of Geomorphology*. Volume 1. *Geomorphology Before Davis*. New York: Wiley, 1964. xvi + 678 pp.

Discusses the work of A. Werner, J. Hutton, J. Playfair, W. Buckland, C. Lyell, J. Dana, G. Greenwood, and others. For volume 2, *The Life and Work of William Morris Davis*, see above (item BD8); Davis (1850-1934) is "the *only* man to whom in our history of landforms we would devote a whole volume" (Introduction to Volume 2).

* Cloud, Preston, ed. *Adventures in Earth History*. Cited above as item A21.

H13 Cuvier, Georges. *Discours sur les révolutions de la surface du globe, et sur les changements qu'elles ont produits dans le règne animal*. Trois édition française. Brussels: Culture et Civilisation, 1969. ii + 400 pp.

Facsimile reprint of the Paris, 1825 edtion. See next item for English translation.

H14 Cuvier, Georges. *Essay on the Theory of the Earth*. Translation of *Discours sur les révolutions*, with Mineralogical Notes, and an Account of Cuvier's Geological Discoveries, by Robert Jameson. Reprint of the third edition (Edinburgh: Blackwood, 1817). New York: Arno Press, 1978. xviii + 348 pp.

A convenient source of a classic exposition of "catastrophism." Cuvier examines four major causes of geological change that can be presently observed (rains and thaws, rivers, seas, volcanoes) and concludes that they are not adequate to account for the formation of the major features of the earth's surface. He also argues that "revolutions" of the surface have taken place in a fairly short time.

H15 Davies, Gordon L. "From Flood and Fire to Rivers and Ice--Three Hundred Years of Irish Geomorphology." *Irish Geography* 5 (1964): 1-16.

Survey of the views of R. Kirwan and other catastrophists; responses to the glacial hypothesis; J.B. Jukes.

H16 Davies, Gordon L. "The Concept of Denudation in Seventeenth-Century England." *Journal of the History of Ideas* 27 (1966): 278-284.

"The concept of denudation is one of the most fundamental tenets of modern geology, and since the concept was so firmly established in XVIIth-century England, it might reasonably have been expected to survive intact until the emergence of scientific geology during the second half of the following century. It failed to do so, however [for two reasons].... First, there was the widespread rejection of the belief in a decaying universe. The Calvinism of earlier decades had been replaced by Arminianism.... The second reason [was that] the new appreciation of the deity encouraged a teleological approach to nature.... No longer were mountains seen as useless warts and tumors; instead they came to be regarded as part of the adornment of a wonderful creation" (pp. 280-281). How, then, could they be subject to decay?

H17 Davies, Gordon L. *The Earth in Decay: A History of British Geomorphology, 1578-1878*. New York: American Elsevier Pub. Co., 1969. xvi + 390 pp.

A comprehensive survey of 17th-century "theories of the earth" (N. Steno, T. Burnet, J. Woodward, R. Hooke), the 18th-century "denudation dilemma," revival of theory in the 1770s (J. Whitehurst, J.A. de Luc, J. Williams, R. Kirwan, R. Jameson), the "Huttonian earth-machine," uniformitarianism and the rejection of pluralism, the glacial theory (L. Agassiz), revival of pluralism (J.G. Jukes, A. Ramsay, A. Geikie). See Carozzi's essay review (item H10).

H18 Dott, Robert H., Jr. "James Hutton and the Concept of a Dynamic Earth." *Toward a History of Geology* (item A89), pp. 122-141.

"It is the purpose of this paper to show that Hutton provided a revolutionary unifying synthesis of geology in the late 18th century, the full significance of which has not been sufficiently appreciated and which seems to have been somewhat misinterpreted. I feel that the theory provided by him, though less concise and nonmathematical, was of the same order of importance to geology as were the synthesis of Newton and Lavoisier.... Hutton's theory postulated a continued, cyclic decay and resurrection of land.... He argued that the earth's interior had always been structurally dynamic."

H19 Ellenberger, François. "Le dilemme des montagnes au XVIIIe siècle: Vers une réhabilitation des diluvianistes? *Revue d'Histoire des Sciences* 31 (1978): 43-52.

"The theories of the Earth of the great British and Swiss catastrophists of the late 17th and early 18th centuries were very harshly judged by the partisans of uniformitarianism. An impartial examination, however, shows that the latter eluded numerous problems and, in particular, too often eliminated tectonics with the Flood; moreover, they provided a poor explanation for the process of sedimentation by the mere natural shifting of the sea over land. Where the catastrophists went wrong was in accelerating Time beyond all measure. With that proviso, some of their visions were often ahead of their time; such is the case with [John] Woodward in stratigraphy or Louis Bourguet for orogenesis" (Author's summary).

H20 Ellenberger, François. "De l'influence de l'environment sur les concepts: L'exemple des théories géodynamiques au XVIIIe siècle en France." *Revue d'Histoire des Sciences* 33 (1980): 33-68.

"France presents three kinds of geological and geomorphic landscapes, each of which is quite distinct from the others: the plains of the North and West; the mountain of the Massif Central and the Mediterranean; the Alpine chains. This article surveys the work of the principal French naturalists of the 18th century who studied the earth in each of these fields (especially the first two), in order to determine the extent to which the environment influenced their views. One finds that it played an important role. The southern French naturalists realized much more clearly than their colleagues elsewhere the extent of secular work accomplished by running water and the reality of tectonic deformations. Those who studied the North of France, or more especially the Alps, were often misled, albeit in good faith, by a model which was in fact derived from the glacial epoch. All the same the natural environment served rather to confirm these writers in their theories than to engender new doctrines" (Author's summary).

H21 Engelhardt, Wolf v. "Neptunismus und Plutonismus." *Fortschritte der Mineralogie* 60 (1982): 21-43.

Reviews the ideas of 18th-century naturalists in France, Italy, and Germany on precipitation of rocks from a primeval ocean, as a background for the system of A.G. Werner. The discovery of the volcanic origin of basalts inspired the volcanist school which ultimately triumphed over Neptunism. Hutton and Playfair replaced the historical neptunist model of the receding ocean by an ahistorical pattern of endlessly repeating cycles. "During the 19th century two branches of geodisciplines developed from the concepts of neptunism and plutonism: of neptunistic origin are geohistoric disciplines like Historical Geology, Palaeontology; from the plutonistic-actualistic standpoint disciplines started with more physical or chemical backgrounds, such as mineralogy, petrology, geochemistry and geophysics.... The unity of geoscience ... consists in the dialectic synthesis of [these] two opposite approaches."

* Frängsmyr, Tore. "The Geological Ideas of J.J. Berzelius." Cited above as item BB24.

H22 Gerstner, Patsy A. "James Hutton's Theory of the Earth and His Theory of Matter." *Isis* 59 (1968): 26-31.

"... one of the most unusual aspects of Hutton's whole geological theory is his conception of the nature of the heat that served as the basic mechanism for fusing loose debris elevating continents. Unlike his vulcanist contemporaries and predecessors who discussed geological formations in terms of heat resulting from fire, Hutton conceived of a heat distinct from fire."

H23 Gerstner, Patsy A. "A Dynamic Theory of Mountain Building: Henry Darwin Rogers, 1842." *Isis* 66 (1975): 26-37.

H.D. Rogers (1809-1866) "argued that mountains were built during paroxysmal upheavals of the land caused by a wave-like undulation of the crust that occurred because molten matter beneath was pulsating or moving with a wave-like motion. The structure of the Appalachians suggested the validity of this theory to Rogers."

H24 Gohau, Gabriel. "Aux commencements de le géologie." *Raison Présenté* 11 (1969): 27-40.

An essay on actualism in the theories of Buffon and Elie de Beaumont on the evolution of the earth's surface.

H25 Greene, Mott T. "Major Developments in Geotectonic Theory Between 1800 and 1912." Ph.D. dissertation, University of Washington, 1978. v + 505 pp.

In this important study Greene shows that "The idea that the works of James Hutton and Charles Lyell represent the mainstream of geological thought in the nineteenth century is a misconception based on the underevaluation of their European contemporaries.... Not only did the geology of A.G. Werner and later the catastrophists survive their celebrated contests with Huttonian and Lyellian geology ... but they were the foundation of a robust and often predominant tradition of nonuniformitarian geology." The theory of Leonce Elie de Beaumont was extended by J.D. Dana and Eduard Suess. Seuss's theory of mountain ranges and oceans, based on the contraction hypothesis, was rejected when the latter hypothesis was successfully challenged on geophysical grounds in the early 20th century. But Wegener tried to reconcile the geology of Suess with the new geophysical theories of continental drift. Thus the "revolution" of Wegener against a Lyellian orthodoxy is an artifact of historical misinterpretation; instead, Wegener was drawing on a strong non-uniformitarian tradition. (Based on author's summary in *Dissertation Abstracts International* 39 [1978]: 3106-A.)

A revised version of this dissertation was published by Cornell University Press in 1982 under the title *Geology in the Nineteenth Century: Changing Views of a Changing World*.

H26 Hawkes, Leonard. "Some Aspects of the Progress of Geology in the Last Fifty Years. I. Anniversary Address Delivered at the Annual General Meeting of the Society on 1 May 1957." *Quarterly Journal of the Geological Society of London* 113 (1957, pub. 1958): 309-321.

"The interpretation of the record of the rocks current fifty years ago has been modified as a result of new evidence and the realization of the immensity of geological time. There is as yet no good reason to doubt that conditions under which the rocks available for examination were formed have remained the same throughout geological time. The hypothesis of a world-wide rhythm in the intensity of crustal movement remains unproved. On present evidence, Lyellian uniformitarianism is applicable to the whole of geological history" (Author's summary).

H27 Hooykaas, R. "Catastrophism in Geology, Its Scientific Character in Relation to Actualism and Uniformitarianism." *Mededelingen der Koninkijke Nederlandse Akademie van Wetenschappen, Afd. Letterkunde*, 33, no. 7. 50 pp. Amsterdam: North Holland Publishing Co., 1970. Reprinted in *Philosophy of Geohistory 1785-1970* (item A7), pp. 310-350.

Surveys the views of Buffon, G. Razumovsky, D. Dolomieu, G. Cuvier, L. Elie de Beaumont, L. Frapolli, Ch. Sainte Claire Deville, W. Conybeare, B. Cotta, H.G. Brown, J. Prestwich.

H28 [Hutton, J.]. Craig, G.Y., D.B. McIntyre, and C.D. Waterston. *James Hutton's "Theory of the Earth": The Lost Drawings*. Edited by G.Y. Craig. Edinburgh: Scottish Academic Press in association with the Royal Society of Edinburgh and the Geological Society of London, 1978. 67 pp. + portfolio.

H29 Lamarck, J.B. *Hydrogeology*. Translated with an introduction and index by Albert V. Carozzi. Urbana: University of Illinois Press, 1964. vi + 152 pp.

Translation of *Hydrogéologie* (1802), a work "concerned with basic geological problems such as the action of terrestrial waters, the existence and possible displacement of the ocean basin, the immensity of geological time, the origin of mountains, the significance of fossils, and the action of living organisms on the earth's crust" (p. 3).

* Lawrence, Philip. "Heaven and Earth--The Relation of the Nebular Hypothesis to Geology." Cited above as item D44.

H30 Lees, George Martin. "The Evolution of a Shrinking Earth." *Quarterly Journal of the Geological Society of London* 109 (1953): 217-257.

While not historical in nature, this review called attention to a number of fundamental problems in geotectonics and stimulated a lively discussion, reported in *Nature* 175 (1955): 575-576.

H31 Manten, A.A. "Uniformitarianism, Catastrophism and Christian Faith." *Atlas: News Supplement to Earth-Science Reviews* 3 (1967): A253-256.

Review of D.W. Patten, *The Biblical Flood and the Ice Epoch* and discussion of similar works by G.M. Price, I. Velikovsky, and others who present a catastrophist view of earth history.

H32 Margerie, Emmanuel de. *Critique et géologie: Contribution à l'histoire des sciences de la terre (1882-1942)*. 4 vols. Paris: Colin, 1943-1948. 2108 pp., 768 figs.

According to G. Sarton, this massive work is essentially a critical autobibliography preceded by a short biography of the author (1862-1953), one of the creators of tectonics It reviews Margerie's own work on tectonics and provides an extensive bibliography of each topic. Almost as remarkable as the work itself are the circumstances of its publication with a subsidy from the Pontificia Accademià delle Scienze at a time when Italy was at war with France. Sarton's enthusiastic reviews are worth reading as an appetizer; see *Isis* 36 (1945): 74-75; 38 (1948): 263-264; 40 (1949): 390.

* Mather, Kirtley F., ed. *Source Book in Geology, 1900-1950*. Cited above as item A65.

H33 Mayo, Dwight Eugene. "The Development of the Idea of the Geosyncline." Ph.D. dissertation, University of Oklahoma, 1968. 292 pp.

Since the middle of the 19th century, the concept of the geosyncline has been an important part of most theories of mountain building. Though rooted in the speculations of Charles Babbage and John F.W. Herschel, it is largely the creation of James Hall. The concept was born in a uniformitarian background but applied by J.D. Dana in a catastrophist-oriented contraction theory. The idea also became associated with C.E. Dutton's isostatic hypothesis.

(Based on the author's summary in *Dissertation Abstracts International* 28 [1968]: 4995-A.)

H34 Page, Leroy E. "Diluvialism and Its Critics in Great Britain in the Early Nineteenth Century." *Toward a History of Geology* (item A89), pp. 257-271.

On the theories of Werner and Cuvier and their discussion by W. Buckland and J. Fleming. "... there were strong uniformitarian tendencies in Wernerianism that may have facilitated the later acceptance of Lyell's more comprehensive uniformitarian views." The diluvial theory was not supported, primarily because of a desire to confirm the Bible, and most geologists agreed with Lyell that their science should be kept free from involvement with religion.

H35 [Prévost, C.]. Laurent, Goulven. "Actualisme et anti-transformisme chez Constant Prévost." *Histoire et Nature: Cahiers de l'Association pour l'Histoire des Sciences de la Nature* 8 (1976): 33-51.

H36 Pyne, Stephen J. "Certain Allied Problems in Mechanics: Grove Karl Gilbert at the Henry Mountains." *Two Hundred Years of Geology in America* (item A90), pp. 225-238.

Gilbert "was a pure Newtonian who sought to make geological systems analogous to mechanical systems." For further discussion see item BG13.

H37 Rappaport, Rhoda. "Geology and Orthodoxy: The Case of Noah's Flood in 18th-Century Thought." *British Journal for the History of Science* 11 (1978): 1-18.

Naturalists were not (at least in France) forced to believe in the Flood by religious orthodoxy, but most of them accepted a flood as one of several natural upheavals that had formed the earth's surface, without adhering literally to the Genesis account. In some cases they thought the earth had once been covered by a universal ocean, different from the flood.

H38 Raspe, Rudolf Erich. *An Introduction to the Natural History of the Terrestrial Sphere principally concerning New Islands born from the Sea and Hooke's Hypothesis of the Earth on the Origin of Mountains and Petrified Bodies to be further established from Accurate Descriptions and Observations*. Translated and edited by Audrey Notvik Iversen and Albert V. Carozzi, including a facsimile of the 1763 edition. New York: Hafner Publishing Co., 1970. cxvii + 190 + 191 pp.

Raspe's work was inspired by the Lisbon earthquake of 1755 and his reading of Robert Hooke's *Lectures and Discourses of Earthquakes and Subterraneous Eruptions*. He developed Hooke's ideas into a general theory of the origin of islands and continents.

H39 Rezvoi, D.P. "K istokam ucheniia o glubinnykh razlomakh" [On the Sources of Deep-Fault Study] (in Russian with English abstract). *Geologischeskoe Obrazovanie i istoriia geologii*. Edited by G.P. Gorshkov. Moskva: Nauka, 1976, pp. 44-59.

"Deep-fault study began at the end of the XIX c. but could not make any progress as long as the contraction theory was accepted. The idea of extensive dislocation belts of non-folding origin was clear only to eminent intellects (the 'rudiment range' of A.P. Karpinsky, the 'lineaments' of W. Hobbs, R. Sonder, H. Steele). In the thirties I.G. Kuznetsov divided the Caucasus into mobile blocks and indicated that the existence of the separating faults was of great duration. In 1933 V.A. Mikolayer described the 'main structure line' of the Tien-Shau, and in 1938 V.I. Popov--the 'discordant lines.' N.S. Shatsky assumed the great faults as the basis of the platform and geosynctive development (1936-1948). A.V. Peive was the first to use the term 'deep fault' in its modern sense. A number of explorers (A.V. Peive, V.E. Khain, V.A. Kosygin) undertook attempts to classify and systematize the deep faults. V.I. Smirnov, E.A. Radkevich and others underlined the metallogenic importance of the deep-faults more than once. The distinguishing features of the deep-faults: their extraordinary length, the great depth of origin and longevity allow us to suppose them to be tectonical elements of great stability. And this comes into conflict with the ideas of mobilism" (Author's abstract, slightly edited).

H40 Rudwick, M.J.S. "A Critique of Uniformitarian Geology: A Letter from W.D. Conybeare to Charles Lyell, 1841." *Proceedings of the American Philosophical Society* 111(2) (1967): 272-287.

"Five aspects of the letter are especially noteworthy: Conybeare's insistence that rival models of earth history--'historical' and 'steady-state'--lay at the root of the debate; his criticism of Lyell's denial of organic progression; his argument for a similar 'progression' in the diminishing intensity of igneous and tectonic processes; his explicit acceptance of Lyell's vast time-scale but implicit failure to grasp its implications; and his belief that Lyell's emphasis on uniformity derived from his fear of supporting the arguments of natural theology" (pp. 273-274). In the letter Conybeare also expressed his views on the hypothesis that the earth has a hot fluid interior.

H41 Rudwick, Martin J.S. "Uniformity and Progression: Reflections on the Structure of Geological Theory in the Age of Lyell." *Perspectives in the History of Science and Technology*. Edited by Duane H.D. Roller. Norman: University of Oklahoma Press, 1971, pp. 209-227.

"Lyell's work ... must be seen in the context of a rival synthesis of great scientific power and sophistication. Directionalism [based on the theory of a cooling earth] gave meaning and coherence to a vast range of empirical observations, and it was based on the best physics and chemistry of the time."

H42 Rudwick, Martin J.S. "Lyell on Etna and the Antiquity of the Earth." *Toward a History of Geology* (item A89), pp. 288-304.

Lyell's analysis of volcanic action at Etna played a key role in the strategy of the *Principles*. He "had expounded the sufficiency of actual causes to interpret the phenomena of the past if only a vast enough time scale were conceded." Etna seemed to have been built up by a series of small flows, yet the flows recorded in historic times did not amount to even a small portion of what was needed to produce the result. Thus the geological time scale must be much longer than the scale of human history.

H43 Scheidegger, Adrian E. "Recent Advances in Geodynamics." *Earth-Science Reviews* 1 (1966): 133-153.

A review of research on the theory of the earth's rotation, epeirogenesis, geotectonic hypotheses, theory of faulting, earthquake origination, folding, etc. Approximately 150 references.

H44 Sweet, Jessie M., and Charles D. Waterston. "Robert Jameson's Approach to the Wernerian Theory of the Earth, 1796." *Annals of Science* 23 (1967): 81-95.

Jameson objected to James Hutton's explanation of the formation of geological strata by consolidation, compression, and slow cooling.

H45 Taylor, Kenneth L. "Natural Law in 18th-Century Geology: The Case of Louis Bourguet." *Actes du XIIIe Congrès International d'Histoire des Sciences* 8 (1971): 72-80.

Bourguet (1678-1742) proposed a "principle of corresponding angles" for mountains on opposite sides of a valley and tried to relate it to a physical theory of the development of the earth's surface.

* Thams, J.C., ed. *The Development of Geodesy and Geophysics in Switzerland*. Cited above as item A103.

H46 Vysotskii, B.P. "Geologicheskie idei P.S. Pallasa i teoria katastrof Zh. Kiuv'e" [The Geological Ideas of P.S. Pallas and Cuvier's Theory of Catastrophies]. *Voprosy Istorii Estestvoznaniia i Tekhniki* 52 (1976): 59-63.

H47 Webby, B.D. "Some Early Ideas Attributing Easterly Dipping Strata to the Rotation of the Earth." *Proceedings of the Geologists' Association* 80 (1969): 91-97.

On the suggestions of John Ray (1692), William Stukeley (1724), John Strachey (1725), John Williams (1789), and William Smith (1801).

H48 White, George W. "William Maclure Was a Uniformitarian and Not a Real Wernerian." *Journal of Geological Education* 18 (1970): 127-128.

H49 Wilson, Leonard G. "The Development of the Concept of Uniformitarianism in the Mind of Charles Lyell." *Proceedings of the 10th International Congress of the History of Science 1962* (pub. 1964) 2: 992-996.

On the influence of J. Playfair's *Illustrations of the Huttonian Theory* and Scrope's *Geology of Central France* on Lyell's thinking in the 1820s. Following Scrope, Lyell showed that valleys in volcanic regions had been produced by gradual erosion rather than a single catastrophic deluge.

For additional information on this topic see works cited in the biography sections on the following scientists:

Abich, O.H.W.
Anuchin, D.N.
Bertrand, M.
Berzelius, J.J.
Boullanger, N.
Bucher, W.H.
Burnet, T.
Dana, J.D.
Davis, W.M.
Desmarest, N.

HA. Glaciology

HA1 Agassiz, Louis. *Études sur les glaciers*. London: Dawsons of Pall Mall, 1966. 347 pp. + 32-plate atlas.

Reprint of the first edition (1840). For English translation see next item.

HA2 Agassiz, Louis. *Studies of Glaciers, preceded by the Discourse of Neuchatel*. Translated and edited by Albert V. Carozzi. New York: Hafner, 1967. lxxi + 213 pp., plates.

In the introduction, Carozzi discusses "the circumstances in which Agassiz became interested in glaciology, delivered his famous *Discours de Neuchatel* of 1837, investigated Alpine glaciers, published the *Études sur les Glaciers* translated here, and, after a bitter struggle, eventually succeeded in obtaining a general acceptance of his concept of an Ice-age."

HA3 Alexander, David. "Leonardo Da Vinci and Fluvial Geomorphology." *American Journal of Science* 282 (1982): 735-755.

"Following Pliny and other classical authors, Leonardo believed that the hydrological cycle is based on circulation of water inside the Earth, through which rainwater is raised up inside mountains by capillary action aided by the heat of the sun.... Leonardo was a Uniformitarian, not a Diluvialist, and supported the theory of slow uplift proposed by Albert of Saxony and others."

HA4 Beck, Hanno. "Alexander von Humboldt und die Eiszeit." *Gesnerus* 30 (1973): 105-121.

HA5 Cameron, Dorothy. "Goethe--Discoverer of the Ice Age." *Journal of Glaciology* 5 (1965): 751-754.

J.W. von Goethe "was one of the first to attribute the transport of erratic blocks to glaciers, and to believe that an ice sheet covered northern Germany; furthermore, he was the very first to believe in an ice age."

HA6 Campbell, Ian and David Hutchison. "A Question of Priorities: Forbes, Agassiz, and Their Disputes on Glacier Observations." *Isis* 69 (1978): 388-399.

In 1841 J.D. Forbes and L. Agassiz studied *bandes bleues* in the Unteraar glacier in Switzerland. In independent later publications they established the importance of the *bandes* in the theory of glacial ice and its movement. Unpublished correspondence throws some light on their priority dispute.

HA7 Carozzi, Albert V. "Agassiz's Amazing Geological Speculation: The Ice-age." *Studies in Romanticism* 5 (1966): 57-83.

An account of the years 1836-42 during which L. Agassiz conceived, developed, published, and defended his Ice Age theory.

HA8 Carozzi, Albert V. "Agassiz's Influence on Geological Thinking in the Americas." *Archives de Sciences* (Geneva) 27(1) (1973): 5-38.

Louis Agassiz's concept of Pleistocene continental glaciation was discussed by T.A. Conrad, E. Hitchcock, W.W. Mather, and S. St. John. Charles Whittlesey (1859) was the first to confirm continental glaciers in North America as proposed by Agassiz. Because of his opposition to Darwinian evolution, Agassiz misinterpreted his geological observations in Brazil as evidence for an Amazonian glaciation; if the Ice Age had been worldwide, it would have severed all genetic relationships between past and present life.

HA9 Davies, Gordon L. "Another Forgotten Pioneer of the Glacial Theory: James Hutton, 1726-97." *Journal of Glaciology* 7 (1968): 115-116.

In his *Theory of the Earth* (1795), Hutton postulated transport of Alpine boulders by "Valleys of ice sliding down in all directions."

HA10 Flint, Richard Foster. "Introduction: Historical Perspectives." *The Quaternary of the United States: A Review Volume for the VII Congress of the International Association for Quaternary Research.* Edited by H.E. Wright, Jr., and David G. Frey. Princeton, N.J.: Princeton University Press, 1965, pp. 3-11.

Survey of the history of glacial geology from B. Silliman (1821) to R.A. Daly (1910).

HA11 Frängsmyr, Tore. *Upptäckten av istiden: Studier i den moderna geologins framväxt* [The Discovery of the Ice Age]. Stockholm: Almqvist & Wiksell, 1976. 188 pp. + bibliography and index.

With English-language summary, "The Discovery of the Ice Age": "The aim of this book is to analyse the geological discussion on the quaternary phenomena and the glacial theory, with special regard to the contributions of Swedish geologists.... Nils Gustaf Sefström ... thought that a great inundation, a 'petridelaunic' flood (i.e., of rolled stones), had swept over the country, but he did not in any way suggest that this flood was identical with the biblical deluge.... The famous chemist Jöns Jacob Berzelius ... was a vulcanist, and he meant that the earth had been the object of violent revolutions ... he accepted Élie de Beaumont's theory of a cooling earth. He explained the land elevation with the help of Élie de Beaumont's theory, and he defended Sefström's idea of a petridelaunic flood against the glacial theory of Agassiz.... Otto Torell ... applied the glacial theory on the whole of Scandinavia; so he was convinced not only of the real existence of an ice age but also of glacial erosion and its dominant influence on the landscape in Scandinavia.... the very famous Swedish geologist Gerard De Geer ... was pupil of Torell and in many ways fulfilled his work."

HA12 Fristrup, Boerge. *The Greenland Ice Cap.* Translated by David Stoner. Copenhagen: Rhodos, 1966. 312 pp. Also Seattle: University of Washington Press, 1966. Translation of *Indlansisen.*

Includes the history of scientific study of the ice sheet; many illustrations in color.

HA13 Gellert, Johannes F. "100 Jahre Glazialtheorie und das quartäre Erdbild von heute." *Petermanns Geographische Mitteilungen* 119 (1975): 241-252.

Brief historical survey, mentioning many names but no references.

HA14 Hansen, Bert. "The Early History of Glacial Theory in British Geology." *Journal of Glaciology* 9 (1970): 135-141.

Reviews British geological thought in the second quarter of the 19th century as background to the British response to Louis Agassiz's theory; suggests that the delay in its acceptance was due to its merger with the older drift theory of Charles Lyell.

HA15 Herneck, Friedrich, et al. *100 Jahre Glazialtheorie im Gebiet der skandinavischen Vereisungen*. Berlin: Akademie-Verlag, 1978. 368 pp.

A collection of 22 articles on glaciology. Includes; F. Herneck, "Entwicklung und Bedeutung der Glazialtheorie," pp. 9-20; T. Frängsmyr and L.-K. Königsson, "Otto Torell und die Entwicklung der Eiszeitforschung in Schweden," pp. 21-33; E.V. Sancer, "Russische Wissenschaft und der Herausbildung der Glazialtheorie," pp. 35-42 (noting the role of P.A. Kropotkin).

HA16 Imbrie, John, and Katherine Palmer Imbrie. *Ice Ages: Solving the Mystery*. Short Hills: N.J. Enslow Publishers, 1979. 224 pp.

Discusses theories of the cause of Ice Ages, especially the ideas of James Croll and Milutin Milankovich, and concludes with a review of the recent research in which J. Imbrie has been deeply involved.

HA17 Imbrie, John. "Astronomical Theory of the Pleistocene Ice Ages: A Brief Historical Review." *Icarus* 50 (1982): 408-422.

The article "traces the evolution of the astronomical theory of the Pleistocene ice ages from its origin in the works of [J.A.] Adhémar and [J.] Croll, through its quantitative transformation in the hands of [M.] Milankovitch, to its current renaissance." The author summarizes the history presented in his 1979 book and presents results of more recent research.

HA18 Keyes, Charles. "Glacial Concept Before Agassiz." *The Pan-American Geologist* 62, no. 4 (1934): 283-300.

On the writings of J. Scheuchzer (1708), P. Martel, Besson (1780), W. Coxe (1779), F.J. Hugi (1830), I. Venetz (1833), J. de Charpentier (1836), K. Schimper, and J. Esmarch (1824). The cause of the ice age was associated by J. Adhémar (1832) with the eccentricity of the earth's orbit.

* Mather, Kirtley F., ed. *Source Book in Geology, 1900-1950*. Cited above as item A65.

HA19 Mayer-Oakes, William J., ed. *Life, Land and Water. Proceedings of the 1966 Conference on Environmental Studies of the Glacial Lake Agassiz Region*. Occasional Papers, Dept. of Anthropology, University of Manitoba, No. 1. Winnipeg: University of Manitoba Press, 1967. xvi + 416 pp.

Includes some material on the history of glaciology.

HA20 Oldroyd, D.R. "Haast's Glacial Theories and the Opinions of His European Contemporaries." *Journal of the Royal Society of New Zealand* 3 (1973): 544.

The glacial theories of Julius Haast (published 1862-79) were derived from those of Europeans, not based only on his own observations.

* Patten, Donald W. *The Biblical Flood and the Ice Epoch*. Cited above as item A76.

HA21 Rowlinson, J.S. "The Theory of Glaciers." *Notes and Records of the Royal Society of London* 26 (1971): 189-204.

Discusses the theories of J.D. Forbes and J. Tyndall in the 1840s and 1850s on the motion of glaciers.

HA22 Rudwick, Martin J.S. "The Glacial Theory." *History of Science* 8 (1969): 136-157.

Essay review of Albert V. Carozzi's translation of *Studies on Glaciers* by Louis Agassiz, including a discussion of the relation between the glacial theory and geological thinking in the 19th century.

* Runcorn, S.K., ed. *Earth Sciences*. Cited above as item A86.

HA23 Seylaz, Louis. "A Forgotten Pioneer of the Glacial Theory: John Playfair (1748-1819)." *Journal of Glaciology* 4 (1962): 124-126.

In his *Illustrations of the Huttonian Theory of the Earth* (1802) Playfair postulated the glacial origin of wandering boulders in the Alps. (According to Davies, item HA9, the hypothesis was due to James Hutton.)

* Thams, J.C., ed. *The Development of Geodesy and Geophysics in Switzerland*. Cited above as item A103.

HA24 White, George W. "Announcement of Glaciation in Scotland: William Buckland (1784-1856)." *Journal of Glaciology* 9 (1970): 143-145.

Reproduces a letter from Buckland to J. Fleming, 1840.

HA25 Wichmann, Arthur. "Aus den Kindheitstagen der Glazialgeologie." *Der Geologie* 12 (1914): 223-229.

Views various 19th-century scientists on the Ice Age.

For additional information on this topic see works cited in the biographical sections on the following scientists:

Agassiz, L.
Charpentier, J. de
Croll,, J.
Dana, J.D.
Forbes, J.D.
Geikie, J.
Gilbert, G.K.
Goethe, J.W.
Grabau, A.M.
Hamberg, A.
Kropotkin, P.A.
Lamplugh, G.W.
Penck, A.
Ramsay, A.C.
Reid, H.F.
Schimper, K.F.
Shokalsky, Y.M.
Simpson, G.
Venetz, I.
Wright, G.F.

HB. Vulcanology

HB1 Bullard, Fred M. *Volcanoes: In History, in Theory, in Eruption*. Austin: University of Texas Press, 1962. xvi + 441 pp. Revised edition: *Volcanoes of the Earth*. Austin: University of Texas Press, 1976. 579 pp.

Includes brief historical remarks. The revised edition includes a chapter sketching the development of plate tectonics in relation to vulcanism.

HB2 Dean, Dennis R. "Graham Island, Charles Lyell, and the Craters of Elevation Controversy." *Isis* 71 (1980): 571-588.

Account of an episode in the 1830s when Lyell, supported by Constant Prévost, refuted the "craters of elevation" theory of Leopold von Buch.

HB3 De Beer, Gavin. "The Volcanoes of Auvergne." *Annals of Science* 18 (1962; pub. 1964): 49-61.

On the work of Jean-Étienne Guettard (1715-1786) and its reception by other scientists.

HB4 Hooker, Marjorie. "The Origin of the Volcanological Concept *nuée ardente*." *Isis* 56 (1965): 401-407.

In his report on the 1902 eruption of Mt. Pelée, Alfred Lacroix revived the term *nuée ardente* (glowing cloud), introduced earlier by Ferdinand Fouqué. It was derived from the Portuguese words *nuvem ardente* used by Azoreans in eyewitness accounts of earlier eruptions.

HB5 Kugler, Ernst. *Philipp Friedrich von Dietrich, Ein Beitrag zur Geschichte der Vulkanologie*. Münchener Geographische Studien, Achtes Stück. München: Theodor Ackermann, 1899. 88 pp.

On the contributions to vulcanology of P.F. Dietrich (1748-1793).

* Mather, Kirtley F., ed. *Source Book in Geology, 1900-1950*. Cited above as item A65.

HB6 Middleton, William E. Knowles. "Borelli and the Eruption of Etna in 1669: Some Unpublished Papers." *Physis* 15 (1973): 111-130.

Manuscripts of G.A. Borelli indicate his views on meteorology and meteorological optics as well as on vulcanology. He took a thermometer and a Torricelli barometer to the top of the mountain, and commented on the difficulty of breathing at high altitudes. Also presented is a letter from F. d'Arezzo on vulcanism.

HB7 Paisley, P.B., and D.R. Oldroyd. "Science in the Silver Age: *Aetna*, a Classical Theory of Volcanic Activity." *Centaurus* 23 (1979): 1-20.

According to this poem from Roman antiquity, earthquakes and volcanoes arise from the action of winds within the Earth; liquid sulphur, alum, and oily bitumen are possible fuels for the flames. The author may have been the Procurator of Sicily, Lucilius Junior.

* Rudwick, Martin J.S. "Lyell on Etna and the Antiquity of the Earth." Cited above as item H42.

HB8 Rudwick, Martin J.S. "Poulett Scrope on the Volcanoes of Auvergne: Lyellian Time and Political Economy." *British Journal for the History of Science* 7 (1974): 205-242.

In his *Memoir on the Geology of Central France* (1827) and other writings on volcanoes, George Poulett Scrope rejected the distinction between "ancient" and "modern," lava flows, proposing instead a continuous temporal series. His gradualistic description was praised by Charles Lyell, though Scrope also favored a (non-Lyellian) directionalism based on the cooling of the earth. Contrary to the tendency of Charles Daubeney and others to "periodize" geological time on the basis of biblical chronology (e.g., before and after the Flood), Scrope insisted on a vast unbroken scale of time.

Rudwick suggests an analogy between the concept of money developed in Scrope's writings on economics and his concept of time in geology; in neither case is the total amount rigidly fixed, rather one can always "draw" more when needed.

HB9 Sapper, Karl. *Vulkankunde*. Stuttgart: Engelhorn, 1927. viii + 424 pp.

Includes "Geschichte der Vulkanologische Auschauungen," pp. 356-402.

HB10 Taylor, Kenneth L. "Nicolas Desmarest and Geology in the Eighteenth Century." *Toward a History of Geology* (item A89), pp. 339-356.

Desmarest (1725-1815) is known for his work on vulcanism, and for first stating the volcanic origin of basalt. But he did not try to explain geological change and orogeny by volcanic action; he was actually a neptunist. He attributed volcanic energy to chemical reactions rather than central heat.

* Thams, J.C., ed. *The Development of Geodesy and Geophysics in Switzerland*. Cited above as item A103.

For additional information on this topic see works cited in the biographical sections on the following scientists:

Anderson, E.M.
Borelli, G.
Buch, L. v.
Cordier, P.
Dana, J.D.
Daubeny, C.G.B.
Day, A.L.
Desmarest, N.
Dutton, C.E.
Escholt, M.P.
Fenner, C.N.
Fouqué, F.A.
Hamilton, W.
Harker, A.
Jaggar, T.A., Jr.
King, C.R.
Koto, B.
Lacroix, A.
Leibniz, G.W. v.
Levinson-Lessing, F.
Mallet, R.
Omori, F.
Rames, J.B.
Raspe, R.C.
Reck, H.
Scrope, G.J.P.

HC. Continental Drift and Plate Tectonics

HC1 Argand, Emile. *Tectonics of Asia*. Translated and edited by Albert V. Carozzi. New York: Hafner, 1977. xxvi + 218 pp.

Translation of *La tectonique de l'Asie*, an address given at the XIII International Congress (1922) by Emile Argand (1879-1940). It was "a unique synthesis of global tectonics in the light of Wegener's hypothesis of continental drift." Carozzi's introduction reviews Argand's career and gives a brief analysis of the text.

HC2 Batiushkova, Irina V. *Istoriia problemy proiskhozhdeniia materikov i okeanov*. Moscow: Nauka, 1975. 137 pp.

History of the problem of the origin of continents and oceans, since the 17th century; includes an account of 20th-century debates on mobility vs. permanence.

HC3 Bird, John M., ed. *Plate Tectonics, Selected Papers from Publications of the American Geophysical Union*. Second enlarged edition. Washington, D.C.: American Geophysical Union, 1980, 1982. 1021 pp.

Reprint of papers originally published 1968-79; an extensive bibliography covers the period 1963-79. Includes papers by L.R. Sykes, J. Oliver, B. Isacks, D.P. McKenzie, J.R. Heirtzler, G.O. Dickson, E.M. Herron, W.C. Pitman III, X. Le Pichon, W.J. Morgan, E. Irving, W.A. Robertson, D.E. Karig, R. Kay, N.J. Hubbard, P.W. Gast, J.F. Dewey, J.M. Bird, R.I. Walcott, R.S. Dietz, J.C. Holden, W.R. Jacoby, M. Talwani, C.C. Windisch, M.G. Langseth, Jr., R.G. Coleman, W.M. Elsasser, M. Nafi Toksöz, J.W. Minear, B.R. Julian, K. Jinghwa Hsü, E.R. Oxburgh, D.L. Turcotte, P.J. Fox, D.E. Hayes, J.S. Milsom, J.G. Sclater, R.N. Anderson, M.L. Bell, G. Plafker, G. Schubert.

HC4 Bishop, A.C. "The Development of the Concept of Continental Drift." *The Evolving Earth (Chance, Change and Challenge)*. Edited by L.R.M. Cocks. London: British Museum & Cambridge University Press, 1981, pp. 155-164.

HC5 Carozzi, Albert V. "A propos de l'origine de la théorie des dérives continentales: Francis Bacon (1620), François Placet (1668), A. von Humboldt (1801) et A[ntonio] Snider [-Pellegrini] (1858)." *Compte Rendu des Séances de la Société Physique et d'Histoire Naturelle de Geneve*; nouvelle séries, 4 (1969): 171-179.

Discussion of some alleged precursors of the theory of continental drift, with extensive quotations. The first true precursor was the American Antonio Snider-Pellegrini, in his book *Création et ses mystères dévoilés* (1858) whose theory resembles that of Taylor (1908) and Wegener (1910).

HC6 Carozzi, Albert V. "New Historical Data on the Origin of the Theory of Continental Drift." *Geological Society of America Bulletin* 81 (1970): 283-286.

"Examination of the original texts of Francis Bacon's *Novum Organum* (1620) and of Francois Placet's *La corruption du grand et petit monde* (1668) shows that, contrary to common belief, these two authors should not be considered as forerunners of the theory of continental drift; this conclusion also applies to A. von Humboldt (1801). A. Snider, in *La création et ses mystères dévoilés* (1858), is the first naturalist who has unequivocally postulated and illustrated a juxtaposition and drifting of the continents as Taylor and Wegener did in the 20th century."

* Cowen, Richard, and Jere H. Lipps, eds. *Controversies in the Earth Sciences. A Reader*. Cited above as item D17.

HC7 Cox, Allan, ed. *Plate Tectonics and Geomagnetic Reversals.* San Francisco: Freeman, 1973. ix + 702 pp.

Reprints the major research papers from the 1960s, with editorial commentary. Cox, himself one of the leaders in the study of geomagnetic reversals, has obtained from other scientists accounts of how they became interested in the subject of their discoveries. Includes papers by J.T. Wilson, D.P. McKenzie, R.L. Parker, W.J. Morgan, A. Cox, R.R. Doell, G.B. Dalrymple, F.J. Vine, D.H. Matthews, J.R. Heirtzler, G.O. Dickson, E.M. Herron, W.C. Pitman III, X. Le Pichon, H. Benioff, G. Plafker, L.R. Sykes, B. Isacks, J. Oliver, P. Molnar, H.W. Menard, T.M. Atwater, D.E. Hayes, J.G. Sclater, J. Francheteau, D.T. Griggs, A.E. Maxwell, J.F. Dewey, J.M. Bird, T. Matsuda, S. Uyeda, J. Gilluly.

HC8 Du Toit, A.L. *Our Wandering Continents. An Hypothesis of Continental Drifting.* Edinburgh and London: Oliver & Boyd, 1937. Reprint, Westport, Conn.: Greenwood Press, 1972. xiii + 366 pp.

Chapter 2, "Historical," begins with F.B. Taylor, H.B. Baker, and A. Wegener, and gives critical summaries of several later works.

HC9 Englehardt, W. v. "Das Erdmodell der Plattentektonik--ein Beispiel für Theorienwandel in der neueren Geowissenschaft." *Die Struktur wissenschaftlicher Revolution und die Geschichte der Wissenschaften.* Edited by Alwin Diemer. Meisenheim am Glan: Hain, 1977, pp. 91-109.

Argues that the description of I. Lakatos fits better than those of K. Popper or T.S. Kuhn.

HC10 Frankel, Henry. "Alfred Wegener and the Specialists." *Centaurus* 20 (1976): 305-324.

"... I suggest that the great range of Wegener's theory over so many seemingly diverse fields, when coupled with the fact that his theory had no distinct advantage over the competition within any specific area, *was a main reason why* it was given perfunctory treatment by the community of earth scientists. The corroborating facts were so diverse that their great variety was not appreciated by the specialists working in only their restricted field." The response of the specialists can be explained with the help of Imre Lakatos' theory of research programs, as the expected defense against an attack on the "hard core" of one's program.

HC11 Frankel, Henry. "Arthur Holmes and Continental Drift." *British Journal for the History of Science* 11 (1978): 130-150.

"In 1978, Arthur Holmes provided proponents of continental drift theory with an auxiliary hypothesis [convection currents in a fluid substratum] which afforded them a badly needed account of the forces responsible for continental drift."

HC12 Frankel, Henry. "The Career of Continental Drift Theory: An Application of Imre Lakatos' Analysis of Scientific Growth to the Rise of Drift Theory." *Studies in History and Philosophy of Science* 10 (1979): 21-66.

The development of the continental drift research program "is partially elucidated by the account of scientific growth and change as put forth by Imre Lakatos. However, at least two alterations must be made in Lakatos' analysis. One concerns his analysis of 'novel fact,' and the other is concerned with his thesis that the hard core of a research programme remains the same throughout the programme's lifetime."

HC13 Frankel, Henry. "The Reception and Acceptance of Continental Drift Theory as a Rational Episode in the History of Science." *The Reception of Unconventional Science.* Edited by Seymour H. Mauskopf. Boulder, Colo.: Westview Press for the American Association for the Advancement of Science, 1979, pp. 51-90.

HC14 Frankel, Henry. "Why Drift Theory Was Accepted with the Confirmation of Harry Hess's Concept of Sea Floor Spreading." *Two Hundred Years of Geology in America* (item A90), pp. 337-353.

"The acceptance of continental drift theory ... nicely illustrates the reluctance of the scientific community to accept a theory until it predicts novel facts which turn out to be true.... It was only after confirmation of the F. Vine, D. Matthews and J.T. Wilson transform fault hypothesis, both of which were virtually corollaries of Hess's proposal, that the geological community threw its endorsement to drift theory."

HC15 Frankel, Hank. "Hess's Development of His Seafloor Spreading Hypothesis." *Scientific Discovery: Case Studies* (item A73), pp. 345-366.

Argues that Harry Hess's hypothesis "arose out of his long time interests in solving a nest of problems in oceanography and geophysics"; his "major epistemic aim in theorizing was to present hypotheses which effectively solved problems ..."; his adoption of continental drift with the grafting of his own seafloor spreading hypothesis "was undertaken because it increased the problem-solving effectiveness of his own hypothesis and made [it] a much better problem-solver through elimination of its most serious deficiency"; his "major epistemic aim in theorizing, presenting hypotheses which were good problem-solvers, when coupled with his apparent lack of fear in making mistakes, gave him the freedom for inventing, evaluating, rejecting and salvaging potential solutions."

HC16 Frankel, Henry. "The Development, Reception and Acceptance of the Vine-Matthews-Morley Hypothesis." *Historical Studies in the Physical Sciences* 13 (1982): 1-39.

Provides a detailed account, based partly on personal interviews, of the hypothesis about seafloor spreading proposed in 1963 by F. Vine and D. Matthews, and independently by L.W. Morley.

HC17 Glen, William. *The Road to Jaramillo: Critical Years of the Revolution in Earth Science*. Stanford, Calif: Stanford University Press, 1982. xvi + 459.

"Among the diverse research programs that contributed to the modern revolution in earth science, three, during the decade ending in 1966, appear decisive in launching it. Initially each had been undertaken with modest prospects only, and with no thought of serving the others. The dating of young rocks by the potassium-argon method was implemented by the development of the static-mode mass spectrometer and the evolution of new dating techniques at Berkeley. That dating capability was coupled to magnetic studies of young rocks by a Berkeley-trained group working in Menlo Park, CA, and later by competitors in Canberra, Australia. They showed that worldwide, contemporaneous rocks have the same magnetic polarity and thus proved that the earth's magnetic field has reversed polarity repeatedly. By 1965 they had formulated a series of increasingly refined time scales dating those reversals. The 11th such scale, containing the Jaramillo reversal, in conjunction with a uniquely detailed magnetic anomaly profile from the Pacific seafloor, became the key to a store of newly acquired, oceanic magnetic data at Lamont-Doherty Geological Observatory. The suddenly deciphered magnetic evidence proved the highly speculative Vine-Matthews-Morley hypothesis which joined seafloor spreading theory to reversals of the earth's magnetic field in order to explain a puzzling, zebra-striped magnetic pattern on the seafloor. The overnight acceptance of the Vine-Matthews-Morley hypothesis in 1966 triggered the plate tectonics revolution" (Author's summary).

HC18 Hallam, A. *A Revolution in the Earth Sciences. From Continental Drift to Plate Tectonics*. Oxford: Clarendon Press, 1972. vii + 217 pp.

A good historical exposition of the basic concepts, with selective bibliography. The last chapter applies the ideas of T.S. Kuhn to the "revolution."

HC19 Hallam, A. "Alfred Wegener and the Hypothesis of Continental Drift." *Scientific American* 232, no. 2 (February 1975): 88-97.

Popular historical survey with several illustrations.

HC20 Heirtzler, James R. "This Week's Citation Classic: Heirtzler, J.R., Dickson, G.O., Herron, E.M., Pitman, W.C., III & Le Pichon, X. Marine Magnetic Anomalies, Geomagnetic Field Reversals, and Motions of the Ocean Floor and Continents. *J. Geophys. Res.* 73: 2119-36, 1968." *Current Contents, Physical, Chemical & Earth Sciences* 20, no. 8 (February 25, 1980): 14.

Heirtzler discusses the historic significance of this paper, "one of those that ushered in the era of 'plate tectonics'"; it has been cited over 560 times since 1968.

HC21 Holland, Thomas H. *The Permanence of Oceanic Basins and Continental Masses*. London: Macmillan, 1937. 22 pp.

Historical survey, from J.D. Dana and W.T. Blanford to A. Wegener.

HC22 Jones, Barrie. "Plate Tectonics: A Kuhnian Case?" *New Scientist* 63 (1974): 536-538.

Argues that the example of plate tectonics supports T.S. Kuhn's theory of scientific revolutions.

HC23 Kasbeer, Tina. *Bibliography of Continental Drift and Plate Tectonics*. 2 vols. Boulder, Colo.: Geological Society of America, 1973, 1975. (Special Papers 142, 164).

HC24 Laudan, Rachel. "The Method of Multiple Working Hypotheses and the Development of Plate Tectonic Theory." *Scientific Discovery: Case Studies* (item A73), pp. 331-343.

The methodology of the geologist T.C. Chamberlin provides a rationale for the simultaneous development of several rival theories. The author argues that "a methodology of this type was instrumental in generating the plate tectonic revolution in geology in the 1950s and 1960s," particularly as used by J. Tuzo Wilson.

HC25 Lear, John. "Canada's Unappreciated Role as Scientific Innovator." *Saturday Review* 50, no. 35 (September 2, 1967): 45-50.

On J. Tuzo Wilson's and L.W. Morley's contributions to the development of plate tectonics; followed by an essay by Wilson, "Advice for the Establishment," pp. 50-51.

* Lemke, J.L., M.H. Nitecki, and H. Pullman. "Studies of the Acceptance of Plate Tectonics." See item L52.

HC26 Marvin, Ursula B. *Continental Drift: The Evolution of a Concept*. Washington, D.C.: Smithsonian Institution Press, 1973. 239 pp., illus., bibl.

Contents: Geographical speculations (Ptolemy, Roger Bacon, Columbus); terrestrial motion and magnetism; geological speculations on the origins of continents and oceans (17th-19th centuries); catastrophic displacement of continents (Richard Owen's tetrahedral earth; Antonio Snider); geophysical speculations on the nature of the earth's interior--Pratt vs. Airy, geosynclinals, isostasy, effect of lunar fission origin, sunken continents (E. Suess), core and mantle, radioactivity; drift theories of F.B. Taylor, H.B. Baker, A. Wegener; reaction to Wegener's theory; B. Gutenberg's "Fliesstheorie" of continental spreading; Arthur Holmes' convection theory; ideas of W. Bucher, A. du Toit, L. Joleaud; radiometric geochronology (C. Patterson); researches on ocean basins, paleomagnetism, sea-floor spreading; plate tectonics--acceptance and dissent.

HC27 McArthur, Robert P., and Harold R. Pestana. "Is Continental Drift/Plate Tectonics a Paradigm-Theory?" *Proceedings of the XIVth International Congress for the History of Science, 1974* (pub. 1975) 3: 105-108.

The authors argue that T.S. Kuhn's theory successfully accounts for the recent revolution in geology.

HC28 McKenzie, D.P. "Plate Tectonics and Its Relationship to the Evolution of Ideas in the Geological Sciences." *Daedalus: Journal of the American Academy of Arts and Sciences* 106(3) (1977): 97-124.

After summarizing the modern theory of plate tectonics, the author considers reasons why the basic concepts of theory could not have been established much earlier. The significance of isostasy was understood by 1900 but not the relationship between geological faults and earthquakes. R.D. Oldham urged the Geological Survey of India to retriangulate the region in which a large earthquake occurred in 1897, but could not get enough data; similar work by H.F. Reid after the 1906 California earthquake could have produced a major advance in understanding but did not. Wegener's theory of continental drift, though now seen to be largely correct, went too far in some respects and failed to attract the support of the theorists like H. Jeffreys who would have been able to develop it in a more satisfactory way. Wegener's geological supporters did not appreciate the significance of rigidity of continents when displaced. The almost complete ignorance of marine geology before World War II was a major reason for lack of belief in continental drift. Paleomagnetism was too complicated to be persuasive. The crucial event was the acceptance of sea-floor spreading in 1966.

HC29 McKenzie, D.P. "This Week's Citation Classic: McKenzie, D.P., Robert, J.M. & Weiss, N.O. Convection in the Earth's Mantle: Towards a Numerical Simulation. *J. Fluid Mech.* 62: 465-538, 1974." *Current Contents, Physical, Chemical & Earth Sciences* 21, no. 47 (November 23, 1981): 18.

McKenzie recalls the circumstances of the original research leading to the writing of this paper, which has been cited over 140 times since 1974.

HC30 Meyerhoff, A.A. "Arthur Holmes: Originator of Spreading Ocean Floor Hypothesis." *Journal of Geophysical Research* 73 (1968): 6563-6565.

Urges that the credit for the hypothesis be given to Holmes on the basis of his 1931 paper, rather than to H.H. Hess or R.S. Dietz. Other ideas such as mantle convection, continental drift, polar wandering, and expansion were also proposed by earlier writers who do not receive proper credit. See also the replies by R.S. Dietz and H. Hess, both of whom state that Holmes' concept was different from that now known as sea-floor spreading (*ibid.*, pp. 6567, 6569).

HC31 Palmer, Allison R. "Serendipity and the Changing Face of the Earth." *American Scholar* 44 (1975): 242-254.

The author asserts that the recent establishment of plate tectonics "is a dramatic example of serendipity in slow motion." She offers no evidence for this assertion other than to point out that techniques in several areas of the earth sciences had all been developed to the point where they could answer crucial questions raised by the theory. No references.

HC32 Robb, Alfred A. "Anticipation of Wegener's Hypothesis." *Nature* 126 (1930): 841.

On A. Snider's book *La Création et ses mystères dévoilés* (1858), with diagrams showing displacement of continents.

HC33 Rupke, N.A. "Continental Drift Before 1900." *Nature* 227 (1970): 349-350.

"The idea that Francis Bacon and other 17th and 18th century thinkers first conceived the notion of continental drift does not stand up to close scrutiny. The few authors who expressed the idea viewed the process as a catastrophic event" (Author's summary). Views of T.C. Lilienthal (1756), A. von Humboldt (1845), and A. Snider-Pellegrini (1858) are mentioned. It was Taylor (1910) and Wegener (1912) who broke the connection with catastrophism, and proposed that drift is slow and still operates today.

HC34 Stucchi, Max. "Chi ha spostato i continenti? Derive e congressi nelle scienze della terra" [Who Is Responsible for Continental Drift? Drifts and Congresses in the Earth Sciences]. *Testi e Contesti* 1 (1979). Edited by Angelo Baraca, pp. 57-67.

Mentions the importance of oceanographic research sponsored by the U.S. Office of Naval Research.

HC35 Sullivan, Walter. *Continents in Motion: The New Earth Debate*. New York: McGraw-Hill, 1974. xiv + 399 pp.

A detailed account of ideas and observations relating to polar wandering, planetary collisions (Velikovsky), mantle convection, paleomagnetism, the "Mohole" project, sea-floor spreading, hot spots and plumes, mountain formation, earthquakes, geothermal energy, etc., dealing mostly with the period since 1950.

* Takeuchi, H., S. Uyeda, and H. Kanamori. *Debate About the Earth. Approach to Geophysics Through Analysis of Continental Drift*. Cited below as item U45.

HC36 Tasch, Paul. "Search for the Germ of Wegener's Concept of Continental Drift." *Osiris* 11 (1954): 157-167.

Discusses the tradition of Plato's Atlantis and its conjectured association with America; Ptolemy's world map; Columbus and "the narrow Atlantic."

HC37 Van Waterschoot van der Gracht, W.A.J.M., et al. *Theory of Continental Drift: A Symposium on the Origin and Movement of Land Masses, both Intercontinental and Intra-Continental, as Proposed by Alfred Wegener*. Tulsa: The American Association of Petroleum Geologists, 1928. x + 240 pp.

One of the major documents in the history of continental drift theories, in which leading geologists debated Wegener's hypothesis.

HC38 Wegener, Alfred. *The Origin of Continents and Oceans*. Translated from the Fourth Revised German Edition (1929) by John Biram. New York: Dover Pubs., 1966. viii + 246 pp.

The classic presentation of the theory of continental drift.

HC39 Wilson, J. Tuzo. "Static or Mobile Earth: The Current Scientific Revolution." *Proceedings of the American Philosophical Society* 112 (1968): 309-320.

Survey of the recent revival of continental drift theories and the development of plate tectonics. The "revolution" is comparable to those described by T.S. Kuhn, involving a shift of paradigms. Earlier failure to adopt a mobile earth picture "could explain why geology has deteriorated from its original intention of being the study of the earth to being a study instead of rocks, minerals, and fossils and why geophysics has never been integrated but remains fragmented with studies of earthquakes by one group, of geomagnetism by another, and so on" (p. 317).

HC40 Wood, Robert Muir. "Coming Apart at the Seams." *New Scientist* 85 (1980): 252-254.

Celebrating the centennial of Alfred Wegener's birth, the author also calls attention to another pioneer of continental drift, F.B. Taylor.

For additional information on this topic see works cited in the biography sections on the following scientists:

Argand, E.
Benioff, V.H.
Du Toit, A.L.
Ewing, W.M.
Gagnebin, E.
Grabau, A.M.
Holmes, A.
Taylor, F.B.
Wegener, A.

J. SEISMOLOGY

J1 Andrews, William D. "The Literature of the 1727 New England Earthquake." *Early American Literature* 7 (1973): 281-294.

Most ministers adopted a scientific attitude in searching for causes of the earthquake. Thomas Foxcroft postulated an explosion in a subterranean cavern. Thomas Prince suggested sulphurous particles. John Barnard and Benjamin Coleman recognized the wavelike character of earthquakes. Others invoked theological arguments (New England was being punished for its sins). Thus the earthquake literature demonstrates the interaction between science and religion.

J2 Batiushkova, I.V. "K istorii izucheniia zemletriasenii v Chechoslovakii" [The History of the Study of Earthquakes in Czechoslovakia]. *Voprosy istorii estestvoznaniia i tekhniki* 8 (1959): 134-137.

J3 Batiushkova, I.V. "Iz istorii izucheniia zemletriasenii v Pol'she" [From the History of the Study of Earthquakes in Poland]. *Trudy, Institut istorii estestvoznaniia i tekhniki, Akademiia Nauk SSSR* 42 (1962): 228-246.

Begins with Yana Dlugosha (1415-1480) and includes the work of F. Kreutz (1846-1910), Vaclav Laska (1862-1943), M.P. Rudzki (1862-1916), M. Smoluchowski (1872-1917), and E.W. Janczewski.

J4 Carozzi, Albert V. "Robert Hooke, Rudolf Erich Raspe, and the Concept of 'Earthquakes.'" *Isis* 61 (1970): 85-91.

Hooke's "Lectures and Discourses of Earthquakes and Subterraneous Eruptions" (1668) "introduced into structural geology the dynamic concept of vertical movements of the crust ... explaining by means of a simple and spectacular hypothesis the origin of islands, continents, mountains and valleys within continents, and ocean basins." Though he did not have a clear idea about volcanoes and earthquakes he considered both to be effects of "the general congregation of sulphureous underground subterraneous vapors

Hooke's theory was ignored until it was revived in 1763 by Raspe, who combined it with ideas of J.G. Lehmann.

J5 Cremonesi, Arduino. "Potresi v Zgodovini furlanije." *Kronika* (Yugoslavia) 26(2) (1978): 71-82.

"A history of earthquakes which originated in or affected Friuli province of the Habsburg empire since 1000." (*Historical Abstracts* 26A: 6530).

* Davies, Gordon L. "Robert Hooke and His Conception of Earth History." Cited above as item BH27.

J6 Davison, Charles. *A History of British Earthquakes*. Cambridge, Eng.: Cambridge University Press, 1924. xviii + 416 pp.

Primarily a catalog of records and observations; the last two chapters present statistical and theoretical generalizations.

J7 Davison, Charles. *The Founders of Seismology*. Cambridge, Eng.: Cambridge University Press, 1927. Reprint, New York: Arno Press, 1977. xiv + 240 pp.

Includes chapters on John Michell (1724/25-1793), Alexis Perrey (1807-1882), Robert Mallet (1810-1881), Fernand de Montessus de Ballore (1851-1923), John Milne (1850-1913), and Fusakichi Omori (1868-1923).

J8 Davison, Charles. *Great Earthquakes*. London: Thomas Murby, 1936. xii + 286 pp.

An account of 18 earthquakes, beginning with Lisbon (1755) and ending with Hawke's Bay (1931), with 122 illustrations and extensive bibliographies.

J9 Dewey, James, and Perry Byerly. "The Early History of Seismometry (to 1900)." *Bulletin of the Seismological Society of America* 59 (1969): 183-227.

This account jumps rather quickly to the inverted-pendulum seismometer of James Forbes (1844) and the explosion research of Robert Mallet (1851-1862). Italians then take the lead: P.G.M. Cavalleri (1858, 1860), Luigi Palmieri (1856-1874), and P.F. Cecchi (1876) developed effective instruments. But the credit for the invention of the seismograph as an essential tool in seismology belongs to a group of British professors teaching in Japan in the late 19th century: J. Milne, J. Ewing, T. Gray,

J. Perry, W.E. Ayrton. E. von Rebeur-Paschwitz was the first to record an earthquake that occurred on the other side of the world. E. Wiechert's inverted-pendulum seismometer is probably the earliest seismograph which is still used in its original form.

Aside from a discussion of the controversy about whether a seismograph records the tilt of the earth, the authors have nothing to say about the theory of seismic wave propagation or the applications of seismology to geophysics.

J10 Fujii, Yoichiro. "Seismology, History of Seismology, and Philosophy of Science in the Scientific Works of Mishio Ishimoto, 1893-1940" (in Japanese). *Kagakusi Kenkyu* 7 (1968): 22-31; 8 (1968): 146-154.

J11 Gils, J.M. van. "La genèse des stations seismologiques belges." *IIIe Congrès National des Sciences, organisé par la Fédération belge des sociétés scientifiques, Bruxelles, 30 mai-3 juin, 1950*. Volume 1. Published in Liège, 1951.

On the role played by E. van den Broeck and E. Lagrange in the years 1895-98.

J12 Hobbs, W. "The Evolution and the Outlook of Seismic Geology." *Proceedings of the American Philosophical Society* 48 (1909): 259-302.

The natural development of seismology in the 19th century was blocked, according to the author, by Mallet's "false [centrum] theory." Mallet's insistence on attributing earthquakes to a disturbance in a "focal cavity" no more than 3 miles below the surface was not supported by observations, but led subsequent workers to ignore significant spatial and temporal variations in averaging their data. An alternative "fault-block" theory was proposed by Edward Suess in 1872. As a result of the success of this theory, seismology must now be considered part of tectonic or structural geology. The publication of *La Science seismologique* by Montessus de Ballore in 1907 signalled the emancipation of seismology from Mallet's centrum theory. Adjustments in position or attitude of sections of the earth's crust are now regarded as the cause and not the effect of the earthquake shocks.

While vulcanism has long been thought to be connected with earthquakes, both are really indications of a more fundamental geological process, mountain formation, which the author attributes to the contraction of the entire earth. Since the author believes that the earth is solid throughout, he can explain lava formation only by temporary release of pressure in a high temperature region, due to folding of strata.

While presenting a useful review of research in seismology, the author is clearly more interested in advancing his own views than in furnishing an objective historical account of the subject.

J13 Hobbs, William H. "The Cause of Earthquakes, Especially Those of the Eastern United States." [Reprinted from *Papers of the Michigan Academy of Science, Arts and Letters* 5, 1925.] *Annual Report of the Smithsonian Institution for 1926*. Washington, D.C., 1927, pp. 257-277, 5 figs.

Includes a brief historical survey of the earthquake theories of Aristotle, Humboldt, Mallet, Montessus de Ballore, F. Omori, H.F. Reid, and Hobbs.

J14 Kanamori, Hiroo, and Katsuyuki Abe. "Reevaluation of the Turn-of-the-Century Seismicity Peak." *Journal of Geophysical Research* 84 (B11) (1979): 6131-6139.

"According to currently available seismicity catalogues, seismicity ... around the turn of the century, from 1897 to 1906, was significantly higher than in recent years. However, the magnitudes of the earthquakes which occurred during this period were determined by Gutenberg, who used the records obtained by the undamped Milne seismograph with the assumption that the effective magnification is 5. Because of saturation of the Milne seismogram for very large events used by Gutenberg for calibration, the gain (= 5) used by Gutenberg could have been underestimated, and therefore the magnitude overestimated ..." (from Author's abstract).

J15 Kawasumi, Hirosi. "An Historical Sketch of the Development of Knowledge Concerning the Initial Motion of an Earthquake." *Publications du Bureau Central Seismologique International*, Série A; *Travaux Scientifiques*, Fasc. No. 15, 2e Partie (1937): 258-330.

On the work of J.A. Ewing, R.D. Oldham, F. Omori, C.G. Knott, B. Galitzin, K. Sezawa, T. Shida, M. Ishimoto, H. Honda, H. Kawasumi, and others. 256 references.

J16 Kendrick, T.D. *The Lisbon Earthquake*. London: Methuen, 1956. ix + 170 pp.

Includes brief discussion of earthquake theories proposed by B.J. Feyzóv y Montenegro (1755-56) and J.J. Moreira de Mendonça (1758).

J17 Khattri, Kailash. "Earthquake Focal Mechanism Studies--A Review." *Earth-Science Reviews* 9 (1973): 19-63.

"The representation of focus by point and finite sources are reviewed. The use of initial motion and spectral amplitude of P-waves, the polarization of S-waves and the application of numerical methods in determining optimum solutions are discussed. A survey of the recent advances in the application of surface waves, free oscillations, and static dislocations for the determination of focal mechanisms is given. A brief outline of the major results of the interpretation of the focal mechanism solution of earthquakes in terms of regional stress distributions and the geodynamic processes that are currently taking place is also included" (Author's abstract). Approx. 280 references, mostly from the 1960s.

* Mather, Kirtley F., ed. *Source Book in Geology, 1900-1950*. Cited above as item A65.

J18 Melton, Ben S. "Earthquake Seismograph Development--A Modern History." *EOS* 62 (May 26, 1981): 505-510; (June 23, 1981): 545-548.

"The years from 1948 through 1976 saw numerous changes in electrodynamic-type earthquake seismographs."

J19 Milyutina, E.H. *Seismicheskie issledovaniya verkhnei mantii: Evolutsiya metodologii* [Seismic Studies of the Upper Mantle: Evolution of Methodology]. Moscow: Izdatel'stvo "Nauka," 1976. 135 pp.

J20 Montessus de Ballore, [F.] de. "Un point d'histoire de la sismologie: Alexis Perrey." *Cosmos* 67, no. 1432 (July 4, 1912): 22-24.

On the relation between the earthquake catalogues of Perrey and Mallet.

J21 Montessus de Ballore, Fernand de. *Bibliografía General de Temblores y Terremotos*. Santiago, Chile: Imprenta Universitaria, 1915-19. 1515 pp.

The last item is numbered 9140.

J22 Montessus de Ballore, F[ernand] de. "Histoire de la sismologie." *Revue des Questions Scientifiques*; série 3, 29 (1921): 29-57.

Contents: I. La sismologie des primitifs. II. Les philosophes Grecs avant et après Aristote. III. La théorie d'Aristote et ses origines. IV. Les philosophes Latins et leurs successeurs au moyen age. V. Relations des tremblements de terre avec d'autres phénomènes naturels. VI. Volcans et tremblements de terre. VII. Sismologie générale et dynamique. VIII. Sismologie géologique. IX. Constructions antisismiques. X. Les organisations sismologiques.

J23 Montessus de Ballore, [Fernand]. *Ethnographie sismique et volcanique, ou les tremblements de terre et les volcans dans la religion, la morale, la mythologie et le folklore de tous les peuples*. Paris: Champion, 1923. viii + 206 pp.

J24 Needham, Joseph, and Wang Ling. *Science and Civilization in China*. Vol. 3: *Mathematics and the Sciences of the Heavens and the Earth*. Cambridge, Eng.: Cambridge University Press, 1959. xlvii + 877 pp.

"Seismology," pp. 624-635.

J25 Oldroyd, David R. "Robert Hooke's Methodology of Science as Exemplified in His *Discourse of Earthquakes*. *British Journal for the History of Science* 6 (1972): 109-130.

"Hooke's methodology was essentially Baconian, except that he openly urged that hypotheses should be proposed." To explain the existence of marine fossils in strata far removed from the sea, he proposed that the poles of the rotation axis have moved relative to the earth's body, and that the resultant variation in gravitational forces produced internal rearrangements which caused earthquakes. Hooke suggested astronomical tests of his hypothesis but did not succeed in carrying them out.

J26 Pastor, Alfonso Rey. "Las teorias sismogénicas a través de la historia." *Ibérica, El Progreso de las Ciencias y de Sus Aplicaciones* Año XI, Vol. XXI (1924): 234-238, 248-252.

A rapid survey of seismological theories, from Thales to E. Seuss, Montessus de Ballore, W.H. Hobbs, and Alfred Wegener.

J27 Perrey, Alexis. "Bibliographie seismique." *Mémoirs de l'Académie Impériale des Sciences, Arts et Belles-Lettres de Dijon*. Deuxième série, Section des Sciences 4 (1855): 1-112; 5 (1857): 183-253; 9 (1861): 87-192; 10 (1862): 1-53; 13 (1865): 33-102.

J28 Perrey, Alexis. "Documents sur les tremblements de terre et les phénomènes volcaniques dans l'Archipel des Philippines." *Mémoires de l'Academie Impériale des Sciences, Arts et Belles-Lettres de Dijon*, Deuxième Série, Section des Sciences 8 (1860): 85-194.

Includes extensive quotations from various accounts.

J29 Perrey, Alexis. "Documents sur les tremblements de terre et les phénomènes volcaniques au Japon." *Mémoires de l'Academie Impériale des Sciences, Belles-Lettres, et Arts de Lyon* (Classe des Sciences) 12 (1862): 281-390.

Includes extensive quotations from various sources.

J30 Perrey, Alexis. "Documents sur les tremblements de terre et les phénomènes volcaniques dans l'Archipel des Kouriles et au Kamtschatka." *Annales des Sciences Physiques et Naturelles, d'Agriculture et d'Industrie, publiées par la Société Imperiale d'Agriculture, etc. de Lyon*, Troisième Série 8 (1864): 209-374.

J31 Perrey, Alexis. "Documents sur les tremblements de terre et les phénomènes volcaniques des îles Aleutiennes, de la péninsule d'Aljaska et de la côte no. d'Amerique." *Mémoires de l'Academie Impériale des Sciences, Arts et Belles-Lettres de Dijon*, Deuxième Série, Section des Sciences 13 (1865): 121-251.

Includes extensive quotations from various sources.

* Raspe, Rudolf Erich. *An Introduction to the Natural History of the Terrestrial Sphere....* Cited above as item H38.

J32 Rousseau, G[eorge] S. "The London Earthquakes of 1750." *Cahiers d'Histoire Mondiale* 11 (1969): 436-451.

Includes brief discussion of various contemporary explanations of the cause of earthquakes.

J33 Shute, Michael N. "Earthquakes and Early American Imagination: Decline and Renewal in 18th Century Puritan

Culture." Ph.D. dissertation, University of California at Berkeley, 1977. 255 pp.

Chapter I "raises salient points in ancient and renaissance culture concerning the metaphoric basis of early seismic insight...." Chapter II discusses "the impact of the 1727 Boston earthquake. It treats tensions in [Cotton] Mather between Puritan covenantal tradition on the one hand and apocalyptic embrace of the seismic rupture on the other." Chapter III, "The Theology of Rupture: Thomas Prince, 1727" evaluates Prince's reaction to dislocation in 1727. Chapter IV depicts John Winthrop's response to the 1755 earthquake. His confrontation with Prince "exposed tensions inherent in a split between creative activity and pre-evangelical religious unities." A brief conclusion touches on 19th-century culture and Melville, through links from the earthquake controversy itself" (from *Dissertation Abstracts International* 38 [1978]: 5010A).

* Sleep, Mark C.W. "Sir William Hamilton (1730-1803): His Work and Influence in Geology." Cited above as item BH8.

J34 Snare, Gerald. "Satire, Logic and Rhetoric in [Gabriel] Harvey's Earthquake Letter to Spenser." *Tulane Studies in English* 18 (1970): 17-33.

Gabriel Harvey used the 1580 earthquake as an occasion to satirize Cambridge University as well as to apply the philosophical methods of P. Ramus to scientific inquiry.

J35 Stauder, William. "The Focal Mechanism of Earthquakes." *Advances in Geophysics* 9 (1962): 1-76.

The elastic rebound theory of H.F. Reid (1910) is generally accepted by Western seismologists. Perry Byerly (1926) initiated a program of determining the direction of the fault plane along which the earthquake occurred (in accordance with Reid's theory) by tabulating the direction of ground motion in the initial P wave at many seismographic stations. H. Nakano (1923) developed a mathematical theory of the displacements produced by a dipole source, but almost every copy of his paper was destroyed by the Tokyo earthquake and fire of 1923, so few seismologists had a chance to read it. Since 1931, H. Honda has been the central figure among Japanese seismologists working in this field. The Dutch school of focal mechanism study began in the late 1930s with the work of L.P.G. Koning. Since 1948, there has been an active group of researchers in Moscow (V.I. Keylis-Borok, et al.). The review also discusses the work of A.E. Scheidegger, L. Knopoff, F. Gilbert, J.H. Hodgson, and S.D. Kogan.

J36 Stoneley, R. "History of Modern Seismology." *International Dictionary of Geophysics*. Edited by S.K. Runcorn. Oxford: Pergamon Press, 1967, pp. 724-729.

J37 Taylor, John G., Jr. "Eighteenth-Century Earthquake Theories: A Case-History Investigation into the Character of the Study of the Earth in the Enlightenment." Ph.D. dissertation, the University of Oklahoma, 1975. vi + 340 pp.

At the beginning of the 18th century the major theories of the cause of earthquakes were based on (1) subterranean fresh sulphurous vapors (Martin Lister, 1684; Robert Plot, 1686; John Ray, 1693); (2) water (John Woodward, 1695); (3) compressed air (Guillaume Amontons, 1703). The first, supported by experiments of Nicolas Leméry (1700), was dominant in the first half of the century. Other writers attributed earthquakes to underground disturbances, giving more or less importance to the role of heat, iron-sulphur reactions, water; and air; but in 1750 Stephen Hales proposed that the source of earthquakes is a gaseous reaction in the atmosphere ("air-quake"). A similar theory of John Flamsteed, developed in 1693, was first made public in 1750. Also in 1750 William Stukeley, proposed an electrical theory, rejecting subterranean causes.

Stukeley's electrical theory, though not universally accepted, provided the basis for many discussions of earthquakes in the second half of the 18th century. The major alternative was John Michell's theory (1760) that the explosive creation of steam when large quantities of water are brought into contact with internal fires is responsible for both earthquakes and volcanoes. But Michell does not deserve the title "father of seismology."

The author concludes that after 1750 earthquake theories became more complex, and that the field came to be dominated by scientists with more specialized interests in earth science. Theorists paid more attention to new observations and began to discuss and criticize each other's ideas. Yet theories continued to proliferate, and no single cause of earthquakes was generally accepted. Instead, they were attributed to causes as diverse as divine judgment and lunar tides. A final section deals with ideas about the effects of earthquakes in altering the topography of the earth, from Robert Hooke (1668) to Charles Lyell. (Based on the author's summary in *Dissertation Abstracts International* 36 [1975]: 2398-A.)

* Thams, J.C., ed. *The Development of Geodesy and Geophysics in Switzerland.* Cited above as item A103.

J38 Van Gils, J.M. "La chronique séismologique en Belgique (du début de l'ère chrétienne jusqu'au XXe siècle)." *Ciel et Terre* 66 (1950): 218-221.

A miscellaneous collection of reports on earthquakes, mostly from the 19th century.

J39 Wartnaby, John. "John Milne (1850-1913) and the Development of the Seismograph." *Proceedings of the XIVth International Congress of the History of Science, 1974* 3 (pub. 1975): 109-112.

In contrast to the earlier "seismoscopes" which merely indicated that an earthquake had occurred in the area, Milne developed in the 1890s a simple instrument to register "earth tremors" (microseisms) resulting from shocks of moderate intensity anywhere in the world.

J40 Willis, Bailey. "Earthquakes in the Holy Land." *Bulletin of the Seismological Society of America* 18 (1828): 73-103.

Summary of the record of shocks in Palestine and Syria since 1606 B.C. For corrections to this paper, see *Nature* 131 (1933): 550 and *Science* 77 (1933): 351.

J41 Wilson, Arnold T. "Earthquakes in Persia." *Bulletin of the School of Oriental Studies, University of London* 6 (1930): 103-131, 1 plate.

A review of descriptions of earthquakes from A.D. 550 to 1930.

For additional information on this topic see works cited in the biography sections on the following scientists:

Benioff, V.H.
Bullen, K.E.
Chang Heng
Day, A.L.
Dutton, C.E.
Escholt, M.P.
Etzold, Fr.
Gilbert, G.K.

Golitsyn, B.B.
Gutenberg, B.
Hamilton, W.
Heck, N.H.
Hooke, R.
Imamura, A.
Jeffreys, H.
Knott, C.G.
Koto, B.
Lamb, H.
Lawson, A.C.
Lehmann, J.G.
Love, A.E.H.
Macelwane, J.B.
Mallet, R.
Michell, J.
Milne, J.
Mohorovičić, A.
Mushketov, I.V.
Naumann, K.F.
Oldham, R.D.
Omori, F.
Orlov, A.Y.
Raspe, R.E.
Reid, H.F.
Rubey, W.W.
Schmidt, C.A. Von
See, T.J.J.
Sezawa, K.
Stoneley, R.
Suess, E.
Turner, H.H.
Wiechert, E.
Willis, B.

JA. Prospecting and Other Applications of Geophysics

JA1 Allaud, Louis A., and Maurice H. Martin. *Schlumberger. The History of a Technique* [translation of *Schlumberger: Histoire d'une technique*. Paris: Berger-Levrault, 1976. 343 pp.] New York: Wiley-Interscience, 1977. xxv + 333 pp.

In 1927, Conrad Schlumberger invented "electrical coring" (now known as "electrical logging"), a method of analyzing rock samples useful in prospecting for oil. The book provides an account of the techniques developed by Conrad and his brother, Marcel Schlumberger, including some discussion of other geophysical prospecting methods.

JA2 Bates, Charles C., Thomas F. Gaskell, and Robert B. Rice. *Geophysics in the Affairs of Man. A Personalized History of Exploration Geophysics and Its Allied Sciences of Seismology and Oceanography*. New York: Pergamon Press, 1982. xx + 492 pp.

A comprehensive account of applied geophysics in the 20th century, with much valuable information about individuals and organizations. The authors obtained from several scientists summaries of what they considered their most important contributions, including personal vignettes from M. Ewing (on the Lamont Geological Observatory), N.A. Ostenso (on G.P. Woollard), C.H. Green, W.H. Mayne, J.M. Crawford, J. Holmes, W.M. Chapman.

JA3 Johnson, Hamilton M. "A History of Well Logging." *Geophysics* 27 (1962): 507-527.

The science of well logging was begun by Conrad Schlumberger in 1927 as an application of his work on resistivity measurements of the earth in surface exploration. Recent techniques have utilized radioactive and acoustic as well as electrical properties of rocks. Includes chronology, 1869-1960.

JA4 Silverman, Daniel J., S. Norman Domenico, and Mrs. David B. Darden, eds. "SEG 50th Anniversary Issue." *Geophysics* 45, no. 11 (November 1980).

Includes S.N. Domenico and D. Silverman, "Our Society" (Society of Exploration Geophysicists), pp. 1600-1618; S.J. Allen, "Seismic Method," pp. 1619-1633; T.R. LaFehr, "Gravity Method," pp. 1634-1639; M.S. Reford, "Magnetic Method," pp. 1640-1658; S.H. Ward, "Electrical, Electromagnetic, and Magnetotelluric Methods," pp. 1659-1666; F.F. Segesman, "Well-logging Method," pp. 1667-1684; R.D. Regan, "Remote Sensing Method," pp. 1685-1689; J.S. Duval, "Radioactivity Method," pp. 1690-1694.

For additional information on this topic see works cited in the biography sections on the following scientists:

De Golyer, E.L.
Lazarev, P.P.
Numerov, B.V.

K. HYDROLOGY

K1 Adams, Frank Dawson. "The Origin of Springs and Rivers: An Historical Review." *Fennia* 50, no. 1 (1928). 16 pp., 2 plates.

A survey of the theories of G.C. Massei (1564), N. Papin (1674), and A. Vallisnieri (1715).

K2 Antrei, Albert. "A Western Phenomenon. The Origin and Development of Watershed Research: Manti, Utah, 1889." *The American West* 8, no. 2 (March 1971): 42-47, 59.

Following a series of disastrous floods at Manti between 1889 and 1903, the National Forest Service established a Range Experiment Station. The first director was Arthur W. Sampson (from 1912 to 1922), who started a program of range research.

K3 Baker, Victor R. "The Spokane Flood Controversy and the Martian Outflow Channels." *Science* 202 (1978): 1249-1256.

"In a series of papers published between 1923 and 1932, J. Harlen Bretz described an enormous plexus of proglacial stream channels eroded into the loess and basalt of the Columbia Plateau, eastern Washington. He argued that this region, ... was the product of a cataclysmic flood ... considering the nature and vehemence of the opposition to his hypothesis, which was considered outrageous, its eventual scientific verification constitutes one of the most fascinating episodes in the history of modern science. The discovery of probable catastrophic flood channels on Mars has given new relevance to Bretz's insights" (Author's summary).

K4 Biswas, Asit K. "Experiments on Atmospheric Evaporation Until the End of the Eighteenth Century." *Technology and Culture* 10 (1969): 49-58.

On the researches by Hippocrates, E. Halley, J. Dalton, and others.

K5 Biswas, Asit K. *History of Hydrology*. New York: Elsevier, 1970. xii + 336 pp.

A systematic account, from ancient times through mid-19th century; includes among others the contributions of Plato, Aristotle, Vitruvius, Leonardo da Vinci, Bernard Palissy, Edmé Mariotte, Edmund Halley, Paolo Frisi, G.B. Venturi. Extensive bibliography.

K6 Biswas, Asit K. "Edmund Halley, F.R.S., Hydrologist Extraordinary." *Notes and Records of the Royal Society of London* 25 (1970): 47-57.

In 1687, Halley "proved that the amount of water which evaporated from the oceans and watercourses and which came down in the form of rainfall adequately replenished the flow of rivers." He should be considered (with Pierre Perrault and Edmé Mariotte) as co-founder of experimental hydrology.

* Carozzi, Albert V. "Lamarck's Theory of the Earth: *Hydrogéologie*." Cited above as item H7.

K7 Chebotarev, A.I., and I.V. Popov. "Pyatidesyatiletie Gosudarstvennogo gidrologicheskogo instituta" [Fiftieth Anniversary of the State Hydrological Institute]. *Meteorologiya i Gidrologiya* No. 10 (1969): 3-10.

The State Hydrological Institute was established in 1919. A history of its activities, principal investigators, and accomplishments is presented.

K8 Chow, Ven-te, ed. *Handbook of Applied Hydrology*. New York: McGraw-Hill, 1964. xiv + 1454 pp.

Includes a 22-page introduction by the editor on "Hydrology and Its Development," with 137 references.

* Federov, E.K., ed. *Meteorologii i gidrologii za 50 let Sovietskoi Vlasti i Sbornik Statei*. Cited below as item NB15.

K9 Fedoseev, I[van] A[ndreevich]. *Razvitie gidrologii sushi v Rossii*. Moscow: Izd-vo Akademii Nauk SSSR, 1960. 303 pp.

History of hydrology in Russia up to 1917.

K10 Fedoseyev, I.A. "Development of Hydrology of the Land in Pre-Revolutionary Russia." *Proceedings of the 10th International Congress on the History of Sciences, Ithaca, 1962* (pub. 1964), pp. 1023-1026.

Discusses the work of A.I. Voyeykov [Voeikov], N.S. Lelyavsky, V.M. Lokhtin, V.J. Altberg, and others.

K11 Fedoseev, I[van] A[ndreevich]. *Razvitie znanii o proiskhozhdenii, kolichestve i krugovorote vody na zemla* [Development of Conceptions of the Origin, Quantity and Rotation of Water on the Earth]. Moscow: Nauka, 1967. 133 pp.

K12 Fedoseyev, Ivan. "Growth of Knowledge about the Earth's Hydrosphere." *Soviet Studies in the History of Science*. Moscow: "Social Sciences Today" Editorial Board & Institute of History of Natural Science and Technology of the USSR Academy of Sciences, 1977, pp. 227-238.

K13 Frazier, Arthur H. "Daniel Farrand Henry's Cup Type 'Telegraphic' River Current Meter." *Technology and Culture* 5 (1964): 541-565.

Henry (1833-1907) constructed his meter in 1867-68 to measure the volume of water flowing out of the Great Lakes.

* Lamarck, J.B. *Hydrogeology*. Cited above as item H29.

* McDonald, J.E. "James Espy and the Beginnings of Cloud Thermodynamics." Cited below as item PA24.

K14 Meinzer, Oscar E., ed. *Hydrology*. Physics of the Earth--IX. New York and London: McGraw-Hill, 1942. xi + 712 pp. Reprint, New York: Dover, 1949.

Chapter I, Introduction, by Meinzer includes a brief history of hydrology.

* *Meteorology and Hydrology in Czechoslovak Socialist Republic*. Cited below as item NB45.

K15 Mikulski, Zdzislaw. *Zarys hydrografi Polski*. Warsaw: Panstwowe Wydawnictwo Naukowe, 1962. 287 pp.

* Moore, Gerald K. "What Is a Picture Worth? A History of Remote Sensing." Cited below as item OB10.

* Thams, J.C., ed. *The Development of Geodesy and Geophysics in Switzerland*. Cited above as item A103.

K16 Tuan, Yi-Fu. *The Hydrologic Cycle and the Wisdom of God. A Theme in Geoteleology*. Toronto: University of Toronto Press, 1968. xiii + 160 pp.

Surveys the writings of John Ray (1691), William Caxton, William Fulke (1563), Thomas Burnet, George Hakewill, John Keill, Edmund Halley, James Hutton.

K17 Weikinn, Curt. *Quellentexte zur Witterungsgeschichte Europas von der Zeitwende bis zum Jahre 1850*. Deutsche Akademie der Wissenschaften zu Berlin. Institut für Physikalische Hydrographie. Quellensammlung zur Hydrographie und Meteorologie, Bd. 1. Hydrographie. Teil 1 (Zeitwende-1500). Berlin: Akademie-Verlag, 1958-63. Vii + 531 pp.

The four parts published so far (1958-63) cover hydrography to 1750.

K18 White, George W. "John Keill's View of the Hydrologic Cycle, 1698." *Water Resources Research* 4 (1968): 1371-1374.

Keill rejected Thomas Burnet's theory of the origin of the waters of the deluge, and presented quantitative estimates of the water flowing in rivers.

K19 Zaikov, B.D. *Ocherki gidrologicheckikh issledovanii v Rosii*. Edited by A.P. Domanitskogo. Leningrad: Gidrometeoizdat, 1973. 325 pp.

Essays on the development of hydrology in Russia since the 12th century, with a 95-page appendix giving one-paragraph biographies of the scientists mentioned.

For additional information on this topic see works cited in the biographical sections on the following scientists:

Anuchin, D.N.
Birge, E.A.
Dalton, J.
Deryugin, K.M.
Hamberg, A.
Leibniz, G.W. v.
Lokhtin, V.M.
Mariotte, E.
Meinzer, O.E.
Nikitin, S.N.
Palissy, B.
Perrault, P.
Rubey, W.W.
Schardt, H.
Schimper, K.F.
Shokalsky, Y.M.
Woltman, R.

L. OCEANOGRAPHY (General and Physical)

* Akademiia Nauk SSSR. "Fizika atmosfery i okeana v SSSR za 50 let." Cited below as item M1.

L1 Arx, William S. von, and Ann J. Martin. "Some Events That Have Influenced Thought in the Marine Sciences." *An Introduction to Physical Oceanography* by William S. von Arx. Reading, Mass.: Addison-Wesley Pub. Co., 1962, pp. 351-396.

A very useful annotated chronology with references to primary and secondary sources for each event.

L2 Bache, F. "Where Is Franklin's First Chart of the Gulf Stream?" *Proceedings of the American Philosophical Society* 76 (1936): 731-741.

Reviews the history of Franklin's interest in the Gulf Stream. No copies of his original chart can be found in any American library. Three versions are reproduced in facsimile.

L3 Bailey, Herbert S., Jr. "The Background of the *Challenger* Expedition." *American Scientist* 60 (1972): 550-560.

Charles Wyville Thomson and William Benjamin Carpenter discussed the need for an oceanographic expedition and conducted preliminary research during the four years before the start of the *Challenger* voyage in December 1872. Matthew Fontaine Maury and Edward Forbes had some influence on their thinking.

L4 Bencker, H. "The Bathymetric Soundings of the Oceans." *Hydrographical Review* 7, no. 2 (November 1930): 64-97.

Consists primarily of a list of ocean expeditions since 1900 and the results of depth measurements in various locations.

L5 Burstyn, Harold L., and Susan B. Schlee. "The Study of Ocean Currents in America Before 1930." *Two Hundred Years of Geology in America* (item A90), pp. 145-155.

"... Before 1850 three agencies of the federal government were founded to furnish what we now call research and development services to maritime commerce, and the employees of these agencies brought American physical oceanography to its dominant position in world science." These agencies were the Coast Survey, the Depot of Charts and Instruments, and the Nautical Almanac Office. American marine science declined after the Civil War, while the study of ocean currents was taken up in Scandinavia. Alexander Agassiz formed the link between European scientists and the revival of American oceanography in the 20th century.

L6 Carpine-Lancre, Jacqueline. "Les expéditions océanographiques françaises du XIXe siècle." *Actes XIIe Congrès International d'Histoire des Sciences, Paris 1968* (pub. 1971) 7: 61-65.

A review of French activity in the period 1880-1910, which was rather meagre compared to that of other countries.

L7 Carruthers, J.N. "Some Oceanography from the Past." *The Journal of the Institute of Navigation* 16 (1963): 180-188.

Miscellaneous remarks on Count Marsilli (or Marsigli) and his *Histoire Physique de la Mer* (1725), Georges Aime and his measurements of temperature in the Mediterranean, and ideas about subterranean channels; based on "the excellent writings of Commandant J. Rouch," and a "recent paper" by M. Rodewald.

L8 "*Challenger* Expedition Centenary: Proceedings of the Second International Congress on the History of Oceanography, Edinburgh, September 12-20, 1972." *Proceedings of the Royal Society of Edinburgh*, Section B (Biology), 1972-1973. 2 vols. Edinburgh: Royal Society, 1972; vol. 1: viii + 462 pp.; vol. 2: viii + 435 pp., illus., bibl.

Part I includes papers on the *Challenger* expedition by Maurice Yonge, Harold L. Burstyn, Margaret Deacon, Bernard L. Gordon, and others; by J. Thomson and A. Walton and by G.N. Baturin, T.D. Ilyina, and N.I. Popov on studies of radioactive elements; by M.N.A. Peterson and A.J. Field comparing the H.M.S. *Challenger* with D/V Glomar *Challenger*; by L. Dangeard on submarine geology since 1868; by Peter H. Klystra and Arend Meerburg on Jules Verne and M.F. Maury; by Gordon Lill on activities of the Office of Naval Research, 1950-59; by N.K. Panikkar and T.M. Srinivasan on "Early Concepts of Oceanographic Phenomena of the Indian Ocean"; by D.E. Cartwright, C.R. Mann, L.M. Brekhovskikh, and others on ocean circulation; by John M. Edmond and John Lyman on the CO_2 system in sea water; by A. Preston and others on artificial radioactivity; by E.C. LaFond and B.K. Couper on the bathythermograph; and several other short papers. Part 2 contains papers on navigation, tidal predictions, cartography, and biological topics.

L9 Charlier, Roger H. "The Transatlantic Telegraphic Cable and the *Physical Geography of the Sea*. *Actes, XII*[e] *Congrès International d'Histoire des Sciences, Paris 1968* (pub. 1971) 7: 73-76.

An assessment of the book by M.F. Maury (1855) and its relation to the submarine cable project of Cyrus West Field.

L10 Charnock, H. "*H.M.S. Challenger* and the Development of Marine Science." *Journal of the Institute of Navigation* 26 (1973): 1-12.

In retrospect, after all the reports had been published, the author concludes that despite the great achievements of the *Challenger* expedition, there was "one serious gap--the physics of the sea had been relatively neglected. None of the scientific staff was a physicist and the observations had been left to the ship's officers, to whom they were an extra obligation ... it was left to German scientists, decades later, to make good use of the temperature and density observations." Moreover, "British marine science proved relatively static after the *Challenger* Expedition.... It was during the life of the most recent H.M.S. *Challenger* (launched at Chatham in 1931) that Marine Science, both in the United Kingdom and abroad, became reactivated" (p. 10).

L11 Cox, Donald W. *Explorers of the Deep. Pioneer Oceanography*. Maplewood, N.J.: Hammond, 1968. 93 pp.

Chapters on B. Franklin, C. Wilkes, M.F. Maury, A. Agassiz, J.P. Holland, W. Beeke, H.G. Rickover, W.R. Anderson, A. and J. Piccard, R. Carson, M. Ewing, J.Y. Cousteau, E.A. Link, G.F. Bond, M.S. Carpenter, W. Bascom, R. Revelle. Popular book with several color illustrations.

L12 Daugherty, Charles M. *Searchers of the Sea. Pioneers in Oceanography*. New York: Viking Press, 1961, 160 pp.

A popular history, with no references to original sources.

L13 Deacon, George E.R. "Scientific Exploration of the Oceans." *Memoirs and Proceedings of the Manchester Literary and Philosophical Society* 102 (1960): 41-64.

"A hundred years ago advances in navigational instruments, charts, and wind and current atlases made Admiral [W.H.] Smyth feel that the union of science and practical seamenship was bringing the elements into subjection. Modern scientific exploration and research can no longer shorten ocean passages by two or three weeks but it is providing new information about waves, currents, and other sea conditions that is just as exciting and may mean as much to future use of the sea and protection against it" (Author's summary).

L14 Deacon, George E.R., ed. *Oceans: An Atlas History of Man's Exploration of the Deep*. London: Hamlyn, 1962, 297 pp., illus., maps, charts. The American edition (Garden City, N.Y.: Doubleday) has title: *Seas, Maps, and Men: An Atlas-History of Man's Exploration of the Oceans*.

This is a picture-book with brief essays on famous expeditions, capsule biographies, etc. The development of the modern science of oceanography is surveyed rather superficially in the chapters by Deacon, A.S. Laughton, J.C. Swallow, and K.F. Bowden.

L15 Deacon, G.E.R., and Margaret B. Deacon, eds. *Modern Concepts of Oceanography*. Stroudsburg, Pa.: Hutchinson Ross Pub. Co., 1982. xiv + 386 pp.

Reprints 41 research papers including H. Stommel (1951, 1963), I. Langmuir (1938), W. Brennecke (1911, 1914, 1918, 1921), L.H.N. Cooper (1961, 1967), J.W. Cooper and H. Stommel (1968), R.D. Pingree (1972), M.S. Longuet-Higgins (1949, 1953), M.S. Longuet-Higgins and R.W. Stewart (1963), H. Charnock and J. Crease (1957), A.H.W. Robinson (1961), H.F. Baird and C.J. Banwell (1940), B.C. Heezen and M. Ewing (1952), W.G. Metcalf, A.D. Voorhis, and M.C. Stalcup (1962), A.E. Gill (1973), T.D. Foster and E.C. Carmack (1976), and A.C. Redfield (1960).

L16 Deacon, Margaret. "Founders of Marine Science in Britain: The Work of the Early Fellows of the Royal Society." *Notes and Records of the Royal Society of London* 20 (1965): 28-50.

On the observations and instruments of Robert Boyle, Robert Hooke, and others.

L17 Deacon, Margaret. *Scientists and the Sea, 1650-1900: A Study of Marine Science*. London: Academic Press, 1971. xvi + 445 pp., illus., maps, bibl., index.

After the Renaissance there were three main periods of activity in the marine sciences: (1) in the 1660s; (2) in the late 18th century; (3) from the 1860s onwards. (We are in the middle of a fourth such period today.) Progress depended heavily on advances in other sciences. The first period is illuminated by studying the writings of Robert Boyle, Robert Hooke, and others. In the 18th century interest in the sea was reawakened by the work of L.F. Marsiglio, Stephen Hales, and Benjamin Franklin. The search for regularities in the tides challenged the mathematical powers and physical understanding of scientists such as Galileo, John Wallis, Newton, Laplace, and William Whewell. The third period was dominated by the *Challenger* expedition, organized by Charles Wyville Thomson and W.B. Carpenter. The author succeeds in presenting a comprehensive and readable account of the origins of modern oceanography, based on extensive study of primary and secondary sources; her book deserves a careful reading by anyone with a serious interest in the history of the earth sciences.

L18 Deacon, Margaret B., ed. *Oceanography, Concepts and History*. Benchmark Papers in Geology/35. Stroudsburg, Pa.: Dowden, Hutchinson & Ross, Inc., 1978. xvii & 394 pp.

A very useful anthology of extracts from original sources with brief editorial comments. Contents: Part I: "Some Seventeenth Century Instructions for Making Scientific Observations at Sea"; Part II: "Circulation of the Ocean"--L.F. Marsigli (1681), J. Rennell (1832), E. Lenz (1832), notes in *Nature* (1871, 1874), G. Wüst (1928), G.E.R. Deacon (1934), J.C. Swallow (1955), J. Crease (1962), J. Namias (1963); Part III: "Tides"--R. Moray (1666), E. Halley (1797), J. Fleming (1818), note in *Nautical Magazine* (1832), W. Hewett (1841), J. Proudman (1944), D.E. Cartwright (1967); Part IV: "Waves"--G.E.R. Deacon (1946), N.F. Barber, et al. (1946, 1947), F.P. Shepard (1949), D.E. Cartwright (1967); Part V: "Chemistry of Sea Water"; Part VI: "Depths of the Ocean and the Sea Bed"--S. Hales (1754), J.D. Dana (1843), M.F. Maury (1853), W. Thomson, Lord Kelvin (1873-75), J. Murray and A. Renard (1882-84), R.S. Dietz (1961); Part VII: "Marine Biology."

L19 DeVorsey, Louis, Jr. "Hydrography: A Note on the Equipage of 18th-Century Survey Vessels." *Mariner's Mirror* 58 (1972): 173-182.

On William Gerard De Brahm's hydrographic survey vessel.

L20 Drubba, Helmut. "On the First Echo-Sounding Experiment." *Annals of Science* 10 (1954): 28-32, 1 plate.

On an experiment performed in 1838 by Charles Bonnycastle (1792-1840); transcription of MS at American Philosophical Society. The experiment failed because of poor equipment and lack of experimental data on underwater acoustics.

L21 Drubba, Helmut, and Hans Heinrich Rust. "Die Entwicklung der akustischen Meerestiefenmessung." *Zeitschrift für angewandte Physik* 5 (1953): 388-400.

History of acoustic measurements of the depth of the sea, beginning with J.D. Colladon (1826) and C. Bonnycastle (1838).

L22 Drubba, Helmut, and Hans Heinrich Rust. "Historische Geräte für die Meerestiefenbestimmung." *Hansa* 92 (1955): 1601-1605, 5 figs.

A brief survey of depth-measuring instruments, starting with the proposal by Nicolaus of Cusa. The devices invented by Robert Hooke (1665), C.W. Siemens (1876), and others are illustrated.

L23 Ekman, V.W. *On the Influence of the Earth's Rotation on Ocean Currents*. Uppsala: Royal Swedish Academy of Sciences, 1962.

Reprint of Ekman's 1905 paper, with a biography by B. Kullenberg and bibliography of Ekman's publications.

* Federov, K.K., ed. *Meteorologii i gidrologii za 50 let Sovietskoi Vlasti i Sbornik Statei*. Cited below as item NB15.

L24 Fedoseev, I.A. *Istoriia izucheniia osnovnykh problem gidrosfery* [History of the Study of the Fundamental Problems of the Hydrosphere]. Moscow: Nauka, 1975. 208 pp.

L25 Fedoseyev, Ivan. "Growth of Knowledge about the Earth's Hydrosphere." *Soviet Studies in the History of Science*.

Moscow: "Social Sciences Today" Editorial Board, 1977, pp. 227-238.

Mostly 20th century, including work of V. Vernadsky.

L26 Gibson, George H. "Ben Franklin and the Gulf Stream." *Sea Frontiers* 11, no. 3 (1965): 172-177.

After British bureaucrats complained in 1768 about delays in mail deliveries to New York City, Benjamin Franklin investigated the effect of the Gulf Stream in retarding ships. The results of his research, with a theoretical explanation, were presented to the American Philosophical Society in 1786.

L27 Gordon, Bernard L., ed. *Man and the Sea: Classic Accounts of Marine Explorations*. Garden City, N.Y.: The Doubleday Natural History Press, 1972. xxv + 498 pp.

An anthology of 71 short extracts on various topics related to oceanography.

L28 Great Britain, Challenger Office. *Report on the Scientific Results of the Voyage of H.M.S. Challenger During the Years 1873-1876*. 50 vols. New York: Johnson Reprint, 1972 (reprint of the 1880-95 edition).

L29 Guberlet, Muriel L. *Explorers of the Sea. Famous Oceanographic Expeditions*. New York: Ronald, 1964. v + 226 pp.

Includes chapters on M.F. Maury, C.W. Thomson, Willard Bascom, and the Mohole Project, etc. Primarily biographical and anecdotal.

L30 Herdman, William A. *Founders of Oceanography and Their Work: An Introduction to the Science of the Sea*. New York: Longmans, Green, 1923. xii + 340 pp.

Includes chapters on E. Forbes, C.W. Thomson, J. Murray, L. Agassiz, A. Agassiz, and on hydrography and ocean currents.

* Herschel, John, ed. *Admiralty Manual of Scientific Enquiry*. Cited above as item A55.

L31 Idyll, C.P., ed. *Exploring the Ocean World: A History of Oceanography*. New York: Crowell, 1969. viii + 280 pp. Revised edition, 1972, viii + 296 pp.

Includes: C.P. Idyll, "The Science of the Sea," pp. 2-21; R.S. Dietz, "The Underwater Landscape," pp. 22-41; J.B.

Rucker, "Physics of the Sea," pp. 62-97. Many illustrations.

* *International Congress on the History of Oceanography, First, Proceedings*. Cited below as item L43.

* *International Congress on the History of Oceanography, Second, Proceedings*. Cited above as L8.

* *International Congress on the History of Oceanography, Third, Proceedings*. Cited below as item L52.

L32 Jones, A.G.E. "Lieutenant T.E.L. Moore, R.N. and the Voyage of the Pagoda, 1845." *Mariner's Mirror* 56 (1970): 33-40.

"The magnetic, meteorological and hygrometric observations were published in Philosophical Transactions as part of a paper by Lieut.-Col. [Edward] Sabine, who had written the instructions.... Whether any use was made of all the observations is anybody's guess."

L33 Lewthwaite, Gordon R. "Maury to Schoolcraft: Correspondence on Ocean Currents and Pacific Migrations." *Professional Geographer* 22 (1970): 128-131.

Letter from Henry R. Schoolcraft to M.F. Maury asking about the possibility that early peoples could have navigated the Pacific and Polynesian waters, and Maury's reply (1850).

L34 Long, E. John, ed. *Ocean Sciences*. Annapolis, Md.: U.S. Naval Institute, 1964. xii + 304 pp.

Contains a chapter on the history of oceanography by John Lyman, pp. 13-25, and brief historical notes in other chapters devoted to related sciences, instrumentation, and institutions.

L35 Manten, A.A. "The Origin of Marine Geology." *Marine Geology* 2 (1964): 1-28.

Following a survey of ideas about the oceans held in ancient and medieval times, the author places the origin of marine geology in the 18th century with the writings of Philippe Buache, J.F. Henkel, M.P. Colonne, and L.F. Comte de Marsilli. Later events noted are Charles Darwin's theory of coral reefs, sounding methods developed in the 19th century, and the *Challenger* expedition.

L36 Marmer, H.A. "The Gulf Stream and Its Problems." *Geographical Review* 19 (1929): 457-478. Reprinted in *Annual Report of the Smithsonian Institution for 1929*. Washington, D.C., 1930, pp. 285-307.

Includes a brief historical review.

L37 Maury, Matthew Fontaine. *The Physical Geography of the Sea and Its Meteorology*. Reprint of the 8th edition (1861). Edited with an Introduction by John Leighly. Cambridge, Mass.: The Belknap Press of Harvard University Press, 1963. xxxi + 432 pp., 10 plates.

L38 McEwen, George F. "The Status of Oceanographic Studies of the Pacific." *Proceedings of the 1st Pan-Pacific Scientific Conference*. Bernice P. Bishop Museum, Special Publication, 7. Honolulu: Honolulu Star, 1921, pp. 487-497.

L39 McEwen, G.F. "Modern Dynamical Oceanography: An Achievement of Applications to Ocean Observations of Principles of Mechanics and Heat." *Proceedings of the American Philosophical Society* 79 (1938): 145-166.

Primarily an historical (but undocumented) review of researches on ocean circulation since Maury. "Zöppritz was the first to appreciate the importance of the viscosity or internal friction of water in investigations of ocean currents" (1878). "V. Bjerknes, as a result of collaboration with his father ..., founded the science of physical hydrodynamics.... His student, V.W. Ekman, devoted special attention to the concept of eddy motion or turbulence and its consequences."

L40 Multhauf, Robert P. "The Line-less Sounder: An Episode in the History of Scientific Instruments." *Journal of the History of Medicine and Allied Sciences* 15 (1960): 390-398.

Cardinal Nicolaus Cusanus (1476) suggested the feasibility of such an instrument; illustrations of instruments developed by R. Hooke (1667, 1691), Luiscius (1805), J.T. Desaguliers, and S. Hales (1728).

* Panzram, Heinz. "Einhundert Jahre im Dienste der maritimen Meteorologie und der Ozeanographie." Cited below as item NB54.

L41 Plakhotnik, A.F. "Istoriia izucheniia okeanov i morei. Predystoriia okeanograficheskikh issledovanii (ot drevneishikh vremen do poslednikh desiatiletii XIX v.)." *Trudy Instituta Istorii Estestvoznaniia i Tekhniki* 37 (1961): 52-79.

History of oceanography from antiquity to the late 19th century.

L42 Plakhotnik, A.F. "Istochniki i printsipy periodizatsii istorii okeanografii." *Trudy Instituta Istorii Estestvoznaniia i Tekniki* 42 (1962): 103-129.

"The origins and principles of the periodization of the history of oceanography." The author proposes four periods: (1) to 15th century; (2) 15th-17th centuries; (3) 18th-late 19th centuries; (4) starting in the 1870s with the *Challenger* expedition. The bibliography contains 183 items, mostly in Russian.

L43 "Premier Congrès International d'Histoire de l'Océanographie, Monaco, 1966." *Bulletin de l'Institut Océanographique*, special no. 2. 3 vols. Monaco: Musée Océanographique, 1968. xlii + 807 pp.

Contents, vol. 1: Papers on the historical influence of tides by G. Sager; on theories of tides by M. Kasumovic and Z. Dadic; on early studies of currents in the Strait of Gibraltar, by M. Deacon; on water-level measurements of antiquity, by W. Matthäus; on B. de Maillet's precursor of the bathyscaphe, by T. Monod; on measurements of depth-temperatures in the sea, by W. Matthäus; on the Gulf Stream, by T.F. Gaskell; on the work of [Urbain] Dortet de Tessan and Anatole Bougnet de la Grye on ocean circulation, by A. Gougenheim; on the quantity of water in the World Ocean, by I.A. Fedoseyev; on longitudinal deep-sea circulation, 1800-1922, by G. Wüst; on contributions to oceanography by the British and Italian navies, by G.S. Ritchie and L. DiPaola, respectively; on M.F. Maury, by J. Leighly; on precursors of submarine optics, by Y. LeGrand; on theoretical oceanography in Sweden 1900-1910, by P. Welander; on P.F. Kendall's marine-geological speculations, by J.N. Carruthers; several papers on cartography and navigation; papers on regional oceanography by V.F. Burkhanov, G.E.R. Deacon, J.B. Tait, R.H. Charlier, E. Leloup, A. Crovetto, M. Buljan, M. Zore-Armanda, and J. Morovic; on Gustave Gilson (1859-1944), by A. Capart.

Vol. 2 contains papers on biological oceanography, marine biology, medical aspects, and expeditions; on the

Challenger expedition, by H.L. Burstyn; on S.O. Makarov (1849-1904), by A.I. Soloviev; on documentary sources, by C.A. Ronan, J. Carpine-Lancre; on historiography, by H.L. Burstyn.

Vol. 3 is a comprehensive index.

L44 Rice, A.L. "The Oceanography of John Ross's Arctic Expedition of 1818: A Re-appraisal." *Journal of the Society for the Bibliography of Natural History* 7 (1975): 291-319.

Discussion of his soundings and temperature measurements in Baffin's Bay.

L45 Ritchie, George S. "Great Britain's Contribution to Hydrography During the Nineteenth Century." The Eva G.R. Taylor lecture, given 26 October 1966. *Journal of the Institute of Navigation* 20 (1967): 1-11.

A discussion of survey voyages, with very brief mention of scientific results.

L46 Ritchie, George S. "Developments in British Hydrography Since the Days of Captain Cook." *Journal of the Royal Society of Arts* 118 (1970): 270-283.

Describes methods and instruments for surveying as developed and used by James Cook, Murdoch Mackenzie, Edward Belcher, and others. Includes discussion following the presentation of the paper on 3 February 1970. Rear Admiral Ritchie is the 19th holder of the position "Hydrographer of the Royal Navy."

L47 Rotschi, H. "Man Discovers the Sea. The Influence of the Sea on the Growth of Human Societies." *Impact of Science on Society* 10 (1960): 79-103.

A brief survey of the history of oeeanography, followed by a discussion of current problems.

L48 Schlee, Susan. *The Edge of an Unfamiliar World: A History of Oceanography.* New York: Dutton, 1973. 398 pp.

This book received the Pfizer Award for the best book on the history of science by an American or Canadian author published in 1973. It begins with oceanography in 19th-century America (Charles Wilkes, M.F. Maury, A.D. Bache) and British oceanography before 1870 (Sir James Clark Ross, T.H. Huxley, W.B. Carpenter) and tells the well-known story of the *Challenger* expedition. A chapter on "The Sea in Motion" comes into the 20th century with the work of Scandinavian scientists (Otto Pettersson, Gustav Ekman, Fridtjof Nausen, Vilhelm Bjerknes, Bjorn Helland-Hansen). After an account of several 20th-century American expeditions, the book ends with a chapter on "Geophysical Studies and the New Theory of Sea-Floor Spreading" featuring the discoveries of Maurice Ewing, Harry Hess, Frederick Vine, and D.H. Matthews.

L49 Schlee, Susan. *On Almost Any Wind: The Saga of the Oceanographic Research Vessel "Atlantis."* Ithaca, N.Y.: Cornell University Press, 1978. 301 pp.

The *Atlantis* was a steel-hulled ketch used by the Woods Hole Oceanographic Institution from 1931 to 1966, making 299 cruises over more than a million and a half miles of ocean. "Because for six years she was the only vessel in the country large enough ... to undertake such extensive investigations, and because for another fifteen years she was one of only a few such ships, it was inevitable that scores of important discoveries were made from her decks" (p. 13). These included explorations of the Mid-Atlantic Ridge under the direction of Maurice Ewing.

L50 Schulz, Bruno. "Geschichte und Stand der Entwicklung des Behmlotes unter besonderer Beruchsichtigung der Lotungen auf D.S. "Hansa," Hamburg-Amerika Linie. *Annalen der Hydrographie und Maritimen Hydrologie* 52 (1924): 254-271, 289-300.

On the echo-sounding method of A. Behm.

L51 Schulz, Bruno. "Stand und Bedeutung der Echolotfrage." *Tijdschrift van het Nederlandsch Aardrijkskundig Genootschap*, 2nd Series, 42 (1925); 85-103.

On the methods of echo-sounding developed by A. Behm and H.C. Hayes.

L52 Sears, M., and D. Merriman, eds. *Oceanography: The Past.* Proceedings of the Third International Congress on the History of Oceanography, held at Woods Hole, Mass., 22-26 September 1980. New York: Springer-Verlag, 1980. xx + 812 pp.

Includes several papers on institutions and expeditions; also, "The Role of Instruments in the Development of Physical Oceanography" by M.C. Hendershott, "Some Origins and Perspectives in Deep-Ocean Instrumentation Development," by F.N. Spiess; "The Historical Development of Tidal Science, and the Liverpool Tidal Institute" by D.E. Cartwright; "Six's Thermometer: A Century of Use in Oceanography" by A. McConnell; "North Pacific Surface Temperature Observations: A History" by K.E. Kenyon; "Physical Oceanography of the Chilean Sea: An Historical Study" by Guillermo Barros G.; "Artificial Radionuclides in the Oceans" by W.L. Templeton; "Physical Oceanography in India: An Historical Sketch" by S. Markanday and P.R. Srivastava; "Studies of the Acceptance of Plate Tectonics" by J.L. Lemke, M.H. Nitecki, and H. Pullman; "How Secure Is Plate Tectonics" by A. Hallam; "Oceanography and Geophysical Theory in the First Half of the Twentieth Century: The Dutch School" by R. Laudan; "William De Brahm's 'Continuation of the Atlantic Pilot,' an Empirically Supported Eighteenth-Century Model of North Atlantic Surface Circulation" by L. D. Vorsey, Jr.

L53 Spiess, Fritz, et al. *Das Forschungsschiff und seine Reise*. Wissenschaftliche Ergesnisse der Deutschen atlantische Expedition auf dem Forschungs- und Vermessungsschiff "Meteor," 1925-1927. Band I. Berlin: De Gruyter, 1932. xii + 442 pp.

Account of the expedition planned by Alfred Merz, to make oceanographic, geomagnetic, and meteorological observations.

L54 Stommel, Henry. "The Gulf Stream. A Brief History of the Ideas Concerning Its Cause." *Scientific Monthly* 70 (1950): 242-253.

Includes theories of Peter Martyr (1515), Benjamin Franklin (1786), M.F. Maury (1859), J.E. Pillsbury (1891), William Ferrel (1856), V.W. Ekman (1902), C.G. Rossby (1936), C. O'D. Iselin (1941), H. Stommel (1948), and W.H. Munk (1949).

L55 Sullivan, Timothy J. "The Impact of the First Transatlantic Telegraph Cable Upon Oceanographic Progress." *XIIe Congres International d'Histoire des Sciences, Paris 1968* (pub. 1971), *Actes* 7: 87-91.

Problems involved in laying cables stimulated interest in oceanographic problems and led to innovations such as the Massey Indicator (developed by the American naval officer C. Massey) and the Deep Sea Sounding Apparatus (developed by another American naval officer, J.M. Brooke).

* Thams, J.C., ed. *The Development of Geodesy and Geophysics in Switzerland*. Cited above as item A103.

* Von Arx, William S. *An Introduction to Physical Oceanography*. See item L1.

L56 Warren, Bruce A., and Carl Wunsch, eds. *Evolution of Physical Oceanography. Scientific Surveys in Honor of Henry Stommel*. Cambridge, Mass.: MIT Press, 1981 [published 1980]. xxxiii + 623 pp.

Contents: essays on Henry Stommel (b. 1920) and list of his publications; review-historical articles on general ocean circulation by B.A. Warren, L.V. Worthington, J.L. Reid, N.P. Fofonoff, G. Veronis, A. Leetmaa, J.P. McCreary, Jr., D.W. Moore, R.C. Beardsley, and W.C. Boicourt, on physical processes in oceanography by J.S. Turner, W. Munk, M.C. Hendershott, C. Wunsch, J.H. Steele, and W.V.R. Malkus; on techniques of investigation by D.J. Baker, Jr., W.S. Broecker, A.J. Faller; on ocean and atmosphere interactions by H. Charnock, J.G. Charney, and G.R. Flierl. The references for all the articles are collected in a joint bibliography of about 1900 items.

L57 Wüst, Georg. "Repräsentative Tiefsee-Expeditionen und Forschungsschiffe 1873-1960." From "Denkschrift zur Lage der Meereskunde" of the *Deutschen Forschungsgemeinschaft* 1962. *Naturwissenschaftliche Rundschau* 16 (1963): 211-214.

Overview of oceanographic research from *Challenger* to IGY.

L58 Wust, G. "The Major Deep-Sea Expeditions and Research Vessels, 1873-1960." *Progress in Oceanography* 2 (1964): 3-52.

"From study of this history of oceanography ... we learn that its progress depends on:

1. The results of the great oceanographic expeditions, i.e. on the research work at sea, and the interpretation of data.
2. The improvement of instruments and methods on board ships and in the laboratories, and
3. The development of theory, particularly in dynamic oceanography."

(The article consists mostly of illustrations, tables, and bibliography.)

For additional information on this topic see works cited in the biography sections on the following scientists:

Albert I.
Buchanan, J.Y.
Charcot, J.-B.
Dalrymple, A.
Deryugin, K.M.
DesBarres, J.F.W.
Duperrey, L.
Eckart, C.H.
Ekman, V.W.
Ewing, W.M.
Heck, N.H.
Knudsen, M.H.C.
Lenz, E.K.
Makarov, S.O.
Maury, M.F.
Mohn, H.
Penck, A.
Pettersson, H.
Proudman, J.
Rennell, J.
Rouille, A.-L.
Shokalsky, Y.M.
Shtokman, V.B.
Thomson, Sir C.W.
Zubov, N.N.

LA. Tides

LA1 Aiton, E.J. "The Contributions of Newton, Bernoulli and Euler to the Theory of the Tides." *Annals of Science* 11 (1955): 206-223.

"Newton's great achievement was the qualitative verification which transformed [Kepler's idea that the moon is responsible for tides] into the principle of universal gravitation.... Newton's work can be divided into two distinct hypotheses ... (1) a kinetic theory in which the tides were explained by an analogy with the motion of satellites under the influence of a disturbing body, and (2) an equilibrium theory in which the ocean was assumed to take up the position of equilibrium under the attractive force of the sun and moon.... Daniel Bernoulli ... developed the equilibrium theory to a point where it could be used in the construction of tide tables.... Euler's most notable contribution, apart from his proof of the fundamental proposition of the equilibrium theory, was his recognition that it is the *horizontal* and not, as Newton had supposed, the *vertical* components of the disturbing forces which are effective in generating the tides."

LA2 Bencker, Henri Lucien Georges. *Les machines à prédire les marées*. Bureau Hydrographique International, Publication Spéciale, 13. Monaco, 1926. 110 pp., illus.

For English version see item LA8.

* Birett, H. "Zur Vorgeschichte der Newtonschen Theorie der Gezeiten." *Zur Geschichte der Geophysik* (item A13), pp. 15-28.

LA3 Brown, Harold I. "Galileo, the Elements, and the Tides." *Studies in History and Philosophy of Science* 7 (1976): 337-351.

"Galileo did hold a [pseudo-Aristotelian] theory of elements, ... this theory plays a central role in the *Dialogue*, and ... once this is recognized the apparent inconsistency between the theory of motion developed on the Second Day and the theory of tides and the remarks about the winds presented on the Fourth Day is dissolved ..." (p. 350). For Galileo, the element water is defined "by its ability to conserve motion while lacking any natural motion" (p. 351).

LA4 Burstyn, Harold L. "Theory and Practice in Man's Knowledge of the Tides." *Proceedings of the 10th International Congress of the History of Science, Ithaca, 1962* (pub. 1964) 2: 1019-1022.

"... the mariners of late medieval and early modern Europe knew that the moon and sun cause the tides, whatever those who were their intellectual betters might have thought." Following a flurry of activity by scientists in the 17th and 18th centuries, most of it of little practical value, Laplace gave the first successful treatment of the tides as a problem in hydrodynamics in 1778-9. But empirical methods ran ahead of theory until the late 19th century when William Thomson introduced modern methods for the calculation of tide tables.

* Cartwright, D.E. "The Historical Development of Tidal Science, and the Liverpool Tidal Institute." See item L52.

LA5 Drake, Stillman. "History of Science and Tide Theories." *Physics* 21 (1979, pub. 1980): 61-69.

"A neglected resemblance between Galileo's tide theory and post-Laplacian explanations is pointed out. Both are based on flows, not bulges, of water. That Kepler's theory resembled Newton's in assuming bulges lifted gravitationally is known, but historians are mistaken in supposing that modern tide theory accepts them. Ernst Mach appears to have been responsible for modern scorn of Galileo's reasoning about tides and their periods" (Author's summary).

LA6 Duhem, Pierre. *Le Système du monde. Histoire des doctrines cosmologiques de Platon à Copernic*. Tome II. Paris: Hermann, 1914. 522 pp.

See Chapter XIII for an account of theories of the tides in Greek antiquity.

* Hobbs, William. *The Earth Generated and Anatomized*. Cited above as item D30.

LA7 Houzeau, J.C., and A. Lancaster. *Bibliographie Générale de l'astronomie jusqu'en 1880*. New edition by D.W. Dewhirst. London: Holland Press, 1964. Volume 2, cols. 626-634, "Théorie des marées"; cols. 634-636, "La Marée et la rotation du globe."

LA8 International Hydrographic Bureau. "Tide Predicting Machines." Special Publication no. 13. Cannes: Robaudy, 1926. 110 pp., several plates.

A report compiled for the Bureau by Lieutenant Bencker, with illustrations and descriptions of machines developed by Sir William Thomson, Lord Kelvin (1872 and later versions), William Ferrel (1882), the U.S. Coast and Geodetic Survey (1896-1910), etc.

LA9 Panikkar, N.K., and T.M. Srinavasan. "The Concept of Tides in Ancient India." *Indiana Journal of History of Science* 6 (1971): 36-50.

"The cumulative evidence from the *Vedas*, *Upaniṣads*, *Saṃhitas*, epics, Purāṇas, and early literary works in Tamil and later works in Sanskrit show that the ancients had not only observed this physical phenomenon but also evolved a causal concept by linking it with the moon."

LA10 Pitt, Joseph C. "The Untrodden Road: Rationality and Galileo's Theory of the Tides." *Nature and System* 4 (1982): 87-99.

LA11 Proudman, Joseph. "Newton's Work on the Theory of Tides." *Isaac Newton, 1642-1727*. A Memorial Volume edited for the Mathematical Association by W.J. Greenstreet. London: Bell, 1927, pp. 87-95.

Presents Newton's theory and evaluates the accuracy of his results.

LA12 Shea, W.R.J. "Galileo's Claim to Fame: The Proof That the Earth Moves from the Evidence of the Tides." *British*

Journal for the History of Science 5 (1970): 111-127.

Author's conclusion: "There can be no doubt that Galileo's theory of the tides opened no new scientific vista to his successor. He neglected to take cognizance of the four well-known periods of the tides, he rode roughshod over the discrepancies between his theory and experience, he did not investigate striking observational consequences entailed by his explanations, and he brushed aside contemptuously any appeal to the influence of the moon.... The fact that Galileo stuck to his idea in the teeth of all these difficulties, ... might be taken as a testimonial to his faith in Copernicanism.... The new cosmology had to have terrestrially observable consequences, and ... he was determined to find them in the sea."

LA13 Thorade, Hermann. "Ebbe und Flut in der Nordsee: ein geschichtlicher Rückblick." *Petermanns Mitteilungen Ergänzungsheft* 209 (1930): 195-206.

LA14 Thorade, Hermann. *Probleme der Wasserwellen*. Hamburg: Grand, 1931. viii + 220 pp. + 11 plates.

Includes historical introduction (pp. 1-11) and a bibliography of more than 300 items, arranged chronologically (starting with Leonardo da Vinci and Newton).

For additional information on this topic see works cited in the biography sections on the following scientists:

Childrey, J.
Fichot, L.
Hassler, F.R.
Lamb, H.
Laplace, P.
Proudman, J.
Van der Stok, J.P.

M. METEOROLOGY: GENERAL HISTORIES

M1 Akademiia Nauk SSSR. "Fizika atmosfery i okeana v SSSR za 50 let" [Atmospheric and Oceanic Physics in the USSR in 50 years]. *Izvestiia Fizika Atmosfery i Okeana* 3 (1967): 1131-1136.

A brief review of the major Russian contributions to atmospheric physics and physical oceanography in the five decades since the October Revolution.

M2 Arakawa, Hidetoshi. *Chronology of World Meteorology*. Tokyo, 1957. 229 pp.

Lists important events in meteorology from 1442 to 1956.

M3 Aristotle. *Meteorologica*. Translated by Erwin Wentworth Webster. Oxford: Clarendon Press, 1923. 140 pp.

The classical treatise on meteorology from Greek antiquity

* [Aristotle]. Fobes, F.H. *Aristotelis Meteorologicum Libri Quattuor*. Cited below as item M15.

* [Aristotle]. Frisinger, H. Howard. "Aristotle's Legacy in Meteorology." Cited below as item M17.

* Bartels, Julius. "Mathematische Methoden der Geophysik." Cited below as item U3.

M4 Bettoni, Pio. "La meteorologia nella sua origine e nel suo sviluppo." *La Meteorologia Pratica* 6, no. 2 (1925): 50-57.

M5 Bjerknes, J. "Half a Century of Change in the 'Meteorological Scene.'" *Bulletin of the American Meteorological Society* 45 (1904): 312-315.

M6 Brunt, Sir David. "A Hundred Years of Meteorology: 1851-1951." *Advancement of Science* 30 (1951).

Traces meteorological work, mainly in Britain, since the founding of the Royal Society (1660). The work of Howard is singled out along with founding of the Meteorological Society (1850) and the establishment of a Weather Service (1855), later called Meteorological Office, which rendered great services in the two World Wars.

M7 Caskey, James E., Jr., ed. *A Century of Weather Progress*. Boston: American Meteorological Society, 1970. 170 pp.

A collection of papers commemorating the 100th anniversary of the U.S. Weather Bureau and the 50th anniversary of the American Meteorological Society, with reviews of progress in those intervals.

M8 [Co-ching Chu]. "Iz istorii Kitaiskoi meteorologii" [From the History of Chinese Meteorology]. *Voprosy Geografi* No. 35 (1954): 278-284.

A translation from the Chinese by IA.M. Berger tracing Chinese meteorology from early folklore and early inventions of instruments to the present system, which shows heavy European influence.

M9 Dettwiller, Jacques. *Chronologie de quelques événements météorologiques, en France et ailleurs*. Direction de la Météorologie, Boulogne-Billancourt, Monographies, nouvelle série, No. 1 (1982). 63 pp.

A chronology of meteorological contributions from antiquity to the present. Richly illustrated, it presents instruments, charts, portraits, book titles, and very brief comments on progress. Although the emphasis is on French affairs, especially in the current century, the most important foreign advances are included. An appendix gives major historic weather events that have influenced military campaigns throughout history.

* Dettwiller, J. "La révolution de 1789 et la météorologie." Cited above as item C5.

M10 Dorsch, Fr. "Kant und die Meteorologie." *Meteorologische Zeitschrift* 41 (1924): 280-282.

A review of Kant's views on meteorology on the occasion of the 200th anniversary of his birth. His speculations on anomalous weather of whole seasons, attributed to exhalations of vapors from the earth's interior caused by earthquakes, were erroneous. But his paper on the theory of winds (1756) correctly interpreted wind deflections as due to the rotation of the earth (earlier published by Hadley, 1735). Kant's philosophy advised adaptation to adverse weather.

M11 Dufour, L. "Les grandes époques de l'histoire de la météorologie." *Ciel et Terre* 59 (1943): 357.

M12 Dufour, Louis. "Quelques considérations sur le developpement de la météorologie." *Actes, Ve Congrès International d'Histoire des Sciences* (Lausanne, 1947), pp. 108-112. Paris: Hermann, 1948. Also in *Archives Internationales d'Histoire des Sciences.*

M13 Dufour, Louis. *Les écrivains français et la météorologie, de l'âge classique à nos jours.* Brussels: Institut Royal Météorologique de Belgique, 1966. 122 pp.

M14 Fassig, Oliver L., ed. *Bibliography of Meteorology.* Washington, D.C.: U.S. Signal Corps, vol. 1: 1889; vols. 2-4; 1891.

A classified list, under the headings temperature, moisture, winds, storms, comprising about 60,000 titles of meteorological literature since the beginning of printing to about 1887.

M15 Fobes, F.H. *Aristotelis Meteorologicum Libri Quattuor.* 1919. Reprint, Hildesheim: Georg Olms Verlagsbuchhandlung, 1967.

An edited version of Aristotle's writings on meteorology with a Latin introduction indicating the sources of manuscripts. Includes an extensive word analysis and an elaborate index of words used. The text material is in classical Greek.

M16 Frisinger, H. Howard. "Meteorology Before Aristotle." *Bulletin of the American Meteorological Society* 52 (1971): 1078-1080.

A brief note on pre-Aristotelian meteorology, principally in Greece, and the hypotheses of Thales, Empedocles, Democritus, and Eudoxus of Cnidos.

M17 Frisinger, H. Howard. "Aristotle's Legacy in Meteorology." *Bulletin of the American Meteorological Society* 54 (1973): 198-204.

Aristotle's "Meteorologica" dominated in atmospheric science for nearly 2000 years. This is an account of concepts from Greek antiquity to the early 17th century when Descartes (1596-1650) ended the era of speculation. A well-annotated paper.

M18 Frisinger, H. Howard. *Early History of Meteorology*. New York: Neale Watson Academic Publications, Inc., 1976.

See also item M19, which overlaps this work considerably.

M19 Frisinger, H. Howard. *The History of Meteorology: To 1800*. Historical Monograph Series, American Meteorological Society. New York: Science History Publications, 1977. 147 pp.

This is a very compact treatment of the history of meteorology. It can serve as a quick orientation for the interval from antiquity to the time when the field became a science. There are fairly good references for those who want to dig deeper.

M20 Gauthier, Henri. "Questions de géophysique contemporaine." *Revue des Questions Scientifiques* 86 (1924): 5-25, 345-368.

The position of problems of practical and theoretical meteorology during the last 20 years.

M21 Gilbert, Otto. *Die Meteorologischen Theorien des Griechischen Altertums*. 1907. Reprint, Hildesheim: Georg Olms Verlagsbuchhandlung, 1967. iv + 746 pp.

Elaborate survey of Greek concepts of the physical world and the influence of this thinking on meteorological phenomena, especially the causes of hydrometeors and winds, according to the various doctrines, especially those of Aristotle. The work contains extensive quotations in Greek and many annotations.

M22 Hanik, Jan. *Dzieje meteorilogii i obserwacji meteorologicznych w Galicji od XVIII do XX wieku*. Monografie z dziejow nauki i techniki, 75. Wroclaw: Ossolineum, 1972. 216 pp.

History of meteorology and meteorological observations in Galicia from the 18th to the 20th century. Includes 4-page English summary and bibliography of 588 items.

M23 Hellmann, Gustav. *Repertorium der deutschen Meteorologie*. Leipzig, 1883. 996 pp.

Lists German works and observations on meteorology and terrestrial magnetism from the earliest known to the end of 1881.

M24 Hellmann, Gustav. "Contribution to the Bibliography of Meteorology and Terrestrial Magnetism in the Fifteenth, Sixteenth and Seventeenth Centuries." *Report of the Chicago Meteorological Congress*, Part II, August 1893, pp. 352-394.

M25 Hellmann, Gustav. "The Dawn of Meteorology." *Quarterly Journal of the Royal Meteorological Society* 34 (1908): 223-227.

M26 Hellmann, G. "Die Meteorologie in den deutschen Flugschriften und Flugblättern des XVI. Jahrhunderts. Ein Beitrag zur Geschichte der Meteorologie." Abhandlungen der Preussische Akademie der Wissenschaften, 1921. Physikalisch-mathematische Klasse, Nr. 1, Berlin. 96 pp.

An account of pamphlets and broadsides published as reports of or reactions to unusual weather events. It includes a list of authors and events, starting with a lightning occurrence in Constantinople in 1490. Thunderstorms, cloudbursts, auroras, and halos are the most common topics. A reproduction of all known texts and the location of copies of the originals is given, ending in 1599.

M27 Hellmann, G., ed. *Neudrucke von Schriften und Karten über Meteorologie und Erdmagnetismus*. Berlin: A. Asher and Co., 1893-1904. Reprint ed., Nendeln, Liechtenstein: Kraus Reprint, 1969.

A series of facsimile reproductions of classical and seminal contributions, all with annotations and explanatory introductions by the editor. Those of greatest importance for meteorology are:

No. 3 (1894). Luke Howard, On the Modifications of Clouds, J. Taylor, London, 1803, 32 pp., 3 plates + 5 pp. notes.

No. 6 (1896). George Hadley, Concerning the Cause of the General Tradewinds, *Philos. Transact.* XXXIX (1735) 58-62 + 12 pp. notes.

No. 7 (1897). Evangelista Torricelli, Esperienza dell' Argento Vivo; together with: Accademia del Cimento, Instrumenti per conoscer l'Alterazioni dell' Aria; published in 1663 in Florence (Firenze), Insegna della Stella, 17 pp. + 18 pp. of notes.

No. 8 (1897). E. Halley, A. von Humboldt, E. Loomis,

U.J. Le Verrier, E. Renou, Meteorologische Karten, 1688, 1817, 1846, 1863, 1864 (five significant facsimile maps, showing the use of charts in meteorology with 9 pp. notes indicating the original sources).

No. 12 (1899). Wetterprognosen und Wetterberichte des XV and XVI Jahrhunderts. Reproductions of weather forecasts and diaries of the 15th and 16th centuries with 25 pp. of notes.

No. 14 (1902). Meteorologische Optik, 1000-1836. Reproductions of texts and figures with rainbows, rays, halos, including material from R. Descartes and I. Newton. 107 pp. + 8 pp. of notes.

No. 15 (1904). Denkmäler Mittelalterlicher Meteorologie. Reproductions and some translations of meteorological concepts held by philosophers from the 7th to the 14th centuries, including treatises by Albertus Magnus and Roger Bacon. 267 pp. + 40 pp. of notes.

M28 Henninger, S.K., Jr. *A Handbook of Renaissance Meteorology with Particular Reference to Elizabethan and Jacobean Literature*. 1960. Reprint, New York: Greenwood Press, 1968. 269 pp.

Contains only very limited references to science but is an extensive compendium of references to weather and weather elements in the general literature, including George Chapman and William Shakespeare, covering primarily the era from 1558 to 1625. Technically most valuable is the index of writers prior to that time with citations of many incunabula.

M29 Hornberger, Theodore, ed. "A Goodly Gallerye: William Fulke's Book of Meteors (1563)." *Memoirs of the American Philosophical Society Held at Philadelphia for Promoting Useful Knowledge*, Vol. 130. Philadelphia, 1979. 121 pp.

An annotated reprint of William Fulke's treatise on meteorological ideas of the 16th century, with a biographical note and a history of the various editions of the book. It deals with thunder, lightning, rainbows, halos, dew, hail, frost, clouds, and rain, mixed with some astrological overtones.

M30 Kassner, Carl. "Meteorologische Geschichtstabellen." *Linke's Meteorologisches Taschenbuch*, Vol. I. Edited by F. Baur. Leipzig: Akademische Verlagsgesellschaft Geest u. Portig, 1951, pp. 330-359.

Chronology of meteorological observations, important meteorological advances, forecasting, and lists of dates of German and international meteorological conferences.

M31 Keil, Karl. "Streiflichter auf die Entwicklung der Meteorologie zu Lebzeiten R. Sürings." *Zeitschrift für Meteorologie* 5 (1951): 133-135.

A brief history of meteorology between 1890 and 1950 with emphasis on the main interests of R. Süring (1866-1950); aerology, thermodynamics, radiation, clouds, and forecasting.

M32 Keil, Karl. "Die Meteorologie im Speculum Majus des Vincenz von Beauvais." *Berichte des Deutschen Wetterdienstes in der U.S. Zone* 38 (1952): 373-376.

The Dominican friar of Beauvais, who died about 1246, prepared a 3-volume encyclopedia, which was printed in 1476. Meteorological material is contained in a number of chapters in books 5 and 16 of this work, based on the views of classical middle ages authorities. All meteorological elements are covered and speculations on causes are given.

M33 Keil, Karl. "Meteorologische Literatur." *Meteorologische Rundschau* 8 (1955): 115-116.

A brief statistical survey of the development and current status of periodical publications on meteorology.

M34 Khrgian, A.Kh. *Ocherki razvitiia meteorologii*, Vol. I, 2nd ed. Edited by Kh. P. Pogosyan. Leningrad: Gidrometeorologicheskoe Izdatelstvo, 1959. 427 pp. English translation: *Meteorology: A Historical Survey*. Published for the Environmental Science Services Administration and the National Science Foundation, Washington, D.C., by the Israel Program for Scientific Translations, Jerusalem, 1970. 387 pp. Report TT69-55106, available from National Technical Information Service, Springfield, Va. 22151.

A fairly comprehensive history of meteorology based on published sources. Only a few pages are devoted to the prescientific era. Instrument developments are well covered. The evolution of synoptic meteorology is well presented. There are chapters on the general circulation and climatology. Some mathematical models are given. Russian contributions to the science are stressed but the presentation of other work is fair and well balanced. The bibliography is rather limited.

M35 Koerber, Hans-Guenther. "Meteorologische Anschauungen bei Immanuel Kant." *NTM, Schriftenreihe fuer Geschichte der Naturwissenschaften, Technik und Medizin* 14, no. 2 (1977): 29-36.

In one work (1756) Kant investigated the theory of winds; in another (1794) he discussed the influence of the moon on the weather. A more comprehensive presentation of his meteorological views is found in his *Vorlesungen über physische Geographie* (1803).

M36 Kopcewicz, Teodor. "O glownych Kierunkach rozwoju meteorologii światowej w ostatnim 50-leciw" [Main Trends in World Meteorological Developments in the Last 50 Years]. *Wiadomosci Sluzby Hydrologicznej i Meteorologicznej* 5(4) (1969): 3-19.

Reviews the development of the state of the art in recent decades, particularly noting use of radiosondes, newer theories, and weather modification.

M37 Lewis, C.L. "Maury--First Meteorologist." *The Southern Literary Messenger* 3, no. 10/11 (1941): 482-483.

An appraisal of the contributions of Matthew Fontaine Maury (1806-1873) to meteorology.

M38 McAllen, P.F. "Brief History of Weather Broadcasting." *Weather* 34 (1979): 436-441.

The BBC began weather reporting in 1922. Airway broadcasts in Britain began in 1946. Article covers only the United Kingdom.

M39 McIntyre, D.P., ed. *Meteorological Challenges: A History.* Ottawa: Information Canada, 1972. 338 pp.

A collection of 10 essays and 2 addresses prepared in honor of the centenary of the Canadian Meteorological Service (now the Atmospheric Environment Service). Contains items M41, OB14, PA21, PD1, QA4, R8, R10, R34, R35, V36.

M40 Mügge, Ratje, et al. *Meteorology and Physics of the Atmosphere.* FIAT Review of German Science, 1939-1946. [Wiesbaden:] Office of Military Government for Germany, Field Information Agencies Technical, British, French, U.S., 1948. 291 pp.

Text in German. Includes articles on climatology by K. Knoch et al.; radiation in the atmosphere by F. Möller and R. Meyer; thermodynamics and cloud physics by F.

Möller; dynamics of the atmosphere by H. Lettau; weather forecasting by R. Scherhag et al.; upper atmosphere by R. Penndorf and D. Stranz; instruments by H.G. Müller; and chemistry of the atmosphere by H. Cauer.

M41 Munn, R.E. "Applied Meteorology and Environmental Utilization." *Meteorological Challenges: A History* (item M39), pp. 283-310.

In the outgoing 19th century, meteorological applications to human health, agriculture, water supplies, air quality, and engineering began to develop. There was some progress in the 20th century, especially for the military aspects, but since World War II a veritable information explosion has taken place. A brief but useful reference section is given.

M42 Neis, Bernhard. *Fortschritte in der Meteorologischen Forschung seit 1900*. Frankfurt a.M.: Akademische Verlagsgesellschaft, 1956. xviii + 238 pp.

A book-length discussion of the developments in meteorology since the beginning of the 20th century, with emphasis on air mass, frontal, and three-dimensional analysis.

M43 Nemoto, Junkichi. "Some Characteristics of the History of Meteorology in Japan." *Japanese Studies in History of Science* 13 (1974): 1-8.

M44 Nezdiurov, D.F. "Meteorologicheskii muzei." *Meteorologiya i Gidrologiya* No. 7 (1961): 57-58.

Description of the meteorological museum at Voeikovo which is devoted to the history of meteorology with notable pictures and a fine collection of instruments used over the years. Also contains some archival material.

M45 Okaela, T. "Sekai Kishoga ku Nenpyo" [Chronology of Weather Science]. Tokyo: Chizin Shokan Co., 1956. 229 pp.

A chronological history of meteorology from 1442 to 1956 with many literature citations and portraits of famous meteorologists.

M46 Oliver, J. "William Borlase's Contribution to Eighteenth-Century Meteorology and Climatology." *Annals of Science* 25 (1969): 275-317.

Borlase gives "a valuable insight into the growing concern with recording techniques and instrumental exposure, and into the extent and nature of speculation on meteorological phenomena amongst his contemporaries."

M47 Paoloni, Barnardo. "I Benedettini e la meteorologia in Italia." *La Meteorologia Pratica* 14 (1933): 198-199.

Review of the contributions of Benedictine monks to Italian meteorology.

M48 Philipps, H. "Historische Betrachtungen zur Meteorologie." *Zeitschrift für Meteorologie* 3 (1949): 230-235.

Philosophical musings about the checkered history of meteorology and the problems of weather forecasting.

M49 Rigby, Malcolm. "Major Sources of Historical Meteorological Information." *Weatherwise* 29 (1976): 86-88.

Calls attention to the principal general histories of meteorology, including the works by Shaw, Khrgian, Schneider-Carius, Frisinger, and Arakawa.

M50 Robinson, A.H., and Helen M. Wallis. "Humboldt's Map of Isothermal Lines: A Milestone in Thematic Cartography." *Cartographic Journal* 4 (1967): 119-123.

In 1817 "Alexander von Humboldt read a long essay to the Paris Academy of Sciences on the subject of isothermal lines and the distribution of heat over the globe.... Although neither the cartographic concept of the isarithm as a graphic analytical and communicative tool nor the employment of the term *iso* was new in 1817, Humboldt's use of the isotherm on a map seems to have been the catalyst for many similar uses of the isarithm in the thematic cartography just beginning to develop."

M51 Rouch, J. "Coup d'oeil sur l'histoire de la météorologie." *Revue Scientifique Illustrée* 69 (1931): 530-533, 563-568.

* Runcorn, S.K., ed. *Earth Sciences*. Cited above as item A86.

M52 Saltzman, B. "Meteorology: A Brief History." *The Encyclopedia of Atmospheric Sciences and Astrogeology*. Edited by R.W. Fairbridge. New York: Reinhold, 1967, pp. 583-591.

A very succinct history is presented. It covers the period of scientific development from the 15th century onward to about 1960. The names of the principal contributors, with dates of publication, are given, but the references are restricted to collective works.

M53 Sansom, H.W. "Meteorology in the Bible." *Weather* 6 (1951): 51-54.

Cites weather events and weather proverbs alluded to in the Bible.

M54 Schlaak, Paul. "300 Jahre Wetterforschung in Berlin. Ihre Geschichte in Persönlichkeitsbildern." *Jahrbuch des Vereins für die Geschichte Berlins* 25. Folge (1976): 84-123.

Presents sketches of the personalities who carried on meteorological research in Berlin, Germany, during the past three centuries.

M55 Schmacke, Erik. "Från meteorologina gryning. Nagra drag i fortidens meteorologi" [From the Dawn of Meteorology. Some Ancient Ideas on Meteorology]. *Wotisec och Poeliminära Rapporter*, Serie Meteorologi No. 29, Meteorologiska och Hydrologiska Institut, Stockholm (1971). 54 pp.

A review of hypotheses of meteorology in classical antiquity with quotations from Greek and Roman writings.

M56 Schneider-Carius, K. *Wetterkunde Wetterforschung: Geschichte ihrer Probleme und Erkenntnisse in Dokumenten aus drei Jahrtausenden*. Freiburg: Verlag Karl Alber, 1955. 423 pp. English translation: *Weather Science, Weather Research*. Published for the National Oceanic and Atmospheric Administration and the National Science Foundation, Washington, D.C., by the Indian National Scientific Documentation Center, New Delhi, 1975. 554 pp.

This is one of the few comprehensive histories of meteorology. It is very brief for the pre-scientific era but then proceeds in chronological order from the invention of meteorological instruments to the 1930s. Sources and literature are cited but there are no references to any but printed material. The physical principles of weather formation are well discussed but no mathematical formulations are given. A biographical register of important meteorologists is appended. Has a good bibliography.

M57 Schneider-Carius, Karl. "Geschichtlicher Überblick über die Entwicklung der Meteorologie." *Meteorologisches Taschenbuch begründet von Franz Linke*. New edition. Edited by Franz Baur. Leipzig: Akademische Verlagsgesellschaft Geest u. Partig K.-G., 1962, pp. 662-708.

Chronological tables listing old meteorologic observations, development of physics and meteorology in antiquity and the Middle Ages, results in meteorology and physics from the 17th century to the beginning of World War II, major texts in meteorology, journals in meteorology, dates of important meteorological scientists, and major congresses.

M58 Schröder, W. "Anregung zum Quellenstudium der Entwicklung von Meteorologie und Geophysik." *Wetter und Leben* 26 (1974): 42-47.

Advocates an international program to sponsor research into the history of meteorology and geophysics and reviews the relatively meager studies available. Suggestions for source studies are made and some works in the field are cited.

M59 Shaw, William Napier. "The Meteorology of Yesterday, Today and Tomorrow." *Scientia* 51 (1932): 393-404.

Brief sketch of the coordination of observations, synoptic charts, pressure measurements, etc., since 1860.

M60 Shaw, William Napier. *The Drama of Weather*. New York: Macmillan, 1933. xiv + 269 pp.

Includes historical remarks but no references.

M61 Shaw, Sir William Napier, and Elaine Austin. *Manual of Meteorology*, Vol. I: *Meteorology in History*. Cambridge, Eng.: Cambridge University Press, 1926. 339 pp. Reprinted with corrections and addenda, 1932, 1942.

An interesting narrative on early meteorological concepts, scientific developments, and associated personalities, with many illustrations of historical significance. There are only very limited references.

M62 Smith, H.T. "Marine Meteorology: History and Progress." *Marine Observer* 2 (1925): 90-92, 173-175.

A brief review of advances in marine meteorology since 1853.

M63 Ströhm, Hans. "Untersuchungen zur Entwicklungsgeschichte der aristotelischen Meteorologie." *Philologus*, Supplement Band 28, Nr. 1 (1935): 1-28.

M64 Ströhm, Hans. "Zur Meteorologie des Theophrast." *Philologus* 92 (1937): 249-268, 401-428.

* Thams, J.C., ed. *The Development of Geodesy and Geophysics in Switzerland*. Cited above as item A103.

M65 Vidales, Carlos Zabaleta. "La mar en la historia de la meteorología." *Revista di Meteorología Maritima* No. 34 (1978): 8-12.

Traces the lessons learned about wind and weather by ancient mariners and the Vikings.

M66 Waterman, A.T. "Remarks Before the American Society, New York City: January 30, 1952." *Bulletin of the American Meteorological Society* 33 (1952): 183-187.

Includes a review of the scope of meteorological research in World War II.

M67 Weikinn, C. *Quellentexte zur Witterungsgeschichte Europas von der Zeitwende bis zum Jahre 1850*. Band I: Hydrographie. Berlin: Akademie-Verlag, 1961. 486 pp.

M68 Wild, Heinrich, ed. *Meteorologicheskii Sbornik: Repertorium fur Meteorologie*. 17 vols. St. Petersburg: Akademiia Nauk, 1870-94.

A collection of important contributions to meteorological research, much in Western languages. These volumes are a good source for gauging Russian contributions to the science in the late 19th century.

M69 Zinszer, Harvey A. "Meteorological Mileposts." *Scientific Monthly* 58 (1944): 261-264.

A rapid survey with no references.

For additional information on this topic see works cited in the biography sections on the following scientists:

Abbe, C.
Bartels, J.
Bjerknes, V.
Brooks, C.E.P.
Brooks, C.F.
Buchan, A.
Buys Ballot, C.H.D.
Chapman, S.
Cotte, L.
Dalton, J.

Dove, H.W.
Ferrel, W.
Ficker, H. von
Fitzroy, R.
Greenwood, I.
Hann, J.
Hellmann, G.
Keil, K.
Loidis, A.P.
Mahlmann, C.H.W.
McAdie, A.
Medici, L. de
Meinardus, W.
Mendel, G.
Milham, W.I.
Palmén, E.H.
Schmauss, A.
Schneider-Carius, K.
Shaw, W.N.
Süring, R.
Sutcliffe, R.C.
Sverdrup, H.U.
Walker, J.
Werner, J.
Wexler, H.

N. METEOROLOGY: INSTITUTIONS

NA. International

NA1 Ashford, O.M. "The First International Meteorological Conference, Brussels 1853." *Weather* 8 (1953): 153-154.

The conference of sea-faring nations, organized by M.F. Maury (1806-1873), dealt with the principles of marine weather observations.

See also: N.L. Canfield. *Mathew Fontaine Maury and the World Meteorological Organization*. Pilot Chart of the Indian Ocean, U.S. Hydrographic Office Publication No. 2603 (1952); backside: Schumacher, Arnold. "Matthew Fontaine Maury und die Brüsseler Konferenz 1853." *Deutsche Hydrographische Zeitschrift* 6 (1953): 87-93.

* Ashford, O.M. "The Past Presidents of the International Meteorological Organization and the World Meteorological Organization." Cited below as item NB3.

NA2 Cannegieter, Hendrik Gerritt. "The History of the International Meteorological Organization: 1872-1951." *Annalen der Meteorologie*, New Series, no. 1 (1963). 280 pp.

An illustrated history of informal international cooperation in meteorology in an organization composed of directors of meteorological services, later succeeded by a specialized agency of the United Nations based on a formal convention (World Meteorological Organization). It lists all meetings, establishment of technical commissions, and agreements; names of members of committees, commissions, and executive council are given. Literature concerning the organization, its publications, and its reports are cited.

NA3 Cehak, Konrad. "100 Jahre organisierte internationale Zusammenarbeit in der Meteorologie: Zum 100. Geburtstag der IMO-WMO." *Wetter und Leben* 25 (1973): 67-75.

Traces the early history of instrumental observations, the beginnings of synoptic meteorology, and the international cooperation in meteorology since the founding of the International Meteorological Organization in 1873.

See also: Heinz Panzram, "Von der lokalen zur globalen Wetterbeobachtung 100 Jahre internationale meteorologische Zusammenarbeit," *Naturwissenschaftliche Rundschau* 26 (1973): 391-394.

NA4 Cehak, Konrad. "100 Jahre Internationale Meteorologische Zusammenarbeit. Hundert Jahre Meteorologische Weltorganisation und die Entwicklung der Meteorologie in Österreich." *Zentralanstalt für Meteorologie und Geodynamik in Wien* 207 (1975): 17-24.

An account of 100 years of developments in cooperative world meteorology since the founding in 1873 of the International Meteorological Organization, which held its first meeting in Vienna in that year.

NA5 Daniel, Howard. *One Hundred Years of International Cooperation in Meteorology (1873-1973): A Historical Review.* World Meteorological Organization publication no. 345. Geneva: World Meteorological Organization, 1973. 53 pp.

A profusely illustrated pamphlet tracing the development of organized efforts by national weather services to develop a world-wide system of meteorological services and research, with milestones of progress during the past hundred years.

NA6 [Daniel, Howard]. "One Hundred Years of International Cooperation in Meteorology (1873-1973): A Historical Review." *World Meteorological Organization Bulletin* 22 (1973): 156-199.

Excerpts from the pamphlet with the same title, but somewhat differently illustrated, giving the highlights of meteorological developments from early beginnings to the centenary of the International Organization.

NA7 Harley, D.G. "British Architects of the International Meteorological System." *Meteorological Magazine* 103 (1973): 249-257.

Extols the contributions of British meteorologists to the International and later World Meteorological Organizations.

NA8 Harley, D.G. "IMO, WMO and 100 Years of Interdependence." *Weather* 28 (1973): 372-377.

Describes history of international cooperation in meteorology under the aegis of the International and later World Meteorological Organizations.

* Landsberg, H.E. "A Bicentenary of International Meteorological Observations." Cited below as item OC2.

NA9 Langlo, Kaare. "Impact of IMO and WMO on Meteorological Research." *World Meteorological Organization Bulletin* 22 (1973): 3-6.

Reviews the efforts of the World Meteorological Organization and its predecessor, the International Meteorological Organization, to promote atmospheric research.

* Lieurance, N.A. "The Development of Aeronautical Meteorology." Cited below as item OB9.

NA10 Maresco, Roberto. "Century of Meteorology in Argentina." *Bulletin of the World Meteorological Organization* 22 (1973): 95-99.

The weather service in Argentina was organized by Benjamin Apthorp Gould, who was director of the Astronomical Observatory in Córdoba, in 1870. An account is given of observational network and activities over the years.

NA11 Rigby, Malcolm. "The Evolution of International Cooperation in Meteorology (1654-1965)." *Bulletin of the American Meteorological Society* 46 (1965): 630-633.

Gives a brief history of international cooperative efforts in meteorology, including some notes on Antarctic exploration, the International Geophysical Year, and the World Weather Watch. No references are given.

NA12 Schlegel, Max. "100 Jahre Internationale Meteorologische Zusammenarbeit." *Beilage zur Berliner Wetterkarte Institut für Meteorologie*, Freie Universität Berlin, 47/73, SO 13/73 (1973). 4 pp.

A brief review of the history of international cooperation in meteorology, with a listing of the aims of the World Meteorological Organization, which by convention of 1951 functions as a specialized agency of the United Nations. Its efforts to establish a World Weather Watch and a Global Atmospheric Research Program in the centennial year are hopeful signs for continuing joint international undertakings.

NA13 Swoboda, G. "The First Four Years." *Bulletin of the World Meteorological Organization* 4 (1955): 46-51.

An account of the activities and accomplishments in the first four years of the World Meteorological Organization.

For additional information on this topic see works cited in the biography sections on the following scientists:

Childrey, J.
Fichot, L.
Hassler, F.R.
Lamb, H.
Laplace, P.
Proudman, J.
Van der Stok, J.P.

NB. National Weather Services; National Histories

NB1 Alter, J. Cecil. "National Weather Service Origins." *Bulletin of the Historical and Philosophical Society of Ohio* 7, no. 3 (1949).

NB2 [Argentina, Servicio Meteorológico Nacional]. "78° aniversario del Servicio Meteorológico Nacional." *Meteoros* 1 (1951): 110-114.

Brief history of meteorology in Argentina with special reference to the history of the National Meteorological Service since 1872.

NB3 Ashford, O.M. "The Past Presidents of the International Meteorological Organization and the World Meteorological Organization." *World Meteorological Organization Bulletin* 22 (1973): 85-93.

On the occasion of the centenary of international cooperation in meteorology the careers of the ten distinguished scientists who guided the international bodies devoted to this effort are reviewed.

NB4 Béll, Bela. "Die neunzig Jahre der Zentralanstalt für Meteorologie von Ungarn." *Idöjaras* 64 (1960): 257-270.

Although observations started in 1781 in Buda, a regular meteorological service was established only in

1870. In addition to routine tasks, its staffs contributed many scholarly works over the years.

NB5 Bench, A.T. "Reminiscences of the Meteorological Office, 1891-1910." *Meteorological Magazine* 110 (1981): 323-329.

A very personal memoir on personalities, surroundings, social relations, and procedures in the British central weather office at the turn of the century.

* [British Meteorological Office]. Brunt, Sir David. "A Hundred Years of Meteorology: 1851-1951." Cited above as item M6.

NB6 [British Meteorological Office]. "Meteorological Office Centenary, 1855-1955." *Meteorological Magazine* 84 (1955): 161-198.

An entire issue of the house organ of the British Meteorological Office is devoted to contributions by their staff from the beginnings under Admiral Robert Fitzroy (1805-1865) to the present.

NB7 Brodskiy, A.V. "[Fiftieth Anniversary of the Main Aviation Meteorological Center]." *Meteorologiya i Gidrologiya* no. 2 (February 1981): 121-124.

A commemorative article tracing the aircraft weather warning and meteorological support service given in the Soviet Union.

NB8 Brunt, Sir David. "The Centenary of the Meteorological Office: Retrospect and Prospect." *Science Progress* 44 (1956): 193-207.

Reviews scientific work at the Meteorological Office.

NB9 Bull, G.A. "Short History of the Meteorological Office." *Meteorological Magazine* 83 (1955): 163-167.

NB10 Czelnai, Rudolf. "A magyar meteorologiai szolgálat 100 eve" [100 Years of Hungarian Meteorological Service]. *Időjáras* 74 (1970): 12-22.

A brief history of the Hungarian meteorological office, with the principal developments and contributions listed. An extensive abstract in English is appended.

NB11 Dixon, F.E. "Meteorology in Ireland." *Weather* 5 (1950): 63-65.

An account giving the names of early Irish weather observers, with the first instrumental pressure observations dating back to 1676. Dublin temperatures were published as early as 1753. The most important contributions were made by Richard Kirwan (1733-1812), whose portrait is included. The Royal Irish Academy issued thermometers to 20 localities in 1787 and periodically published results. In 1824 an observatory was established out of which the Irish Meteorological Service developed.

NB12 Dufour, L. "Sketch History of Meteorology in Belgium." *Weather* 6 (1951): 359-364.

Belgians have made important contributions to meteorology. A. Quetelet (1796-1874) was the first to develop a climatological statistics. J. Jaumotte (1887-1940) designed equipment and explored cyclones three-dimensionally. Reproduces several portraits.

NB13 Dufour, L. "Ésquisse d'une histoire de la Météorologie en Belgique." *Miscellanées Institut Royal Météorologique de Belgique*, Fasc. 40 (1953). 55 pp.

Review of observations, meteorological organizations, and publications in Belgium from about 1600 onward but only sketchily outlined after 1918.

NB14 Estoque, Mariano A. "History of the National Weather Service and Present Meteorological Research in the Philippines." *International Union of Geodesy and Geophysics, Association of Meteorology, Assembly Brussels 1951, Rapports Nationaux* (1952): 94-96.

NB15 Federov, E.K., ed. *Meteorologii i gidrologii za 50 let Sovietskoi Vlasti i Sbornik Statei* [Meteorology and Hydrology During 50 Years of Soviet Rule: Collection of Papers]. Leningrad: Gidrometeoizdat, 1967. 267 pp.

A collection of papers dealing with progress in various phases of meteorology and hydrology by 20 authors, describing the contributions of Soviet scientists since the 1917 revolution.

NB16 Federov, E.K. "The Soviet Hydrometeorological Service on the Fiftieth Anniversary of the U.S.S.R." *Bulletin of the World Meteorological Organization* 17 (1968): 2-14.

Weather and climate have always played a major role in Russia with a fifth of its area beyond the arctic circle. Starting with M.V. Lomonosov (1711-1765) research in meteorology began in Russia and was gradually expanded into a meteorological service. Soviet rule brought a rapid expansion and far-flung practical applications.

NB17 Ficker, Heinrich von. *Die Zentralanstalt für Meteorologie und Geodynamik, 1851-1951*. Vienna: Österreichische Staatsdruckerei, 1951. 32 pp. Also in *Denkschriften, Österreichische Akademie der Wissenschaften, Mathematisch-Naturwissenschaftliche Klasse*, Band 109, 1. Abhandlung.

Gives account of founding and history of the Central Austrian Institute for Meteorology and Geodynamics, its contributions and directors (with portraits), and a list of early publications.

NB18 [France]. Météorologie Nationale. *Ce qu'est la météorologie française*. Paris: Météorologie Nationale, 1952. 30 pp.

Presents the development of the French meteorological service and lists many French "firsts" in the field.

NB19 Gentilli, J. "A History of Meteorological and Climatological Studies in Australia." *University Studies in History* (University of Western Australia Press, Nedlands) 5 (1967): 54-88.

NB20 Gibbs, W.J. "The Origins of Australian Meteorology: Historical Note." Department of Science, Bureau of Meteorology. Canberra: Australian Government Publishing Service, 1975. 32 pp.

After an introductory chapter reviewing the merits of Robert Fitzroy (1805-1865) and Matthew F. Maury (1806-1873) the author attributes the first Australian meteorological observations to Sir Thomas Brisbane (1773-1860), starting near Sidney in 1822. Other pioneers whose careers are reviewed are Charles Todd, Georg von Neumayer, Lewis J. Ellery, Henry C. Russel, and Clement Wragge. A chronological chart is added.

NB21 Gold, Ernest. "The Meteorological Office and the First World War." *Meteorological Magazine* 84 (1955): 173-178.

* Grisollet, H. "Histoire administrative et scientifique du service d'études...." Cited below as item ND7.

NB22 Gross, Walter E. "The American Philosophical Society and the Growth of Meteorology in the United States: 1835-1850." *Annals of Science* 29 (1972): 321-338.

NB23 Harrison, Louis P. *History of Weather Bureau Wind Measurements*. Key to Meteorological Records Documentation No. 3.151. Washington, D.C.: Weather Bureau, U.S. Department of Commerce, 1963. 68 pp.

Illustrated review of use of anemometers and wind vanes by the Weather Bureau, with pertinent instructions and tables. Contains historical notes on anemometer development and an extensive bibliography pertaining to these instruments.

* Havens, James M. *An Annotated Bibliography of Meteorological Observations in the United States, 1731-1818*. Cited below as item O5.

NB24 Havens, James M. "A Note on Early Meteorological Observations in the United States with Reference to the Germantown Temperature Record of 1731-32." *Bulletin of the American Meteorological Society* 39 (1958): 211-216.

Lists the earliest instrumental records made in America and gives a modern interpretation of the earliest known temperature records made from December 1731 to October 1732 in Germantown, Pennsylvania.

NB25 Hiedorn, Keith. "The Legacy of Admiral Beaufort." *Sea Frontiers* 24 (1978): 39-43.

Francis Beaufort (1774-1857) developed a scale for reporting wind force, adopted by the British Navy in 1838.

NB26 Hellmann, Gustav. *Geschichte des Königlich Preussischen Meteorologischen Instituts von seiner Gründung im Jahre 1847 bis zu seiner Reorganisation im Jahre 1885*. Ergebnisse der Meteorologischen Beobachtungen im Jahr 1885, Königlich Preussisches Meteorologisches Institut, Berlin, 1887.

A history of the first 38 years of the Prussian Meteorological Institute.

NB27 Hellmann, G. "Die Entwicklung der meteorologischen Beobachtungen in Deutschland von den ersten Anfängen bis zur Einrichtung staatlicher Beobachtungsnetze." *Abhandlungen der Preussische Akademie der Wissen-*

schaften, 1926. Physikalisch-mathematische Klasse, Nr. 1, Berlin. 25 pp.

An annotated source list of regular meteorological observations in Germany (borders as established after World War I), starting with pre-instrumental diary notations in the 15th century and instrumental data in the late 17th century to 1847 when the Royal Prussian Meteorological Institute was established.

NB28 Hidrometeoroloski Zavod Scojalisticke Republike Hrvaste. *Trideset godina rada i razvoja Hidrometeoroloskog Zavoda SR Hrvatske* [Thirty Years of Research and Development in the Croatian Hydrometeorological Service]. No. M6-6. Zagreb, 1979. 162 pp.

A review of meteorological work from 1947 to 1977 in one of the States of Yugoslavia with a bibliography of 500 publications and reports of the Service. A brief history of observations and work in the era of the Austrian-Hungarian monarchy and between the two world wars is included.

NB29 Hogan, Lawrence J. "The Anniversary of the Weather Service." *Congressional Record 116* (17), February 9, 1970, Extension of Remarks, E 816.

A note commemorating the 100th anniversary of signing of legislation by President Grant to establish the U.S. Weather Bureau. It includes part of the legislative history and an appraisal of weather service contribution to public welfare.

* Hughes, Patrick. "Early American Weathermen." Cited below as item 07.

NB30 Hughes, Patrick. *A Century of Weather Service: A History of the Birth and Growth of the National Weather Service, 1870-1970*. New York: Gordon and Breach, 1970. xii + 212 pp., illus., maps, ports.

A popular history of the U.S. Weather Bureau, established by Act of Congress in 1870, later incorporated as the National Weather Service into the Environmental Science Services Administration and then the National Oceanic and Atmospheric Administration. Includes a chronology and a large number of photographs, including a number of portraits of persons associated with the Service, but no documentation or annotations.

NB31 [Hughes, Patrick]. "The First Century." *ESSA, U.S. Department of Commerce, Environmental Science Services Administration* 5, no. 1 (1970): 4-7.

An excerpt from the author's book (item NB30).

NB32 Hughes, Patrick. "A Century of Cooperation." *Weekly Weather and Crop Bulletin*, Special Centennial Edition. U.S. Department of Commerce and U.S. Department of Agriculture, Washington, D.C., September 1972, pp. 2-4.

Traces the history of relations between weather service and agriculture in the United States, including the efforts of Joseph Henry (1797-1878) at the Smithsonian Institution, who published five papers on meteorology and agriculture. The *Weekly Weather and Crop Bulletin*, started in 1872 as *Weekly Weather Chronicle*, in 1887 continued as *Weather and Crop Bulletin*, and assumed its present title in 1924. Changes in contents took place gradually.

NB33 Hughes, Patrick. "Weather and Crop Service: A Century of Cooperation." Special issue of *Weekly Weather and Crop Bulletin* (1972): 8-11.

Agricultural meteorological services were started by the U.S. Weather Bureau shortly after its establishment. Close cooperation with the U.S. Department of Agriculture in the survey of crop progress and specialized forecasts were primary functions.

NB34 Hume, Edgar Erskine. "The Foundation of American Meteorology by the United States Army Medical Department." *Bulletin of the History of Medicine* 8 (1940): 202-238.

Traces the early development of meteorology in the U.S. Government in response to orders of the Surgeons General of the Army for post surgeons to make weather observations.

NB35 Johnson, Sir Nelson. "Milestones in a Century of Meteorology." *Weather* 5 (1950): 87-90.

A centenary reminiscence of the head of the Meteorological Office, whose portrait is reproduced, as is that of Sir Robert Watson-Watt (1892-), recalls the contributions of Robert FitzRoy (1805-1865), John Glaisher (1809-1903), the discovery of the stratosphere, the work on numerical prediction by L.F. Richardson (1881-1949), and the first radio-direction tracking balloons for upper wind soundings (1937). Urges research on the general global circulation as the next major task in meteorology.

NB36 Keil, Karl. "Ein Beitrag zur Geschichte der Meteorologie in Preussen. Reichsamt für Wetterdienst." *Wissenschaftliche Abhandlungen* 4 (1938). 17 pp.

The Prussian government began to establish meteorological stations in 1817. Its first guide for observers is reproduced. H.W. Brandes (1777-1834) arranged for observations at a number of locations.

* Khmorov, Sergei Petrovich. "Sto let nashey sluzhby pogody." Cited below as item S10.

NB37 Koelsch, William A. "Pioneer: The First American Doctorate in Meteorology." *Bulletin of the American Meteorological Society* 62 (1981): 362-367.

Reviews early educational efforts in meteorology at Johns Hopkins University under strong prodding from Cleveland Abbe (1838-1916). President Daniel C. Gilman and geology professor William B. Clark, in cooperation with the Weather Bureau, established the Maryland State Weather Service. Abbe served as lecturer at the university, and the Weather Bureau assigned a career meteorologist, Oliver L. Fassig (1860-1936), to its Baltimore office for the express purpose of taking graduate work at the university, culminating in 1899 in the award of the Ph.D. degree on the basis of a synoptic dissertation. Meteorology as a course of study at the university lapsed in 1901 because of financial pressure.

NB38 Končeк, Mikulas. "Ungarisch-slowakische Zusammenarbeit auf dem Gebiet der Meteorologie." *Idöjaras* 74 (1970): 103-111.

The history of meteorological activities in Slovakia and Hungary is briefly presented, with beginnings as early as the 15th and 16th centuries. Instrumental observations began in the late 18th century. Organization of Institutes, first under Austrian rule, is outlined and the investigations after independence are presented.

NB39 Kuhn, L. "Stulecie istnienia Krolewskiego Instytutu Meteorologicznego w Holland" [Centennial of the Netherlands Meteorological Institute]. *Przeglad Meteorologiczny i Hydrologiczny* 7 (1954): 79-80.

A brief account of the great contributions of the Dutch meteorologists to the science.

* Landsberg, H.E. "Early Weather Observations in America." Cited below as item OB7.

NB40 Lauscher, F. "Hundert Jahre öffentliches Beobachtungs-netz in Österreich." *Mitteilungen der Beographischen Gesellschaft, Wien* 91 (1949): 1-6.

Although instrumental weather observations in Austria started in 1734 in Vienna, government-supported observation started in 1848 under the direction of Kark Kreil (1798-1862) with 48 stations. Eventually observations became available from 617 localities. In 1948 there were 190 climatological stations.

NB41 Lauscher, Friedrich. "Hundertjahrfeier der Zentralanstalt für Meteorologie und Geodynamik in Wien." *Wetter und Leben* 3 (1951): 193-195.

Lists the names of the important meteorologists who worked at the Central Meteorological and Geodynamical Agency in Vienna during a century of its existence and the most important work performed there.

NB42 Lewis, R.P.W. "The Founding of the Meteorological Office, 1854-55." *The Meteorological Magazine* 110 (1981): 221-227.

Traces the beginning of the Meteorological Office in Britain to the 1853 Brussels Conference organized by Lt. M.F. Maury (USN). Quotations from early correspondence, a parliamentary inquiry, and reports of Robert FitzRoy are included.

NB43 Maenhout, A. "Ontwikkeling van het koninklikjk Meteorologisch Instituut van Belgie sinds 1944." *Hemel en Dampkring* 67 (1969): 97-100.

A history of the post-World War II meteorological work of the Belgian Meteorological Institute with emphasis on the observational and research work.

* Maresco, Roberto. "Century of Meteorology in Argentina." Cited above as item NA10.

NB44 Meigs, Josiah. "Circular to the Registers of the Land Offices of the United States." *Niles Weekly Register for 1817* (May 10, 1817): 167-168.

A circular letter by the Commissioner of the Land Office with instructions for meteorological observation, also transmitting appropriate forms.

NB45 *Meteorology and Hydrology in Czechoslovak Socialist Republic*. Prague: Hydrometeorologicky Ustav, 1978. 45 pp.

Illustrated pamphlet in English, French, and Russian traces the development of meteorology and hydrology in the area of present Czechoslovakia since Kepler (1604) and Comenius (1633). Some of the later principal scientists in the field are listed, and current activities are described. Some old documents are reproduced.

NB46 Miller, Eric R. "The Evolution of Meteorological Institutions in the United States." *Monthly Weather Review* 59 (1931): 1-6.

Gives a chronology of meteorological work in the United States from Benjamin Franklin (1706-1790) onward to 1919 when the American Meteorological Society was founded. This very important source document has an extensive bibliography.

NB47 Miller, Eric R. "New Light on the Beginnings of the Weather Bureau from Papers of Increase A. Lapham." *Monthly Weather Review* 59 (1931): 65-70.

Based on original papers at the Wisconsin Historical Society, this is a biographical note on Lapham, who began meteorological observations in Milwaukee in 1842. After an escapade in the gold rush to California, he returned to Milwaukee, worked with the Smithsonian weather network, and later cooperated with Cleveland Abbe (1838-1916) in the production of the earliest weather forecasts in the United States. His lobbying efforts with Congressman H.E. Paine, who had been a student of Elias Loomis (1811-1889) at Western Reserve College, led to legislation establishing a weather service in 1870.

NB48 Moshenichenko, I.E. *Ocherki Razvitiya Meteorologii na Dalnem Vostoke* [Progress of Meteorological Work in the Far East]. Leningrad: Gidrometeorologicheskoe Izdatelstvo, 1970.

A history of establishment of meteorological stations in the Far East of Russia and a review of scientific work in meteorology performed in that region. With a literature catalogue of over 400 pertinent publications.

NB49 Munzar, Jan. "Počatky meteorologických měřeni v Československu v 18. stoleti" [The Beginnings of Meteorological Measurements in Czechoslovakia in the 18th Century]. *Dějiny Věd a Techniky* 2 (1969): 183-187.

NB50 Nederlands Meteorologisch Instituut. *1854 Koninklijk Nederlands Meteorologisch Instituut 1954.* Gravenhage: Staatsdrukkerijen Uitgeverijbedrijf, 1954. 469 pp.

History of the Royal Netherland Meteorological Institute, profusely illustrated, including portraits and a complete list of publications.

NB51 Nesdyurov, D.G. *Ocherki Razvitiya meteorologicheskich Nablyudenii v Rossii* [Course of Development of Meteorological Observations in Russia]. Leningrad: Gidrometeoizdat, 1969. 221 pp.

A survey of early weather observations in Russia from 867 to the end of the 17th century is followed by an account of the first instrumental observations in 1725 to organization of a regular network in the middle of the 19th century. Illustrations of various generations of equipment, leading personalities, and observatories are included. The account closes with the establishment of the modern hydrometeorological service by V.I. Lenin in 1921.

NB52 [Netherlands, Royal Meteorological Institute]. "Centenary of Royal Meteorological Institute." *Bulletin of the World Meteorological Organization* 3 (1954): 70-71.

Recounts the founding of the famed Netherlands meteorological institution in 1854 under the directorship of C.H.D. Buys Ballot (1817-1891). A picture of the new institute at DeBilt is included.

NB53 Nikandrov, V.Ya. "Pervye shagi 'meteorologicheskogo dela' v SSSR" [First Steps in the "Meteorological Happenings" of the U.S.S.R.]. *Meteorologiya i Gidrologiya* No. 4 (1970): 10-19.

An account of the early decisions of the Soviets under the leadership of V.I. Lenin for the institution of a meteorological service and for the role of the main geophysical observatory.

NB54 Panzram, Heinz. "Einhundert Jahre im Dienste der maritimen Meteorologie und der Ozeanographie." *Naturwissenschaftliche Rundschau* 21 (1968): 341-343.

NB55 Parczewski, Wladyslaw. "Zarys historii meteorologii w Polsce (od 10 do 19 wisku)" [Contributions to the History of Meteorology in Poland (From the 10th to the 19th Century)]. *Przeglad Meteorologiczny i Hydrologiczny* 1 (1948): 66-72; 2 (1948): 62-77.

Summarizes reports on weather and climate in Poland from the 10th to 15th century. Instrumental records started in Warsaw in 1643 but were interrupted. They resumed in 1779 and are continuous since.

NB56 Penzar, Brauk, and Ivan Penzar. "Razvoj mreže meteoroloških stanica Hrvatskoj u 19. stoljecu" [Development of Meteorological Stations in Croatia in the 19th Century]. *Hidrometeorološki Zavod Sociazalisticke Republike Hrvatske* no. M6-4 (1978). 45 pp.

A history of number, equipment, procedures of meteorological stations in Croatia since 1829, with rapid growth between 1860 and 1900. Reference is made to early eye observations since 1587 and to some thermometer observations since 1641. Good, uninterrupted data are available from Zagreb since 1853.

* Philipps, H. "Rückschauende Betrachtungen zur Meteorologie. Cited below as item S22.

* Pinkett, Harold T., Helen T. Finneran, and Katherine H. Davidson. *Climatological and Hydrological Records of the Weather Bureau (Record Group 27)*. Cited below as item O13.

NB57 Popkin, Roy. *The Environmental Science Services Administration*. New York: Frederick A. Praeger, Publishers, 1967. 278 pp.

Contains, especially in Chapter III ("From Thomas Jefferson to Tiros: The Weather Bureau"), historical references on observations and antecedents of the weather service, with some references to congressional actions, operational procedures, and budgetary developments.

* Queney, Paul. "Rapport sur le dévéloppement de la météorologie dynamique aux Etats Unis pendant la période 1937-1947." Cited below as item R32.

NB58 Ramanathan, K.R. "Development of Atmospheric and Related Earth Sciences in the Last Sixty Years." *Proceedings of the Indian National Science Academy* vol. 45, Part

A, no. 1 (January 1979): 1-5.

Primarily an account of his own work in India.

NB59 Rojecki, Ananiasz. "O Traditsiiakh Pol'skoi Meteorlogii v XVI-XIX Vekakh" [Concerning the Tradition of Polish Meteorology from the 16th through the 19th Centuries]. *Actes, XIe Congrès International d'Histoire des Sciences* 4 (1965; pub. 1968): 299-302.

NB60 Schlaak, Paul. "Vor 130 Jahren wurde das Preussische Meteorologische Institut in Berlin gegrundet." *Beilage zur Berliner Wetterkarte des Instituts für Meteorologie der Freien Universität Berlin* 97/77, SO 24/77 (1977). 9 pp.

Created through the urging of Alexander v. Humboldt, the Prussian Meteorological Institute was in 1847 a relative late-comer in Europe. It had five eminent scientists as successive directors until its absorption into the German Reich's Weather Service in 1934. The impact of the directors' work on the science is characterized.

NB61 Schoner, Robert W., and Patrick M. Brady. "National Weather Service Marine Data Collection, Past--Present--Future." *Mariners Weather Log* 20 (1976): 267-270.

Gives a brief historical sketch of marine meteorology since 1697.

NB62 Scrase, F.J. "History of the Meteorological Research Committee." *Meteorological Magazine* 91 (1962): 310-314.

The Committee was formed by the British Meteorological Office to review and sponsor research in the field.

NB63 Seilkopf, Heinrich. "Zur Geschichte der meteorologischen Arbeit an der Deutschen Seewarte, Hamburg." *Annalen der Meteorologie* 3 (1950): 53-56.

The German marine meteorological institute, led by notable meteorologists including Georg von Neumayer (1826-1909) and Wladimir Koppën (1846-1940), carried on much maritime meteorological research, published in monographs and in its journal, *Annalen der Hydrographie*, during its existence from 1867 to 1945. The paper includes 73 references.

NB64 Seilkopf, Heinrich. "Die meteorologische Tätigkeit der deutschen Seewarte für Seeschiffahrt und Seefischerei 1875-1945." *Wetterlotse* 9 (1951): 1-4.

An account of activities of the German Marine Observatory with a reproduction of the first published German synoptic chart, prepared by W. Köppen (1846-1940).

NB65 Seydl, Otto. "K dwoustému výročí pruých měření meteorologických v Čechach" [Two Hundredth Anniversary of Meteorological Observations in Bohemia]. *Meteorologické Zprávy* 5 (1952): 141-145.

Regular meteorological observations in Prague were started by Josef Stepling (1716-1778) in 1752. The article places this in the context of European meteorological developments at that time.

* Simojoki, Heikki. *The History of Geophysics in Finland, 1828-1918*. Cited above as item A99.

NB66 Sinclair, Bruce. "Gustavus A. Hyde, Professor Espy's Volunteers and the Development of Systematic Weather Observations." *Bulletin of the American Meteorological Society* 46 (1965): 779-784.

James Espy (1785-1860) started a voluntary weather observer network in the United States in 1842. Gustavus A. Hyde of Framingham, Massachusetts, became interested in meteorology as a 17 year old. After settling in Cleveland, Ohio, he started regular observations there in 1859 and continued them for over 50 years. Contains much information on and many sources of 19th-century meteorological work in the United States.

NB67 Smith, [Margaret Chase]. "One Hundredth Anniversary of National Weather Services." *Congressional Record*, February 2, 1970, S 995-997.

Senator Smith placed an anonymous brief history of U.S. weather services, entitled "The First Century," into the *Congressional Record*.

NB68 Stagg, J.M. "The Meteorological Office and the Second World War." *Meteorological Magazine* 84 (1955): 178-183.

Brief history of support by Meteorological Office personnel for the war effort.

NB69 Steinhauser, Ferdinand. "Die Wissenschaftlichen Beiträge der Wiener Zentralanstalt für Meteorologie und Geodynamik zur Umweltskontrolle." *Hundert Jahre Meteorologische Weltorganisation und die Entwicklung der Meteorologie in Österreich.* Vienna: Zentralanstalt für Meteorologie und Geodynamik in Wien, Publication no. 207 (1975), pp. 31-41.

A review of Austrian contributions since 1927 in the field of environmental meteorology with emphasis on anthropogenic influences, with references.

NB70 Stewart, Ronald. "Adams on Espy." *Bulletin of the American Meteorological Society* 46 (1965): 812.

A letter to the editor quoting John Quincy Adams (1767-1848), then a member of the House of Representatives, on Espy's lobbying efforts on behalf of a weather service.

NB71 Sutton, Sir Oliver Graham. "The Meteorological Office, 1855-1955." *Nature* 175 (1955): 963-965.

Centenary account of activities.

NB72 Thomas, Morley K. "A Brief History of Meteorological Services in Canada: Part I: 1839-1930." *Atmosphere* 9 (1971): 3-15.

NB73 Thomas, Morley K. *Canadian Meteorological Milestones.* Downsview, Ontario: Atmospheric Environment Service, Department of the Environment, undated (ca. 1972). 13 pp.

A mixture of important events and dates in the development of meteorology in Canada and extreme weather events that affected that country.

NB74 Tolstobrov, B.Ya. "[Fiftieth Anniversary of the Leningrad Hydrometeorological Institute]." *Meteorologiya i Gidrologiya* no. 3 (March 1981): 126-127.

A short account of the educational effort of the Institute since its founding in Moscow with subsequent (1944) transfer to Leningrad and the commemorative assembly.

NB75 Tverkoi, Pavel Nikolaevich. *Razvitie Meteorologii v U.S.S.R.* [Development of Meteorology in the U.S.S.R.]. Leningrad: Gidrometeorologicheskoe Izd-Vo, 1949. 53 pp.

A brief history of Russian meteorology from the time of Peter the Great to 1949. The contributions of Mendeleev (1834-1907) and Voeikov (1842-1916) are stressed. The work of the Hydrometeorological Service of the Soviet Union and its branches is presented.

* [U.S. Weather Bureau]. Caskey, James E., Jr., ed. *A Century of Weather Progress*. Cited above as item M7.

NB76 [U.S. Weather Services]. "Some Meteorological Observations." *ESSA* (Environmental Science Services Administration) 5(1) (1970): 32-37.

Reproduction of archival photographs of the early days of the U.S. Weather Service, and old instruments and present-day equipment. Very little text.

NB77 [U.S. Weather Services]. "Weather Forecast Center USA--1888-1969." *ESSA* (Environmental Science Services Administration) 5(1) (1970): 8-16.

Description of a typical work day in the year 1969 at the National Meteorological Center where the U.S. daily weather charts and forecasts are made and disseminated. This is compared with a description of the activities of the predecessor prognosticators in 1888, taken from a detailed account by a foreign visitor.

NB78 [U.S.S.R. Meteorological Service]. "Leninskie dekrety ob organizatsii Sovetskoi meteorologicheskoi sluzhby" (Lenin Decrees on the Organization of the Soviet Meteorological Service]. *Meteorologiya i Gidrologiya* No. 4 (1970): 5-9.

A decree signed by V.I. Lenin, dated July 21, 1921, ordered the establishment of the Soviet meteorological service. Functional expansion and changes prior to World War II are noted in the form of subsequent ordinances.

NB79 van Eimern, Josef. "Zur Geschichte des Wetterdienstes in Bayern." In: "100 Jahre Wetterdienst in Bayern, 1878-1978." *Annalen der Meteorologie*, New Series 14 (1979): 7-17.

In October 1878, King Ludwig II of Bavaria approved the establishment of a state weather service with a central station in Munich and 34 observing posts, which were later incorporated into the German Federal Weather Service. Details of scientific, administrative, and fiscal developments are reported.

NB80 Watson-Watt, Sir Robert. "The Evolution of Meteorological Institutions in the United Kingdom." *Quarterly Journal of the Royal Meteorological Society* 76 (1950): 115-124.

This presidential address before the Royal Meteorological Society relates many of the milestones in the British Isles, including the notable role played by scientific societies and the governmental Meteorological Office. The contributions of amateur meteorologists and voluntary observers are stressed.

NB81 Weber, Gustavus A. *The Weather Bureau, Its History, Activities and Organization*. New York: D. Appleton & Co., 1922.

NB82 Weightman, R.H. *Establishment of a National Weather Service--Who Was Responsible for It?* Washington, D.C.: U.S. Department of Commerce, Weather Bureau, 1952 (mimeographed). 35 pp.

Reports on early, abortive attempts to establish a weather service in the United States up to the successful legislation of 1870, whose legislative history is presented. The act establishing the Weather Bureau was sponsored by Congressman H.E. Paine of Milwaukee. The important role of Cleveland Abbe and Increase A. Lapham is described. A full set of references and a detailed chronology is given. This is an extremely valuable source document.

NB83 White, Robert M. "A Century of Weather Science and Service." *Monthly Weather Review* 98 (1970): 91.

A brief note on the 100th anniversary of the establishment of the U.S. Weather Bureau and the start of the *Monthly Weather Review* as its house organ.

NB84 Whitnah, Donald R. *A History of the United States Weather Bureau*. Urbana: University of Illinois Press, 1965. 267 pp.

Well annotated and documented treatise, grown out of a dissertation. It includes the growing pains from a military to a civilian organization, internal and external difficulties of later years, the effects of wars and aviation on progress. An extensive bibliography of published and unpublished sources is given, including many references to Congressional hearings.

NB85 Willfarth, Josef. "Die Entwicklung der Meteorologie in Österreich." *Hundert Jahre Meteorologische Welt-organisation und die Entwicklung der Meteorologie in Österreich, Zentralanstalt für Meteorologie und Geodynamik in Wien* 207 (1975): 1-7.

Traces the development of Austrian meteorology from observations in the early 16th century to the establishment of a meteorological network in the Austrian-Hungarian monarchy in 1848 and the formation of a central institution in 1851 in Vienna. Since then many major advances in meteorology have emanated from Vienna, where a galaxy of important meteorologists worked (among them: A. Defant, F. Exner, H. v. Ficker, K. Jelinek, M. Margules, V. Peruter, W. Schmidt).

* Woeikov (Voeikov), A. "Meteorology in Russia." Cited below as item T31.

NB86 World Meteorological Organization. "Beaufort Scale of Wind Force (Technical and Operational Aspects)." *Reports on Marine Science Affaris* No. 3. Geneva: World Meteorological Organization, 1970. 22 pp.

Gives in its introduction a history of wind scales by visual observations, the earliest of which was proposed by Noppen in the Netherlands in 1733. The standard became Admiral Sir Francis Beaufort's (1774-1857) scale of 1805.

For additional information on this topic see works cited in the biography sections on the following scientists:

Fassig, O.L.
Maury, M.F.
Mieghem, J. van
Sutcliffe, R.C.
Swoboda, G.J.H.

NC. Universities

NC1 Bider, Max. "Geschichte der meteorologischen und Klimatologischen Forschung im Baselgebiet (1900-1949)." *Tätigkeitsberichte der Naturforschenden Gesellschaft in Basel* 18 (1950): 56-63.

The first meteorological observations in the Basel region were made in 1879. A number of stations have been established since. A number of important analyses have been performed, especially clarifying soil temperatures under snow covers. Since 1928 observations are made at the University of Basel.

NC2 Hallgren, Elizabeth Lynn. *The University Corporation for Atmospheric Research and the National Center for Atmospheric Research, 1960-1970: An Institutional History*. Boulder, Colo., 1974. 146 pp.

A detailed account of the formation of a corporation by American universities engaged in atmospheric research and the establishment with National Science Foundation funding of a large laboratory complex in Boulder for atmospheric studies, with organizational details and principal personalities involved. Also contains the assumption of responsibility for the High Altitude Solar Observatory and a balloon facility at Palestine, Texas.

NC3 Hare, F. Kenneth. "Vaulting of Intellectual Barriers: The Madison Thrust in Climatology." *Bulletin of the American Meteorological Society* 60 (1979): 1171-1174.

* Koelsch, William A. "Pioneer: The First American Doctorate in Meteorology." Cited above as item NB37.

NC4 Kutzbach, Gisela. "One Hundred Twenty-Five Years of Meteorology at the University of Wisconsin." *Bulletin of the American Meteorological Society* 60 (1979): 1166-1171.

Professor J.W. Sterling started a meteorological record at Madison in 1853. From that time until 1947 there was some meteorological activity at the University. In 1948 the Department of Meteorology was founded, eventually leading to a large and productive research program with an extensive staff and excellent facilities. See also the article by F.K. Hare (item NC3).

NC5 Möller, Fritz, and Hermann Wachter. "50 Jahre meteorologische Tätigkeit des Physikalischen Vereins und der Universität Frankfurt a.M." *Berichte des Instituts für Meteorologie und Geophysik der Universität Frankfurt a.M.* No. 6 (1956): 3-10.

In 1906 a meteorological division was established by the private Physical Association which in 1914 was incorporated into the University of Frankfurt. Pioneering research on aviation and industrial meteorology was undertaken in the Institute which gained world-wide reputation under the leadership of F. Linke (1878-1944).

For additional information on this topic see works cited in the biography sections on the following scientists:

Rossby, C.-G. Scherhag, R.

ND. Observatories

ND1 Attmannspaeher, W. "Hohenpeissenberg--200 Years of Meteorological Observations." *World Meteorological Organization Bulletin* 30 (1981): 104-106.

Gives a brief account of the history of the uninterrupted observations since 1781 when the station started as part of the network of the Societas Meteorologica Palatina.

ND2 Bělohlávek, Vlastimil. "200 Jahre Meteorologisches Observatorium in Prag-Klementinum." *Wetter und Leben* 28 (1976): 117-121.

A brief history of two centuries of meteorological observations at Prague, Czechoslovakia, at the Klementinum, one of the oldest observatories in Central Europe, originally started by the Jesuits in 1752. The early observers are listed.

ND3 "Blue Hill Observatory Fifty Years Old." *Harvard Alumni Bulletin* 37 (1935): 584-594.

Reminiscences of collaborators of the first American meteorological research observatory at its 50th anniversary celebration.

ND4 Borisenkov, E.P. *A.I. Voeikov Main Geophysical Observatory*. 2d ed. Leningrad: Gidrometeorizdat, 1977, 34 pp.

A brief historical essay on the principal Russian meteorological and geophysical research establishment. It presents some notes on the earliest weather observations in Russia and the establishment of the Observatory in 1848; also contains a complete statement on personalities and activities, with some photographs. The Observatory was incorporated in 1929 into the U.S.S.R. Hydrometeorological Service. The later and current activities and leading scientists are described. No references are provided.

ND5 Budyko, M.I. *Glavnaya Geofizicheskaya Observatorii imeni A.I. Voeikova* [The Main Geophysical Observatory Named for A.I. Voeikov]. Leningrad: Gidrometeorizdat, 1969. 24 pp.

A history of the central meteorological research institute of Russia, established in 1849. It includes a list of the principal investigators, the most important contributions made, and the organization after the Soviet revolution.

* Ferreira, H. Amorim. "Observações meteorológicas em Portugal...." Cited below as item OD3.

ND6 Glebov, P.A. "Istoricheskii ocherk" (Historical Note]. *Sverdlovskaya Magnitnaya i Meteorologicheskaya Observatoriya, 1836-1936, Yubileiny Sbornik* 7-94 (1936).

History of the meteorological and magnetic observatory of Sverdlovsk, formerly Ekaterinburg, a key station in the Russian network.

* Grasnick, K.H. "Wärmehaushalts--Turbulenz--und Ozonforschung als Aufgabengebiete des Potsdamer Meteorologischen Observatoriums." Cited below as item R14.

ND7 Grisollet, H. "Histoire administrative et scientifique du service d'études et de statistique climatique de la ville de Paris, et des observatoires de Montsouris et de la Tour Saint-Jacques." *La Météorologie* 4, no. 19 (1950): 129-142.

A history of the meteorological observing points in and near Paris, where observation began in the 17th and 18th centuries, and the personalities involved.

* Gunther, R.T. *Early Science in Oxford*. Cited above as item A47.

ND8 Hagarty, J.H. *History of Weather Bureau Barometric Pressure Measurements*. Key to Meteorological Records Documentation No. 3.021. Washington, D.C.: Weather Bureau, U.S. Department of Commerce, 1964. 16 pp.

Illustrated documentation of barometers used by the Weather Bureau, systems of reduction to sea level, and pertinent instructions for corrections.

ND9 Hallman, E.S. "110 Years Ago." *Weather* 5 (1950): 155-158.

Gives the history of the founding of the Toronto Meteorological Observatory and its eventual conversion into the Canadian Weather Service.

ND10 *Kodaikanal Observatory, 1901-1905*. New Delhi, India: Meteorology Department, Government Publisher, 1951. 44 pp.

Reviews history of Madras and Kodaikanal observatories and gives a summary of observations at Kodaikanal and a list of publications.

ND11 Lukesch, J. "Die Geschichte des meteorologischen Observatoriums auf dem Hochobir, 2041 m." *Jahresbericht des Sonnblick-Vereins, Wien* 48 (1950): 25-30.

Presents a brief history of the mountain observatory on Hochobir, with a summary of data collected since 1848. The station was destroyed in World War II.

ND12 *Magnitnaya i Meteorologicheskaya Observatoriya Pavlovsk--Voeikovo* [Magnetic and Meteorological Observatories at Pavlovsk and Voeikovo]. Leningrad: Gidrometeoizdat, 1978. 22 pp.

An illustrated pamphlet issued in celebration of the centennial of the observatory at Pavlovsk and the Experiment Station named for the famous Russian meteorologist and climatologist Alexander Voeikov (1842-1916). Lists the leading personalities and activities, past and present without references.

ND13 "Old Wooden Observatory on Greylock." *The Berkshire Hills*, July 1903, p. 131.

Brief story of the establishment of the first American mountain observatory on Mt. Greylock, Massachusetts, by Professor James H. Coffin of Williams College, for "meteorological experiments."

A follow-up story (ibid., July 1906, p. 157), "The First Observatory on Greylock," shows a picture of the structure and gives specifications as contained in the building contract.

ND14 Paech, H., J. Werner, F. Becker, and K. Bringmann. *Meteorologisches Observatorium Aachen, 1900 bis 1950*. Aachen: Arend & Ortmann, 1950. 30 pp.

History of meteorological observations at Aachen since 1820 and the scientific work of the observatory over 50 years, especially in radiation and visibility.

ND15 Pejml, Karel. *200 Let Meteorologicke Observatoře v Pražském Klementinu* [200 Years of Meteorological Observations at the Prague Klementinum]. Prague: Hydrometeorologicky Ustav, 1975. 78 pp.

An illustrated history of meteorological work started by the Jesuits at the Klementinum in Prague, later continued by government employees first under the Austrian-Hungarian monarchy and then under the Czech Republic. (In Czech with Russian and German summaries.)

ND16 Petrovic, Stefan. "Štâtne meteorologické a geofyzikálne observatorium v Hurbanova (Stara' Ďala) päidesiatročné" [The National Meteorological and Geophysical Observatory at Hurbanova (Stara Dala) for 50 Years]. *Meteorologické Zprávy* 4 (1950): 100-103.

A commemoration of meteorological observations in Slovakia, where the Hurbanova is the only station with an uninterrupted 50-year record.

ND17 Rossmann, Fritz. "Zur Geschichte der Bergwetterwarten." *Meteorologische Rundschau* 5 (1952): 63-65.

Reviews the contribution of mountain observatories to the exploration of the high atmospheric layers, especially since 1865.

ND18 Rykachev, Mikhail. *Histoire de l'Observatoire Physique Central pour les premières 50 années de son existence, 1848-1899*, Part I. St. Petersburg: Imp. de l'Académie impériale des sciences, 1900. 290 pp. + 87 pp. Appendices.

A detailed presentation of the founding and operation of the Central Physical Observatory (later Main Geophysical Observatory), which was charged with the responsibility of operating a meteorological and magnetic observational network in the Russian Empire. Two decades of difficulties in getting started are described. Biographical data for the three directors prior to the author are given and lists of operating stations and information on equipment and scientific programs are given.

ND19 Skeib, G. "75 Jahre Meteorologisches Observatorium Potsdam (Festvortrag des Direktors)." *75 Jahre Meteorologisches Observatorium Potsdam, 1892-1967, Meteorologischer Dienst* (der Deutschen Demokratischen Republik) [n.d., 1969?]: 7-12.

Address in celebration of the 75th anniversary of the Potsdam Observatory describing the state of meteorological science at the end of the 19th century and highlighting the important investigations made there since.

ND20 Stone, Robert G. "The History of Mountain Meteorology in the United States and the Mount Washington Observatory." *Transactions of the American Geophysical Union* (1934): 124-133.

Starting with the first observations taken on Mount Washington in New Hampshire for the Smithsonian Institution, reviews the U.S. efforts to occupy peaks for meteorological observations, including Pike's Peak by the Signal Corps in 1870. Has a good bibliography with some annotations.

ND21 Szukalscy, Malgorzata, and Jerzy Szukalscy. "O pierwszych stalych obserwacjach meteorologicznych w Gdańsku w XVIII wieku" [First Permanent Meteorological Observations in Gdansk in the 18th Century]. *Przeglad Geofizycny, Warsaw* 16 (1971): 243-252.

Observations in Gdansk (Danzig) are among the earliest made in the territory of present-day Poland. G. Reiher started them in 1730 and continued them for 26 years. They were published in the second volume of the Gdansk Society of Natural Sciences with an introduction justifying the need for systematic meteorological observations and some theoretical considerations.

ND22 Veselý, Emil. "Památka na Gregora Mendela y archive Hydrometeorologického ústavu" [Records of Gregor Mendel in the Archive of the Hydrometeorological Institute]. *Meteorologické Zprávy* 18 (1965): 28-29.

Mendel (1822-1884), better known for his genetic experiments, was an ardent meteorological observer in Brno (Brünn). His data sheets, charts, and notes are in the archive of the Hydrometeorological Institute in Prague.

See also: Milos Nosek. "Meteorologische Tatigkeit von Johann Gregor Mendel." *Wetter und Leben* 17 (1965): 133-140.

ND23 von Muralt, Alexander L. *Fünfzehn Jahre Hochalpine Forschungsstation Jungfraujoch*. Bern: Bern University, 1946. 45 pp.

History of scientific activities on the Jungfrajoch, which include ultra-violet radiation, atmospheric ozone, snow, and glacier observations. A list of 238 publications is included.

For additional information on this topic see works cited in the biography sections on the following scientists:

Humphreys, W.J.
Neumayer, G.B.
Swinden, J.H. Van

NE. Societies

* [American Meteorological Society]. Caskey, James E., Jr., ed. *A Century of Weather Progress*. Cited above as item M7.

NE1 Cappel, Albert. "Societas Meteorologica Palatina (1780-1795)." *Annalen der Meteorologie*, New Series, no. 16 (1980): 10-27; with Annex pp. 255-261.

Commemorates founding of the meteorological class in the Palatinate Academy at Mannheim and relates the efforts of its permanent secretary, Johann Jacob Hemmer. Recalls Hemmer's endeavor to introduce lightning rods of his design in the Palatinate. The annex consists of facsimiles and transcriptions of the founding rescript by the Prince-Elector of the Palatinate, Karl Thevdor, and Hemmer's instructions for additional demographic and phenomenological information to be added to the meteorological observations.

NE2 Cappel, Albert. "Die Pfälzische Meteorologische Gesellschaft (1780-1795)." *Beilage zur Berliner Wetterkarte, Institut für Meteorologie Freie Universität Berlin* 101/80, SO 22/80 (1980). 5 pp.

Lists equipment given to observers of the Palatinate Meteorological Society, the locations of the observations, and the instructions for phenological observations. The merits of Hemmer for introduction of lightning rods are noted. The first scientific balloon flights from Mannheim, one of them a failure, are mentioned.

NE3 Cockrell, P.J. "The Meteorological Society of London: 1823-1873." *Weather* 23 (1968): 357-361.

This Society was founded in 1823 and had a number of distinguished members, including Luke Howard (1772-1864) and Sir John Herschel (1792-1871). Some facsimile reproductions of minutes of the Society from the *Philosophical Magazine and Journal* are shown. Notes on the publications of the Society are given.

NE4 Corless, R. "A Brief History of the Royal Meteorological Society." *Weather* 5 (1950): 78-83.

The British Meteorological Society was established in 1850 with John Glaisher (1809-1903), the famous balloon explorer of the atmosphere, as secretary. In 1883 the society received a Royal charter from Queen Victoria and henceforth became the Royal Meteorological Society. It published many notable papers, and sponsored investigations on thunderstorms, wind pressure, and free atmosphere exploration by kites and balloons. Some of the medalists and the notable presidents of the society are listed.

NE5 [France]. "La Société Météorologique de France 1852-1952." *La Météorologie* No. 27 (July/September 1952): 107-116.

A brief review of the history of the French Meteorological Society.

NE6 [Japan, Meteorological Society]. *Short History of the Meteorological Society of Japan in Commemoration of the 75th Anniversary*. Tokyo, 1957. 68 pp. (In Japanese.)

Founded in 1882, the Society established seven meteorological observatories and began publication of a journal. Its observing activities were expanded in 1920.

NE7 Kington, J.A. "The Societas Meteorologica Palatina: An Eighteenth-Century Meteorological Society." *Weather* 29 (1974): 416-426.

Shows meteorological equipment, abbreviations and symbols used in records, and gives list of stations in the network of Palatinate Meteorological Society founded in 1780.

* Landsberg, H.E. "A Bicentenary of International Meteorological Observations." Cited below as item OC2.

NE8 Lauscher, Friedrich. "Hundert Jahre Österreichische Gesellschaft für Meteorologie." *Wetter und Leben* 17 (1965): 131.

A note on the centenary of the Austrian Meteorological Society.

NE9 Loewe, Fritz. "The First Meteorological Society of the Southern Hemisphere." *Australian Meteorological Magazine* 30 (1960): 52-53.

NE10 Martin, D.C. "The Gassiot Committee of the Royal Society and Meteorological Research." *Nature* 190 (1961): 212-213.

The Committee, founded in 1871, has supported meteorological observations and research, using its funds to initiate important studies.

NE11 Max-Planck-Institut für Aeronomie Katlenburg-Lindau. *Max-Planck-Gesellschaft Berichte und Mitteilungen, Munchen, 4/81.* 107 pp.

Includes "Geschichte, Aufbau und Arbeitsgebiete des Instituts," pp. 9-21.

NE12 Paulus, R. "200 Jahre Pfälzische Meteorologische Gesellschaft." *Deutsche Meteorologische Gesellschaft e. V., Mitteilungen*, December 1979, pp. 57-62.

Comments on the founding and the network of the Palatinate Meteorological Society in 1780.

NE13 Perta, G. "Benemerenze meteorologiche della 'Société Royale de Medecin de Paris' nel corso del XVIII secolo." *Revista di Meteorologica Aeronautica* 33 (1973): 268-271.

Calls attention to the meteorological efforts of the French Royal Society of Medicine by establishing an observing network in 1772 for the purpose of using climatic information for therapeutic purposes.

NE14 Rigby, Malcolm. "Evolution of International Cooperation in Meteorology, 1654-1965." *Bulletin of the American Meteorological Society* 46 (1965): 630-633.

Traces cooperative efforts in meteorology from those among individual scientists of various nations to the highly organized collaboration of governmental weather services, as well as international research efforts such as the Polar Years and the International Geophysical Year.

NE15 Rigby, Malcolm. "Ephemerides of the Meteorological Society of the Palatinate." *Environmental Data Service*. Washington, D.C.: U.S. Department of Commerce, National Oceanic and Atmospheric Administration, February 1973, pp. 10-16.

A review of the publications of the Palatinate Meteorological Society (1783-1795), with pictures of the instruments used.

NE16 Russ, A. "Der Einfluss der Societas Meteorologica Palatina auf unseren heutigen Beobachtungsdienst." *Berichte des deutschen Wetterdienstes in der U.S. Zone* 38 (1952): 376-380.

Gives a brief history of the weather station network established by the Palatine Academy under the guidance of J.J. Hemmer (1733-1790), their instruments and instructions, and their long-lasting influence on meteorological developments.

NE17 Seydl, Otto. "Mannheimská společnost meteorologická, 1780-1799." *Meteorologicke Zprávy* 7 (1954): 4-11.

An account of the activities of Societas Meteorologica Palatina.

NE18 Steinmayr, J. "Das Beobachtungsnetz der Mannheimer Meteorologischen Gesellschaft und die ältesten meteorologischen Beobachtungen in Wien." *Meteorologische Zeitschrift* 52 (1935): 229-231.

Reviews the observations of the Palatinate Meteorological Society in the 18th century and the start of observations in Vienna (1763).

NE19 Symons, George James. "The History of English Meteorological Societies, 1823 to 1880." *Quarterly Journal of the Royal Meteorological Society* 7 (1881): 66-68.

NE20 Traumüller, V. *Die Mannheimer Meteorologische Gesellschaft (1780-1795): Ein Beitrag zur Geschichte der Meteorologie*. Leipzig, 1885. 48 pp.

A brief history of the Palatine meteorological society with illustrations of the instruments distributed by that organization.

NE21 Wedderburn, Ernest. "The Scottish Meteorological Society." *Quarterly Journal of the Royal Meteorological Society* 74 (1948): 232-242.

The Scottish Meteorological Society was founded in 1855 and merged with the Royal Meteorological Society in 1921. It sponsored a station network and in particular fostered the mountain observatory on Ben Nevis. For the period from 1860 to 1907 Alexander Buchan (1829-1907) was the driving force behind the society's activities, which included important contributions for fisheries and public health.

O. METEOROLOGY: OBSERVATIONS

O1 Aldredge, Robert Croom. "Weather Observers and Observations at Charleston, South Carolina, 1670-1871." *Historical Appendix of the Year Book of the City of Charleston for the Year 1940*, pp. 190-257.

Based on original sources, this article indicates the first casual observations and later John Lining's instrumental record, starting in 1738 (probably the third in the British North American colonies). Principal interest was shown by the medical profession because of the prevalent hypothesis linking weather and yellow fever epidemics.

O2 Beaude, Joseph. "Lettre inédite de Picot à Carcavi relative à l'expérience barométrique 5 août 1649." *Revue d'Histoire des Sciences et de Leurs Applications* 24 (1971): 233-246.

O3 Brock, William H. "William Prout and Barometry." *Notes and Records of the Royal Society of London* 24 (1969): 281-294.

O4 Dufour, L. "Voltaire et les thermomètres." *La Météorologie* 5, no. 4 (1967): 551-555.

Cites an abortive attempt of Voltaire (1694-1778) to engage in scientific experiments with thermometers and barometers in 1737 and 1738, as revealed in his letters to an Abbé Moussinot. In them he comments on the relative merits of the Reaumur and Fahrenheit thermometers.

* Fassig, Oliver L., ed. *Bibliography of Meteorology*. Cited above as item M14.

* Hanik, Jan. *Dzieje meteorilogii i obserwacji meteorologicznych w Galicji od XVIII do XX wieku*. Cited above as item M22.

O5 Havens, James M. *An Annotated Bibliography of Meteorological Observations in the United States, 1731-1818*. Tallahassee: Department of Meteorology, Florida State University Technical Report No. 5 [Contract no. Nonr-1600(00)], 1956. 31 pp.

A fairly complete list of instrumental observations in the area of the United States prior to the institution of governmental observation, organized by States, with period of record, observer, elements observed, and location of data.

06 Hellmann, G. "Die Entwicklung der meteorologischen Beobachtungen bis zum Ende des XVIII. Jahrhunderts." *Abhandlungen der Preussische Akademie der Wissenschaften*, 1927. Physikalisch-mathematische Klasse, Nr. 1, Berlin. 48 pp.

Annotated list of sources of early meteorological observations, by countries, giving both manuscript or diary material and printed data collections. Observations in Germany, with borders as established after World War I, are published in another monograph (Hellmann, 1926).

* Hellmann, G. "Die Entwicklung der meteorologischen Beobachtungen in Deutschland...." Cited above as item NB27.

07 Hughes, Patrick. "Early American Weathermen." *ESSA, U.S. Department of Commerce, Environmental Science Services Administration* 5, no. 2 (1970): 12-15.

Account of early American weather observations by John Jeffries, John Winthrop, William Plumer, Samuel Rodman, with some facsimile reproductions.

08 Hughes, Patrick. *American Weather Stories*. Washington, D.C.: U.S. Department of Commerce, National Oceanic and Atmospheric Administration, 1976. 114 pp.

Illustrated collection of weather-related essays, among them early hurricane observations, including studies of Franklin Redfield and the pioneering 1943 reconnaissance flight of Col. Joseph Duckworth into the eye of a hurricane. An essay on early American weathermen gives an account of persons active in making weather observations in the United States in the late 17th and early 19th century with portraits and some facsimile reproductions of pages from weather diaries, including Jefferson's observations in Philadelphia in July of 1776.

09 Humbert, P. "Les deux récits de l'experience du Puy-de-Dome." *La Météorologie* (1943): 222-224.

There are two slightly different accounts of the famous experiment by Florin Perier, who climbed the Puy-de-Dome on 19 September 1648 with a barometer to ascertain the difference in barometric pressure between the valley and the mountain. One account is by Blaise Pascal (1623-1662), the other by Pierre Gassendi (1592-1655).

* Körber, Hans-Günther. "Über Alexander von Humboldts Arbeiten zur Meteorologie und Klimatologie." Cited below as item T13.

O10 Kovačević, Milan. "Historijski pregled meteoroloških opažanya" [Historical Sketch of Meteorological Observations]. *Hidrometeorološki Glasnik* 1 (1968): 35-46.

Meteorological observations date back to antiquity but organized information is available only since the invention of instruments in the 17th century. Also discusses developments in the present area of Yugoslavia.

O11 Loewe, Fritz. "First Series of Meteorological Observations in Australia." *Australian Meteorological Magazine* 18 (1970): 39-42.

The first regular meteorological records in Australia date back to 1821.

O12 Loewe, Fritz. "Early Mountain Observations in Australia." *Australian Meteorological Magazine* 20, no. 1 (1972): 52-55.

O13 Pinkett, Harold T., Helen T. Finneran, and Katherine H. Davidson. *Climatological and Hydrological Records of the Weather Bureau (Record Group 27)*. Preliminary Inventories of the National Archives No. 38. Washington, D.C., 1952. 76 pp.

A topically classified list of observations of the Weather Bureau (many of them starting when the Bureau was part of the Signal Service of the Army), by locality, giving the years stored at the Archives.

* Russ, A. "Der Einfluss der Societas Meteorologica Palatina auf unseren heutigen Beobachtungsdienst." Cited above as item NE16.

* Solov'ev, Yu.I. "Pervy nauchnyi polet na vozdushnom share." Cited below as item OB12.

* Szukalscy, Malgorzata, and Jerzy Szukalscy. "O pierwszych stalych obserwacjach meteorologicznych w Gdańsku w XVIII wieku." Cited above as item ND21.

O14 Walter, Emil J. "Technische Bedingungen in der historischen Entwicklung der Meteorologie." *Gesnerus* 9 (1952): 55-66.

An outline of major events in the history of meteorology, including the development of instruments and reporting techniques.

For additional information on this topic see works cited in the biography sections on the following scientists:

Leibniz, G.W. v.
Meigs, J.
Washington, G.

OA. Instruments

OA1 Appleton, Sir Edward V. "The Thirty-Sixth Kelvin Lecture: The Scientific Principles of Radiolocation." *Journal of the Institution of Electrical Engineers* 92 (1945): 340-353.

Gives, among other things, a brief historical review of radar development and its applications.

OA2 Arbey, Louis. "Le premier thermomètre de Lavoisier de l'observatoire de Paris." *L'Astronomie* 66 (1952): 80-82.

Antoine Laurent Lavoisier (1743-1794) made three thermometers for the Paris Observatory in 1782 and 1783 and designed scales for them and reported on measurements in the deep cellar below the observatory. The temperature there was often used as a fix point.

O43 Archinard, Margarida. "De Luc et la recherce barométrique." *Gesnerus* 32 (1975): 235-247.

OA4 Austin, Jill. "A Forgotten Meteorological Instrument--Rainband Spectroscope." *Weather* 36 (1981): 151-154.

Use of hand-held spectroscopes led the Astronomer Royal Charles Piazzi Smyth (1819-1900) to the discovery in 1872 of a dark band toward the shorter wave lengths on the right of the Fraunhofer D-line in the lower part of the sky, which he attributed to water vapor absorption. Observation of this band was touted as an aid to local rainfall forecasts but never became a standard meteorological procedure.

OA5 B[ilancini], R[aoul]. "Il Padre Castelli e il pluviometro." *Rivista di Meteorologia Aeronautica* 27 (1967): 126.

An account, quoting correspondence with Galileo (1564-1642), of how the drought of 1639 in Italy stimulated an early design of a rain gauge by the Benedictine monk Benedetto Castelli (1577-1644).

OA6 Bolle, Bert. *Barometers*. Watford, Herts.: Argus, 1981. 256 pp.

Written by a Dutch antique dealer; shows the evolution of barometers. Also briefly discusses ancillary instruments, such as thermometers, hygrometers, and polychrometers.

OA7 Burckhardt, F. "Die Erfindung des Thermometers und seine Gestaltung im XVII. Jahrhundert." *Bericht des Pädagogiums, Basel* (1867): 1-48.

An important memoir on the early development of thermometers.

OA8 Burckhardt, F. "Die wichtigsten Thermometer des achtzehnten Jahrhunderts." *Bericht der Gewerbeschule zu Basel* (1871): 1-30.

A review of 18th-century thermometers.

OA9 Conrads, L.A., and P.J. Jonker. "De regenmeter van Krecke wit 1849 en zijn betekenis bij het onderzoek naar de invloed van de stad Utrecht op de chemische samenstelling van het regenwater." *Hemel en Dampkring* 69 (1971): 178-181.

The early rain gauge by Krecke is described and the chemical composition of rainwater near Utrecht indicated. This is one of the earliest papers on rain chemistry.

OA10 Corton, Edward L. "Celsius or Linnaeus?" *Bulletin of the American Meteorological Society* 58 (1977): 1091.

A letter to the editor on the priority of invention of the centesimal thermometer, attributed by some to Linnaeus but generally claimed for Celsius. See item OA26.

OA11 Danlous-Dumesnils, M. "Du thermomètre de Florence au thermomètre de Lyon." *Mesures et Controle Industriels* 26 (1961): 1023-1030.

Recounts the early development of thermometers from the thermoscope of Jean Rey (1632) and the Florentine thermometer (1648) to the instruments in use in the 18th century, with a reproduction of Martine's (1741) comparison of various scales in use. Also includes a reproduction of a thermometer made by Pierre Casati according to the specifications and scale arrangements of P. Christin (1683-1755), who in 1743 suggested a centesimal scale with the freezing point at 0° and the boiling point of water at 100°. Letters and memoranda pertaining to this are still preserved at the Académie de Lyon.

OA12 Gerland, E. "Zur Geschichte des thermometers." *Zeitschrift für Instrumentenkunde* 13 (1893): 341-342.

A note on the history of thermometers.

OA13 Goodison, Nicholas. *English Barometers 1680-1860: A History of Domestic Barometers and Their Makers*. New York: Clarkson N. Potter, Inc., 1968. 353 pp.

A beautifully illustrated volume showing a very complete collection of barometers made by English instrument makers. Many of them were ornately designed works of art made for royal and aristocratic homes. Shows graphically the development of the "weather glass." A list of makers, their time of activity, and addresses is given.

* Harrison, Louis P. *History of Weather Bureau Wind Measurements*. Cited above as item NB23.

* Hellmann, G. "Die ältesten Quecksilber Thermometer." Cited below as item OD5.

OA14 Hellmann, G. "Beiträge zur Erfindungsgeschichte meteorologischer Instrumente." *Abhandlungen der Preussische Akademie der Wissenschaften*, 1920. Physikalisch-mathematische Klasse, Nr. 1, Berlin. 59 pp.

Source of early publications on thermometers, barometers, rain gauges, wind vane, with extensive literature citations and commentary.

OA15 Hellmann, G. "Ursprung der hundertteiligen thermometerskala." *Meteorologische Zeitschrift* 37 (1920): 36.

Short note to call attention to G.E. Melin's 1710 Uppsala dissertation on thermometry using a centesimal thermometer made by a Swedish instrument maker named Palmburg.

* Hellmann, G., ed. *Neudrucke von Schriften und Karten über Meteorologie und Erdmagnetismus*. Cited above as item M27.

* Hellmann, G. "Die ältesten Quecksilber Thermometer." Cited below as item OD5.

* Khrgian, A.Kh. *Ocherki razvitiia meteorologii*. Cited above as item M34.

OA16 King-Hele, Desmond. "Erasmus Darwin, Grandfather of Meteorology." *Weather* 28 (1973): 240-250.

Erasmus Darwin (1731-1802) devised a number of meteorological instruments, including a recording wind vane. He also had a correct interpretation of adiabatic expansion and its role in cloud formation.

OA17 Lambert, L.B. "A History of Humidity Measurement." *Instrument Practice* 19 (1965): 128-137.

OA18 Lambert, L.B. "Five Part History of Instruments, Pt. 4, Mainly Meteorological Instruments." *Instrument Practice* 23 (1969): 729-735.

Description of weather-observing instruments starting with the Galilean era, with sketches and illustrations of early equipment.

OA19 Landsberg, H.E. "A Note on the History of Thermometer Scales." *Weather* 19 (1964): 2-7.

A brief account of the development of fixed points of thermometers and various scales. Attention is called to the first use of a centesimal scale in Sweden three decades before Celsius and the priority of Christin for the presently used centigrade scale.

OA20 Meyer, Kirstine. *Temperaturbegrebets udvikling gennem tiderne samt dets sammenhaeng med vexlende Forestillinger om Varmens Natur*. Copenhagen: Jul. Gjellerups Boghandel, 1909. 179 pp.

Extensive review of early thermometer developments, with documentation from manuscript sources of Olav Rømer's contributions and the evolution of thermodynamic theory leading to the concept of absolute temperature. (Also appeared in German translation: *Die Entwicklung des Temperaturbegriffs im Laufe der Zeiten*. Die Wissenschaft, vol. 48. Braunschweig: Friedr. Vieweg & Sohn, 1913. 160 pp.)

OA21 Middleton, W.E. Knowles. "The Early History of Hygrometry, and the Controversy Between de Saussure and de Luc." *Quarterly Journal of the Royal Meteorological Society* 68 (1942): 247-261.

Traces the development of hygrometry from the Renaissance to about 1800. H.B. de Saussure (1740-1799) and J.A. de Luc (1727-1817) were rivals in design and use of hygrometers.

OA22 Middleton, W.E. Knowles. *The History of the Barometer.* Baltimore: The Johns Hopkins Press, 1964. 489 pp.

A thoroughly documented and annotated history of the most important early meteorological instrument that helped to lift meteorology from conjecture to science. It covers the period from Galileo's days to the beginning of the present century. It is based on literature, original manuscript sources, and personal inspection of older barometers found in museums. The book has many excellent illustrations and is a superb source.

OA23 Middleton, W.E. Knowles. *A History of the Thermometer and Its Use in Meteorology.* Baltimore: The Johns Hopkins Press, 1966. 249 pp.

A very detailed account of the early controversial history of thermometers and subsequent developments to the end of the 19th century. Very well documented and illustrated, the book cites many original sources, including manuscript material. It covers the problems of calibration points and the confusion of scales. It also gives an excellent account of the development of shelters and ventilation to record a representative air temperature.

OA24 Middleton, W.E. Knowles. *Catalog of Meteorological Instruments in the Museum of History and Technology.* Smithsonian Studies in History and Technology, vol. 2. Washington, D.C.: Smithsonian Institution Press, 1969. v + 128 pp.

OA25 Middleton, W.E. Knowles. *Invention of the Meteorological Instruments.* Baltimore: The Johns Hopkins Press, 1969. xiv + 362 pp.

The chapters on barometers and thermometers are based on the more elaborate books by this author on the development of these instruments. Added are chapters on humidity measurements, rain gauge development, wind vanes and anemometers, meteorographs, sunshine recorders, and equipment for upper air soundings. Everything is carefully annotated, illustrated, and documented. An outstanding contribution on the topic.

OA26 Middleton, W.E. Knowles. "More On 'Celsius or Linnaeus.'" *Bulletin of the American Meteorological Society* 59 (1978): 190.

Letter to the editor commenting on one by Edward L. Corton (item OA10) on the priority of the centesimal thermometer scale. Points out that the respective roles of Linnaeus and Celsius are not clear in this matter. It is quite likely that Jean Pierre Christin of Lyons, France, has the greatest claim for the first use (1743) of the scale presently in use.

* Mügge, Ratje, et al. *Meteorology and Physics of the Atmosphere*. Cited above as item M40.

OA27 Multhauf, Robert P. "The Introduction of Self-Registering Meteorlogical Instruments." *U.S. National Museum* (Smithsonian Institution) *Bulletin* 228, *Contributions from the Museum of History and Technology*, Paper 23, Washington, D.C., 1961, pp. 95-106.

Recording instruments were introduced into meteorological practice around the middle of the 19th century. This well-illustrated paper traces the development of such equipment to the end of that century.

OA28 Oliver, J. "An Early Self-Recording Pressure-Plate Anemometer." *Weather* 12 (1957): 16-21.

A brief review of anemometer developments, with the pressure plate type ascribable to Leonardo da Vinci (1452-1519). A recording variety was first designed by William Edmond in 1839, with drawings and a record reproduced here. (It is not mentioned in W.E.K. Middleton, *Invention of the Meteorological Instruments* [item OA25].)

OA29 Paoloni, Bernardo. "La Nostra Revista nel terzo centenario dell'invenzione del Pluviometro." *La Meteorologia Pratica* 20 (1939): 3-7.

Commemoration of the invention of pluviometry by B. Castelli.

OA30 Patterson, Louise Diehl. "Thermometers of the Royal Society, 1663-1768." *American Journal of Physics* 19 (1951): 523-535.

Thermometry was notably influenced by members of the Royal Society. The thermometers made by Francis Hauksbee, Jr. (1678-1763) in London were widely used but later superseded by those of Fahrenheit (1686-1736), Réaumur (1683-1757), and Celsius (1701-1744). Describes various scales and a standard thermometer kept at Gresham College that was used for calibration, 1665-1709. Has many references.

OA31 Patterson, Louise Diehl. "The Royal Society's Standard Thermometer 1663-1709." *Isis* 44 (1953): 51-64.

Review of progress in thermometry made by members of the Royal Society, including Robert Hooke (1635-1703) and Francis Hauksbee, Jr. (1687-1763).

OA32 Renou, E. "Histoire du thermomètre." *Annuaire de la Société météorologique de France* 24 (1876): 19-72.

A review of the history of thermometers, including the merit of Pierre Christin of Lyon (1683-1753) for the present centesimal scale. Renou was inventor of a sling thermometer to obtain proper ventilation and fast adaptation of the instrument (according to W.E.K. Middleton [*A History of the Thermometer* ..., item OA23] there are some errors in this account).

OA33 Rex, Daniel F. "Early Wind Vanes and Anemometers." *Atmospheric Technology* no. 2. Boulder, Colo: National Center for Atmospheric Research, 1973, pp. 4-10.

Gives an account of wind vanes from the time of the Athenian Tower of winds (1st century B.C.) to the beginning of World War II. Dates and names of inventors and their designs are noted.

OA34 Reynolds, Geoffrey. "A History of Rain Gauges." *Weather* 20 (1965): 106-114.

Rain gauges are mentioned as early as 400 B.C. in Indian writings. They were also known in antiquity in the Near East. In Korea they have been used since the 15th century. Sir Christopher Wren (1632-1723), who designed a number of meteorological instruments, developed in 1662 a tipping-bucket rain gauge. In the 19th century a variety of gauges and their exposures were tested. There are problems of comparability and at present there are large sampling errors. Much of the material in the article is drawn from the publication *British Rainfall*, 1861-1960.

OA35 Sealey, Antony F. "A Further Note on the Storm-Glass." *Weather* 22 (1967): 412-416.

Traces the history of the so-called Camphor Glass or chemical weather glass, which purportedly reacted to various weather stages, in vogue during the late 18th and early 19th centuries, and even advocated by Robert FitzRoy.

OA36 Six, James. *The Construction and Use of a Thermometer.* Edited by Jill Austin and Anita McConnell. Prefaced by an Account of his Life and Works and the Use of his Thermometer over Two Hundred Years. London: Nimbus Books, 1980. 123 pp.

James Six (1731-1793) invented the first effective maximum and minimum thermometer in 1780. He used his thermometers to measure diurnal variations, and found that during calm weather the upper air was much warmer that that close to the ground.

OA37 Taylor, F. Sherwood. "The Origin of the Thermometer." *Annals of Science* 5 (1942): 129-156.

Review of the claims of Galileo, Sanctorius, Fludd, and Drebbel.

OA38 von Oettingen, A.J. *Abhandlungen über Thermometrie von Fahrenheit, Réaumur, Celsius (1724, 1730-1733, 1742).* Ostwald's Klassiker der exacten Wissenschaften, no. 57. Leipzig: Wilhelm Engelmann, 1894. 140 pp.

Annotated German translations of five papers by D.G. Fahrenheit, three papers by R.A.F. de Réaumur, and one paper by A. Celsius dealing with the construction and calibration of the respective thermometers invented by these scientists.

For additional information on this topic see works cited in the biography sections on the following scientists:

Daniell, J.F.
Fuess, R.
Hough, G.W.
Maurer, J.M.
Montanari, G.
Patterson, J.
Robitzsch, M.
Saussure, H.B. de
Wild, H.

OB. Methods

OB1 Crutcher, Harold L. "Wind, Numbers, and Beaufort." *Weatherwise* 28 (1978): 260-271.

Admiral Sir Francis Beaufort (1774-1857) designed a wind scale (1806) for estimating force exercised on a sailing vessel. This force, expressed in numbers from 0 to 12, was observed by estimating height of waves at sea and later on land by effects on trees and persons. It was adopted internationally in 1874, and conversion to wind speeds was introduced as discussed in this paper. The relation to flag and light displays, as long used by the U.S. Weather Bureau along coast lines, is shown graphically.

* Heidorn, Keith. "The Legacy of Admiral Beaufort." Cited above as item NB25.

OB2 Hellmann, G. "Die Entwicklung der meteorologischen Beobachtungen bis zum Ende des 17. Jahrhunderts." *Meteorologische Zeitschrift* 16 (1901): 145-157.

Calls attention to the fact that meteorological observations, albeit sporadic, have been made since antiquity but became fairly systematic since the 16th century. A chronology of the earliest regular weather observations in various parts of the world is given to the end of the 17th century. (Information on others has become available since this paper was published.)

OB3 Kinsman, Blair. "Historical Notes on the Original Beaufort Scale." *The Marine Observer* 39 (1969): 116-124.

Admiral Sir Francis Beaufort (1774-1857) introduced early in the 19th century a scale of wind force judged by the agitation of the sea surface, from calm to hurricane force (0-12).

OB4 Kletter, Leopold. "Die Beiträge Österreichs zur Satellitenmeteorologie." *Hundert Jahre meteorologische Weltorganisation und die Entwicklung der Meteorologie in Österreich*. Zentralanstalt für Meteorologie und Geodynamik in Wien, Publication no. 207 (1975): 9-16.

A literature review of Austrian contributions to the development of satellite meteorology.

OB5 Kohler, J.P. *History of Climatological Record Forms 1009 and 612-14*. U.S. Weather Bureau, Key to Meteorological Records Documentation No. 2.11, Washington, D.C. 40 pp.

Traces instructions to Weather Bureau voluntary cooperation observers and forms for recording observations from 1891 to 1962.

OB6 Kurtyka, J.C. "Methods of Measuring Precipitation for Use with the Automatic Weather Station." *Illinois State Water Survey, Report of Investigation* No. 20 (1953). 178 pp.

Presents a comprehensive review of all rain gauges used since the 19th century.

OB7 Landsberg, H.E. "Early Weather Observations in America." *EDIS* 10(4) (1979): 21-23.

A brief note on weather records kept in the British North American colonies with an English translation of the article that stimulated such observations, Jacob Jurin's "Invitatio ad observationes Meteorologicas communi consilio instituendas," which appeared in the *Philosophical Transactions* (London), No. 379 (September-October 1723): 422-427.

OB8 Laughton, J.K. "Historical Sketch of Anemometry and Anemometers." *Quarterly Journal of the Royal Meteorological Society* 8 (1882): 162-164.

OB9 Lieurance, N.A. "The Development of Aeronautical Meteorology." *Proceedings of the Scientific and Technical Conference on Aeronautical Meteorology*. World Meteorological Organization Technical Note No. 95 (1969): 1-8.

Presents the gradual development of procedures for observations, forecasts, and dissemination of meteorological information to aviators. Emphasizes the role in this process of the International and World Meteorological Organizations and the International Civil Aviation Organization.

OB10 Moore, Gerald K. "What Is a Picture Worth? A History of Remote Sensing." *Hydrological Sciences Bulletin* 24 (1979): 477-485.

Traces the various methods of remote sensing, widely used in meteorology and hydrology, since World War II: radar, sonar, thermal, infrared. Since 1960 these have been incorporated into satellites. LANDSAT has been particularly effective for hydrological purposes because of its exceptional resolution.

* Penzar, Brauk, and Ivan Penzar. "Razvoj mreže meteoroloških stanica Hrvatskoj u 19. stoljecu." Cited above as item NB56.

OB11 Peppler, A. "Zur Geschichte der Pilotballonaufstiege." *Das Wetter* 48 (1931): 189-192.

Discusses the development and use of pilot balloons for the exploration of winds aloft.

* Rojecki, Ananiasz. "Kilka uwag a najdawniejszych obserwaejach meteorologicznych...." Cited below as item OD22.

* Schlegel, Max. "Alte meteorologische Karten und Welt-Atlanten." Cited below as item S26.

OB12 Solov'ev, Yu. I. "Pervy nauchnyi polet na vozdushnom share" [The First Scientific Flight in a Balloon]. *Priroda* 40 (1951): 73-75.

Describes 1804 ascent in a balloon by Ya.D. Zakharov for observation of magnetic and meteorological factors, especially intensity of solar radiation aloft.

OB13 Star, Pieter van der. "A Series of Meteorological Observations at the End of the 17th Century." *Actes du XIIe Congrès International d'Histoire des Sciences, 1968* (pub. 1971) 12: 81-86.

OB14 Suomi, Verner E. "The Acquisition of Meteorological Data." *Meteorological Challenges: A History* (item M39), pp. 159-176.

A very brief note on early instrumentation, the development of the radiosonde and the upper air network of observations, the free-floating upper atmospheric balloons, and early weather satellites.

OB15 Weickmann, L. *Über Aerologische Diagrammpapiere*. Internationale Meteorologische Organisation, Internationale Aerologische Kommission. Berlin: Julius Springer, 1938. 123 pp. + 30 plates.

For additional information on this topic see the work cited in the biography section on the following scientist:

Beaufort, F.

OC. Networks

OC1 Best, William H., Jr. "Radars over the Hump: Recollections of the First Weather Radar Network." *Bulletin of the American Meteorological Society* 54 (1973): 205-208.

Brig. Gen. Best, commander of the Air Weather Service, recounts the establishment of the first weather radar system in 1944/45 in the Bay of Bengal (now Bangladesh) and Burma to help flights "over the hump" into China. Another military weather radar network was installed in Panama in 1944.

OC2 Landsberg, H.E. "A Bicentenary of International Meteorological Observations." *World Meteorological Organization Bulletin* 29(4) (1980): 235-238.

The establishment of an international observing network by the Palatinate Academy of Science under the direction of Johann Jacob Hemmer (1733-1790) and financed by the Prince Elector Karl Theodor marked a very important milestone in the history of meteorology. The data collected under a uniform system and published for 12 years stimulated some of the fundamental early meteorological contributions such as Alexander von Humboldt's isothermal chart and Heinrich Wilhelm Brandes' synoptic studies.

* Lauscher, F. "Hundert Jahre öffentliches Beobachtungsnetz in Österreich." Cited above as item NB40.

* Meigs, Josiah. "Circular to the Registers of the Land Offices of the United States." Cited above as item NB44.

OC3 Ogden, R.J. "Co-Operating Observers and the Climatological Network." *Meteorological Magazine* 107 (1978): 209-215.

Traces the origin and development of the climatological observing network in the United Kingdom since its official inception in 1884, listing the original 17 stations and showing the growth (in 5-year intervals) to 1977, the type sponsors for the stations, and a list of individuals who were recognized for long-term service as observers.

OC4 Robertson, N.G. "The Organisation and Development of Weather Observations in New Zealand." *New Zealand Weather and Climate*. Edited by B.J. Garnier. Christchurch: Whitcomb and Tombs, 1950, pp. 7-25.

Traces the history of the New Zealand meteorological network, with observations starting in 1840 in Auckland.

* Sinclair, Bruce. "Gustavus A. Hyde, Professor Espy's Volunteers and the Development of Systematic Weather Observations." Cited above as item NB66.

* Suomi, Verner E. "The Acquisition of Meteorological Data." Cited above as item OB14.

For additional information on this topic see works cited in the biography sections on the following scientists:

Bailey, S.I.
Guyot, A.
Karazin, V.N.
Lamarck, J. de
Le Verrier, U.J.J.
Tilton, J.
Vines, B.
Wild, H.

OD. Early Observational Series (Prior to 1870)

* Aldredge, Robert Croom. "Weather Observers and Observations at Charleston, South Carolina, 1670-1871." Cited above as item O1.

OD1 Brouillette, Benoit. "Quelques observations climatiques en Nouvelle-France au XVIIIe siècle." *Proceedings and Transactions of the Royal Society of Canada* 8 (Trans.) (1970): 93-99.

OD2 Burnaby, Andrew. *Travels Through the Middle Settlements in North America in the Years 1759 and 1760 with Observations upon the State of the Colonies*. 2d edition. London: T. Payne, 1775. Reprint, Ithaca, N.Y.: Great Seal Books, 1960. 154 pp. (with 2 pp. of notes by the republisher).

Contains a diary of the weather with temperature readings, wind, and weather observations at Philadelphia, January 1760-December 1762.

OD3 Ferreira, H. Amorim. "Observações meteorólogicas em Portugal antes da fundação do Observatório do Infante D. Luiz." *Revista Faculdade de Ciências, Lisboa Universidad* 3-9 (1943): 17-29.

Earliest observations were made 1747-1748 at Funchal, Madeira. At Lisbon data were collected 1781-1785. Other early records prior to the establishment of a meteorological observatory in 1856 are noted. The text includes numerous references.

OD4 Frederick, R.H., H.E. Landsberg, and W. Lenke. "A Climatological Analysis of the Basel Weather Manuscript." *Isis* 57 (1966): 99-101.

Comparing frequency distributions of weather events and wind observations as derived from the Basel manuscript from September 1399 to June 1401 with current data, the authors suggest that this diary might have originated in Besançon (France). See item OD9.

* Havens, James M. "A Note on Early Meteorological Observations in the United States...." Cited above as item NB24.

OD5 Hellmann, Gustav. "Die ältesten Quecksilber Thermometer." *Meteorologische Zeitschrift* 14 (1897): 31-32.

A description of the earliest mercury-in-glass thermometer.

OD6 Kington, Beryl. "Searches for Historical Weather Data: Appeals and Responses." *Weather* 35 (1980): 124-134.

References to sources of weather information in Great Britain, including diaries, church records, and printed material.

OD7 Kington, J.A. "A Late Eighteenth-Century Source of Meteorological Data." *Weather* 25 (1970): 169-175.

Presents a parallel effort to the Palatinate Meteorlogical Society in France, where the Société Royale de Medicine of Paris established a network of stations and reporting schemes. In 1778 there were 50 observers in correspondence with the "Société." In France 39 stations were active between 1780 and 1785. Many of the observations were made by physicians and the data were sent on standardized formats to Louis Cotte (1740-1815), who summarized them. The scheme continued partially until

1793 when the revolutionary regime abolished this and other learned societies.

OD8 Kington, J.A. "Meteorological Observing in Scandinavia and Iceland During the Eighteenth Century." *Weather* 27 (1972): 222-233.

Regular observations in Stockholm started in 1783 by Wargentin, in Trondheim in 1762 by Berlin, in Iceland by Lievog in 1779, and in Copenhagen in 1783 by Bugge. Instruments and units used are described and some biographical notes on the observers are given.

OD9 Klemm, Fritz. "Über die Frage des Beobachtungsortes des Baseler Wettermanuskriptes 1399-1406." *Meteorologische Rundschau* 22 (1969): 83-85.

Comments on a deduction by Frederick, Landsberg, and Lenke (item OD4) that the observations reported in a weather diary in Basel Manuscript F III resurrected by Thorndike (item OD26) could have been made in Besançon. While this is not ruled out by the ecclesiastical and temporal history of the region, the author brings forth arguments that Basel itself is more likely to have been the observing point.

OD10 Klemm, Fritz. "Die Entwicklung der meteorologischen Beobachtungen in Franken und Bayern bis 1700." *Annalen der Meteorologie*, New Series, no. 8. Offenbach am Main: Selbstverlag des Deutschen Wetterdienstes, 1973. 50 pp.

Annotated list of weather observations in Franconia and Bavaria since 1331. Extensive citations of literature and location of sources are given. A valuable source guide.

OD11 Klemm, Fritz. "Die Entwicklung der meteorologischen Beobachtungen in Nord- und Mitteldeutschland bis 1700." *Annalen der Meteorologie*, New Series, no. 10 (1976). 73 pp.

Thorough documentation of weather observations from Northern and Central Germany, with location of sources and illustrations of samples. Information starts in 1533, with the most systematic and extensive set of data accumulated by the Landgrave Hermann IV of Hessia from 1621 to 1650. Included are comments on early attempts by Johannes Kepler to prognosticate the weather from constellation of the planets. A valuable source guide.

OD12 Klemm, Fritz. "Die Entwicklung der meteorologischen Beobachtungen in Sudwestdeutschland bis 1700." *Annalen der Meteorologie*, New Series, no. 13 (1979). 66 pp.

Documents casual and systematic observations in Southwest Germany, some in manuscripts, some printed since 1020, with location of sources, quotations, and illustrations. Included are comments on the systematic weather notes at Linz of Johannes Kepler. A valuable source guide.

OD13 Končekk, M. "Über die ersten meteorologischen Aufzeichnungen in Bratislava (Pressburg) aus dem Ende des 18. Jahrhunderts." *Idöjaras* (Budapest) 76 (1972): 51-53.

Discovery of pamphlets describing the weather conditions in Bratislava during the winter of 1783-84 and temperatures for the winter of 1783-84 and 1784-85 made a comparison with observations in the interval 1851-1950 possible. The early data appear to be valid.

OD14 Lawrence, E.N. "The Earliest Known Journal of the Weather." *Weather* 27 (1972): 494-501.

An account of the rediscovery of the manuscript of daily weather observations kept by William Merle at Driby, 1337-1344, with a facsimile page from the diary. Some views on the observations and a comparison with current conditions.

OD15 Lawrence, E.N. "Merle's Weather Observations." *Weather* 28 (1973): 127.

Discusses the place at which William Merle made the observations recorded in the oldest known weather diary and comes to the conclusion that East Lincolnshire is the most likely locality.

OD16 Lenki, W. "Die ältesten Temperaturmessungen von Nürnberg." *Meteorologische Rundschau* 17 (1964): 163-166.

Nürnberg has a number of casual weather observations, dating back to 1487. Fairly reliable thermometer observations there started in 1719, lasting to 1724. Later observations were from 1732 to 1746 by various observers with diverse instruments.

OD17 Ludlum, David M. *Early American Tornadoes: 1586-1870*. Vol. 4: *History of American Weather*. Boston: American Meteorological Society, 1970.

* Ludlum, David M. "Thomas Jefferson and the American Climate." Cited below as item T21.

OD18 Manley, Gordon. "The Weather and Diseases: Some Eighteenth Century Contributions to Observational Meteorology." *Notes and Records of the Royal Society of London* 9 (1952): 300-307.

Early instrumental weather records in England date back to Robert Hooke (1635-1703) and others to 1808 are reviewed with special attention to the merits of Jacob Jurin (1684-1750), referred to in a number of references as "James," whose portrait is included.

OD19 Middleton, W.E. Knowles. "Royal Meteorologist of the Seventeenth Century." *Weather* 28 (1973): 435-438.

A translation of and commentary on correspondence of Leopold de Medici and his 1657 barometer observations with speculations on relations between barometric changes and weather.

* Parczewski, Wladyslaw. "Zarys historii meteorologii w Polsce...." Cited above as item NB55.

* Penzar, Brauk, and Ivan Penzar. "Razvoj mreže meteoroloških stanic Hrvatskoj u 19. stoljecu." Cited above as item NB56.

OD20 Rigby, Malcolm. "Sources of Historical Meteorological Data Available in Reprints." *Weatherwise* 29 (1976): 242-243.

Lists books and sources of early meteorological work; much of it is devoted to the reprints produced by Hellmann. Also calls attention to a 1964 Dover reprint of M.F. Maury's *Physical Geography of the Sea*.

OD21 Rocznik, Karl. "Ildephons Kennedy, Mönch und Meteorologe in Bayern." *Meteorologische Rundschau* 14 (1961): 136.

A biographical note on the Benedictine monk Ildephons Kennedy (1722-1804), who started the first instrumented meteorological station in Bavaria at Regensburg in 1771.

OD22 Rojecki, Ananiasz. "Kilka uwag a najdawniejszych obserwaejach meteorologicznych w Toruniu na tle wynikow jednoczeshnie prowadzonych spostrzezeń w Warsawie." *Przeglad Geofizyczny* 10 (1965): 141-151. English translation by Mygorzata Widymska: *Some Notes Concern-*

ing the Oldest Meteorological Observations at Toruń on the Background of Contemporaneous Observations at Warsaw. Published for U.S. Department of Commerce, ESSA, and the National Science Foundation, Washington, D.C., by the Scientific Publications Foreign Cooperation Center of the Central Institute for Scientific, Technical and Economic Information, Warsaw, Poland, 1967. 8 pp.

A valuable contribution, not only on weather conditions at Torun, 1760-1762, but also on comparisons of early thermometer scales.

OD23 Rojecki, Ananiasz. "Traditions of Meteorology in Poland, from the 15th to 19th Century." *Acta Geophysica Polonica* 14 (1966): 3-10.

Reports on the earliest weather observations in Cracow and Warsaw, an early 19th-century network of stations, and the late 19th-century regular observing system. Many references to the author's archival studies are given.

OD24 Schove, D.J. "Weather in Scotland, 1659-1660: The Diary of Andrew Hay." *Annals of Science* 30 (1973): 165-177.

* Seydl, Otto. "K dwoustému výrocí pruých měření meteorologických v Čechách." Cited above as item NB65.

OD25 Thorndike, Lynn. "A Weather Record for 1399-1400 A.D." *Isis* 32 (1940): 304-323.

An English translation of the Basel Manuscript F III, giving daily weather notes by an unknown observer at an unknown locality from September 1399 to the end of August 1400.

OD26 Thorndike, Lynn. "A Daily Weather Record Continued from 1 September 1400 to 25 June 1401." *Isis* 57 (1966): 90-98.

Continuation of a translation of daily weather notes by an unknown observer at an unknown locality, from the Basel Manuscript F III (this is the second oldest known systematic weather diary from Europe). See items OD4 and OD9.

OD27 Wang, Pao-Kuan. "Meteorological Records from Ancient Chronicles of China." *Bulletin of the American Meteorological Society* 60 (1979): 313-318.

The records cover a time span from 2187 B.C. to A.D. 3. There are indications of long wet and dry periods of wide areal extent.

For additional information on this topic see works cited in the biography section on the following scientist:

Manley, G.

P. METEOROLOGY: PHENOMENA

PA. Clouds, Cloud Physics, Precipitation

PA1 Affronti, Filippo. "La 'Nuvola pendente' discorso meteorologico nel 'Museo di fisica e di esperienze.'" *Revista di Meteorologia Aeronautica* 10 (1950): 17-23.

Calls attention to an early treatise by Giovan Battista Hodierna published 1697 in Venice on formation of clouds, rain, snow, and sleet, showing notable insight into these phenomena.

PA2 Bergeron, T. "L'origine de la théorie des noyaux de glace comme declencheurs de precipitation un cinquantenaire: Quelques notes autobiographiques." *Journal de Recherches Atmospheriques* 6 (1972): 49-53.

Recollections about the discovery of ice nuclei as instigators of evolution of supercooled clouds start with an acknowledgment of the influence of ideas of Alfred Wegener. Observations of low stratus cloud behavior in Norway showed that ice nucleation could produce clearing (a discovery later attributed to V. Schaefer). These observations were published sketchily in 1929 but explicitly at the Lisbon Congress of the International Union of Geodesy and Geophysics (1931).

PA3 Besson, Louis. *Aperçu historique sur la classification des nuages*. Paris: Office National Météorologique, 1923. 20 pp.

Survey of 19th-century writings from Lamarck (1802) to H. Clayton (1896).

PA4 Biswas, Asit K. "Experiments on Atmospheric Evaporation Until the End of the Eighteenth Century." *Technology and Culture* 10 (1969): 49-58.

On the researches by Hippocrates, Edmund Halley, John Dalton, and Benjamin Franklin.

PA5 Blanchard, Duncan C. "Bentley and Lenard: Pioneers in Cloud Physics." *American Scientist* 60 (1972): 746-749.

Almost simultaneously in 1904 Phillip Lenard (1862-1947) and Wilson Bentley (1865-1931), one a Nobel laureate physicist, the other an amateur photographer, published papers indicating the importance of drop-size spectra for the explanation of rain phenomena. Drop size governs the collection efficiency of a drop and also the break-up phenomenon, clearly recognized by these investigators. Bentley also suspected that the formation of rain related to an initial ice stage, a fact firmly established three decades later independently by T. Bergeron and W. Findeisen.

PA6 Byers, Horace R. "History of Weather Modification." *Weather and Climate Modification*. Edited by W.N. Hess. New York: John Wiley & Sons, 1974, pp. 3-44.

A scholarly appraisal of the development of physical understanding of rain formation in the atmosphere and the attempt to induce rain artificially. Also included are sections on dissipation of fog and suppression of hail. Includes 163 references. A very dispassionate discussion of a still controversial area.

PA7 Changnon, Stanley A., Jr., and J. Loreena Ivens. "A History Repeated: The Forgotten Hail Cannons of Europe." *Bulletin of the American Meteorological Society* 62 (1981): 368-375.

A review of the great hey-day of attempted hail suppression, especially in the Alpine countries in the outgoing 19th century. This followed the alleged successes of Albert Stiger in Austria, who pursued a lucrative business by building mortars for this purpose. The contemporary scientific community either doubted or denied the efficacy of hail cannons. A parallel is drawn to the weather modification efforts in the world after Schaefer's discovery of cloud dissipation with dry ice in 1946, which included the abortive National Hail Project in Colorado in the 1970s.

* Charnock, H. "Fitzroy--Meteorological Statist." Cited above as item BF12.

PA8 Chary, Henry A. *History of Air Weather Service Weather Modification 1965-1972*. U.S. Air Weather Service, Technical Report 247 (1974). 25 pp.

The efforts of the Air Weather Service to dissipate fog by artificial means is reviewed, using dry ice, silver iodide, and liquid propane.

PA9 Day, J.H., and F.H. Ludlam. "Luke Howard and His Clouds: A Contribution to the Early History of Cloud Physics." *Weather* 27 (1972): 448-461.

Howard (1772-1864) was one of the most productive meteorologists of the 19th century. His cloud classification, which is presented here in excerpt, has endured. Excerpts from his writings are presented here and several of his plates are reproduced. He gave also a correct version of the hydrological cycle.

PA10 Dessens, Henri, and Jean Dessens. "Les plus anciennes descriptions et de pluis consécutifs à des incendies." *Bulletin Observatoire Puy de Dôme*, Ser. 2, no. 2 (1962): 103-105.

Review of cases of clouds and rains caused by conflagrations in the early literature as far back as 429 B.C.

PA11 Elliot, Robert D. "Experience of the Private Sector." *Weather and Climate Modification*. Edited by W.N. Hess. New York: John Wiley & Sons, 1974, pp. 45-89.

Recounting of the experimentation to control atmospheri precipitation with the aim to commercialize the processes It includes reports on some of the University work, the legislation to both stimulate and regulate "cloud seeding" by state and federal government.

PA12 Fukuta, Norihiko. "Weather Modification Activities in Japan." *Bulletin of the American Meteorological Society* 52 (1971): 4-14.

Reviews Japanese work on weather modification since 1947, which after some attempts at cloud seeding turned to more fundamental research on cloud dynamics and microphysics.

PA13 Gourlay, R.J. "Clement Wragge and the Stiger." *Weather* 28 (1973): 494-499.

Clement Lindley Wragge (1852-1922) investigated on behalf of the Australian government the so-called vortex gun developed by an Austrian vintner, Albert Stiger. This essay, partly based on an unpublished report in the Queensland State Archives by Wragge, reports on the experiments to use mortars to make rain, carried out unsuccessfully in Brisbane in 1905.

PA14 Halacy, Daniel S. *The Weather Changers*. New York: Harper, 1968. viii + 246 pp.

PA15 Havens, Barrington S., James E. Jiusto, and Bernard Vonnegut. *Early History of Cloud Seeding*. Langmuir Laboratory, New Mexico Institute of Mining and Technology, Socorro Atmospheric Sciences Research Center, State University of New York at Albany Research and Development Center, General Electric Company, Schenectady, New York. Socorro: New Mexico Tech Press, 1978. 75 pp.

This is essentially a recounting of Project Cirrus carried out at the General Electric Laboratory under direction of Dr. Irving Langmuir and close collaborators Vincent J. Schaefer and Bernard Vonnegut. It includes the history of discovery of freezing nuclei (dry ice and silver iodide) and the results of cloud modification flights. While some antecedent World War II work is mentioned, the detailed history of Project Cirrus from 1946 to 1950 with a list of 40 reports is presented (134 references are given).

* Hellmann, G. "Die Meteorologie in den deutschen Flugschriften und Flugblättern des XVI. Jahrhunderts...." Cited above as item M26.

PA16 Hellmann, Gustav. "System der Hydrometeore." *Veröffentlichungen des Preussischen Meteorologischen Institus* no. 285, Berlin, 1915. 27 pp. English translation in *Monthly Weather Review* 44 (1916): 385-392; 45 (1917): 13-17.

Gives a brief review of hypotheses of hydrometeor formation from Aristotle (384 B.C.-322 B.C.) onward, then discusses the definitions for various types of hydrometeors. Has valuable footnotes of historical interest.

* Hellmann, G., ed. *Neudrucke von Schriften und Karten über Meteorologie und Erdmagnetismus*. Cited above as item M27.

PA17 Kerr, Richard A. "Cloud Seeding: One Success in 35 Years." *Science* 217 (1982): 519-521.

In 1946, Vincent Schaefer discovered that dry ice could produce snow in a fog of cold water, on a small scale. But attempts to make rain this way have generally been unsuccessful, the exception being an experiment conducted in Israel from 1961 to 1967 under the direction of A. Gagin, J. Neumann, and R. Gabriel.

PA18 Kington, J.A. "A Historical Review of Cloud Study." *Weather* 23 (1968): 349-356.

Clouds have been an object of interest since antiquity. The Old Testament is a notable source of early cloud observations and their association with weather events. Greek literature, too, has many references. Modern cloud study starts with Jean Lamarck (1744-1829), who in 1802 published an attempt to classify them. This was followed in 1804 by the more successful scheme of Luke Howard (1772-1864). International agreement on cloud classification was reached in 1891 and the first international cloud atlas appeared in 1896.

PA19 Kington, J.A. "A Century of Cloud Classification." *Weather* 24 (1969): 84-89.

Reviews the efforts during the 19th century to classify clouds. These include works by Lamarck (1744-1829) published in 1802 and almost simultaneously by the system of Luke Howard (1772-1864), which is still the basis of contemporary cloud description. Aside from some abortive schemes, the International Meteorological Committee arranged for preparation of the first "official" cloud atlas (1895).

PA20 Loewe, Fritz. "Some Forerunners of Rainmaking by Seeding of Clouds." *Weather* 8 (1953): 35-37.

Calls attention to the attempts in 1891 to use liquid carbon dioxide and in 1922 to use liquid air for cloud seeding.

PA21 Mason, B.J. "The Physics of Clouds and Precipitation." *Meteorological Challenges: A History* (item M39), pp. 73-126.

Although a few references are made to 19th-century work, the essay dwells principally on the developments of the last few decades in the microphysics of clouds, with a comprehensive bibliography (except for Russian contributions).

PA22 Mason, B.J. "Personal Reflections on 35 Years of Cloud Seeding." *Contemporary Physics* 23 (1982): 311-327.

"The results of three recent major experiments conducted in Tasmania, Florida and Israel are discussed.... Despite all the claims to the contrary, there is little convincing evidence to suggest that it is possible to produced economically significant increases of precipitation...."

PA23 Maybank, J. "History of Weather Modification." *Weather Modification*. Edited by J. Maybank and W. Baier. Ottawa: Canada Department of Agriculture, Research Branch, 1970, pp. 6-10.

A brief account of research on weather modification since the 1930s. The changes brought about by the use of dry ice and silver iodide to initiate precipitation and suppress hail are presented.

PA24 McDonald, J.E. "James Espy and the Beginnings of Cloud Thermodynamics." *Bulletin of the American Meteorological Society* 44 (1963): 634-641.

James P. Espy, in his famous work *Philosophy of Storms* (Boston: Little, Brown, 1841; 552 pp.), stated clearly the effect of rising, convective air currents on cloud formation by first dry and later moist adiabatic processes. He also recognized the importance of the release of latent heat on the development and maintenance of clouds and convective storm systems.

PA25 Middleton, W.E. Knowles. *A History of the Theories of Rain and Other Forms of Precipitation*. Oldbourne History of Science Library. London: Oldbourne Book Co. Ltd., 1965. 223 pp.

An account of the processes of water vapor transformations in the atmosphere from the speculations of ancient Hebrews and Greeks to the end of the 19th century, with the emerging knowledge of thermodynamics. The association of precipitation with barometric changes was observed soon after the invention of the barometer but remained quite empirical for two centuries, when the physics of condensation and coalescence became clearer. Brief sections dealing with dew, hoar frost, and hail are included. There are extensive citations.

* Mügge, Raje, et al. *Meteorology and Physics of the Atmosphere*. Cited above as item M40.

PA26 Nikitin, N.I. "Ocherk razvitiya classifikatsii oblakov i oblachnykh sistem" [Outline of the Development of Classification of Clouds]. *Meteorologiya i Gidrologiya* Nr. 3 (1947): 19-30.

Reviews the history of efforts to classify clouds from the 18th century to the present. Gives the Russian equivalents for 77 Latin designations of cloud types.

* Pejml, Karyl. "Johann Kepler (1571-1630) a cešká meteorologie v jeho době." Cited below as item S20.

PA27 Schaefer, Vincent J. "Serendipity and the Development of Experimental Meteorology." *Journal of the Irrigation and Drainage Division, American Society of Civil Engineers* 86/1 (1960): 1-16.

Discusses the early stages of cloud seeding.

PA28 Schaefer, V.J. "The Early History of Weather Modification." *Bulletin of the American Meteorological Society* 49 (1968): 337-342.

The discoverer of the effectiveness of dry ice in dissipating supercooled clouds relates here the early efforts at the General Electric Laboratories to modify clouds in general and produce rain. A project named "Cirrus" under the direction of Irving Langmuir (1881-1957) made notable contributions to the knowledge of cloud physics and valiant attempts to cause rain by seeding clouds with a variety of nucleating substances.

PA29 Schwartz, Leonard E. "Artificial Weather Modification: A Case Study in Science Policy." *Technology and Society* 5(1) (1969): 44-48.

PA30 Scott, William T. "The Personal Character of the Discovery of Mechanisms in Cloud Physics." *Scientific Discovery: Case Studies* (item A73), pp. 273-289.

General remarks on the process of discovery, with comments on the contributions of Tor Bergeron, John Hallett, and Stanley Mossop.

PA31 Spence, Clark C. *The Rainmakers: American "Pluviculture" to World War II*. Lincoln: University of Nebraska Press, 1980. x + 181 pp.

In his review in *American Historical Review* 86 (1981): 662-663, Brooke Hindle says historians of meteorology have unjustly neglected this topic, since it looks like a pseudoscience that failed. Yet it must be considered part of the general cultural history of science; "the attitudes and methods of its practitioners were congenial with going empirical technologies of the time." The story begins with a recognized meteorologist, James P. Espy, and ends just before the first successful cloud seeding of 1946.

PA32 Stephan, Johannes. "Das Tauproblem." *Bioklimatische Beiblätter der Meteorologischen Zeitschrift* 5 (1938): 75-81.

A review of the investigations on dew from the time of Aristotle to 1938, with a bibliography of the principal treatises on dew.

PA33 Townsend, Jeff. *Making Rain in America: A History.* Lubbock: Texas Tech University, ICASALS Publication No. Y5-3, 1975. 93 pp.

A well-annotated pamphlet with ample references to the numerous efforts in the United States to alleviate drought by stimulating rainfall from 1839 onward. Charlatans and scientists tried it. Early failures were followed by some alleged successes, especially claimed by Irving Langmuir (1881-1957), but it remained an uncertain art. The interactions of private rainmakers, university scientists, government agencies, and legislative bodies are carefully presented. It is a prime story of unfinished scientific research prematurely and often fraudulently applied. It also illustrates the unjustified belief in the infallibility of Nobel laureates.

PA34 Walker, J.M. "A Meteorological Curio." *Weather* 25 (1970): 30-32.

A note on a 1784 publication, *Meteorological Imagination and Conjectures* by Benjamin Franklin (1706-1790), which comments on formations of rain from frozen hydrometeors and on cold winters succeeding foggy summers, showing unusual perception of meteorological phenomena.

PA35 Weickmann, Helmut. "The Development of Bergeron's Ice Crystal Precipitation Theory." *Pageoph* 119 (1980/81): 538-547.

A review of the theory proposed by Tor Bergeron (1891-1977), formally in 1933, that rain forms through the falling of ice nuclei through a supercooled water cloud. The essential parts of the theory have not changed but essential clarifying observations have been added since.

For additional information on this topic see works cited in the biography sections on the following scientists:

Bentley, W.
Bergeron, H.P.
Espy, J.P.
Guericke, O. v.
Hellmann, G.
Howard, L.
Jesse, O.
Klein, H.J.
Köhler, H.
Langmuir, I.
Le Roy, C.
Mohorovičić, A.
Montanari, G.
Nakaya, U.
Süring, R.
Thornthwaite, C.W.
Wells, W.C.

PB. Thunderstorms (Thunder, Lightning)

* Fassig, Oliver L., ed. *Bibliography of Meteorology.* Cited above as item M14.

PB1 Fraunberger, Fritz. "Gedankenblitze, welche die Donnerwetter betreffen. Ausserungen zumeist aus dem 18ten Jahrhundert." *Physikalische Blätter* 24 (1968): 145-149.

PB2 Frisinger, H. Howard. "Early Theories on the Cause of Thunder and Lightning." *Bulletin of the American Meteorological Society* 46 (1965): 785-787.

A brief note on the concepts of lightning and thunder in antiquity and the Middle Ages. This is followed by presentation of the misconceptions of Rene Descartes (1596-1650) and the mathematician and physicist John Wallis (1616-1703), a co-founder of the Royal Society.

PB3 Prince, H. "Lightning in History." *Lightning*, vol. I. Edited by R.H. Golde. London: Academic Press, 1977, pp. 1-21.

A concise history of lightning phenomena from mythological beginnings to the first scientific experiments in the 18th century to current explanations and laboratory simulation. Protection against lightning is also included as a topic.

PB4 Remillard, Wilfred J. "The History of Thunder Research." *Weather* 16 (1961): 245-253.

Traces the formulation of hypotheses of thunder back to Lucretius (50 B.C.) and through the centuries to the present.

For additional information on this topic see works cited in the biography sections on the following scientists:

Bergman, T.O.
Dove, H.W.
Franklin, B.
Mohn, H.
Schmidt, C.A. Von

PC. Other Severe Storms and Hurricanes

PC1 Anthes, Richard A. *Tropical Cyclones. Their Evolution, Structure and Effects.* Meteorological Monographs, vol. 19, no. 41. Boston: American Meteorological Society, 1982. xviii + 208 pp.

Includes a summary of research from the time of Redfield to about 1980.

PC2 Dampier, William. "A Discourse of the Tradewinds, Breezes, Storms, Seasons of the Year, Tides and Currents of the Torrid Zone Throughout the World." *Dampier's Voyages.* Edited by G.E. Richards. London: John Mace Field, 1906, pp. 229-321.

A book on the explorations of William Dampier (1652-1715). The cited chapter contains Dampier's observations on winds and tropical storms.

PC3 De Angelis, Richard M. "Tropical Cyclones Around the World (1. The Early Pioneers)." *Mariners Weather Log* 14 (1970).

Recites early efforts to understand and map the courses of tropical storms, starting with William Dampier (1652-1715), with a number of now-forgotten contributors, and better-known ones, such as William Redfield (1789-1857) and William Reid (1791-1858). Has reproductions of two portraits and three old charts.

PC4 Dettwiller, J. "Lutte anti-grêle." *Bulletin d'Information Direction de la Météorologie* no. 54 (January 1982): 25-30.

With reference to recent damaging hailstorms in France, some of the cloud-seeding experiments to mitigate the impacts are described. This is followed by an illustrated chronology of attempts to affect hail clouds from antiquity to the present. It includes a reproduction of a 16th-century print showing archers shooting arrows at clouds. In the 19th century, cannons were fired to dispel clouds.

PC5 Hay, R.F.M. "William Dampier's Contribution to Meteorology." *Weather* 10 (1955): 194-199.

Dampier (1651-1715) made many exploratory voyages to the tropics and described the weather encountered, especially the tropical storms.

PC6 Hellmann, G. "Die ältesten Untersuchungen über Windhosen." *Beiträge zur Geschichte der Meteorologie* 2 (1917): 329-334.

References to early accounts of whirlwind, principally in the 17th and 18th centuries, with sources.

PC7 Ludlum, David M. *Early American Tornadoes: 1586-1870.* History of American Weather, Vol. 4. Boston: American Meteorological Society, 1970.

PC8 Lugt, Hans. "Wirbelstürme im Alten Testament." *Biblische Zeitschrift*, New Series, No. 19 (1975): 195-204.

An analysis of the equivalence of Hebrew, Greek, and Latin expressions for rotating storms, principally referring to whirlwinds and tornadoes. A table lists mention of these phenomena in the Septuaginta and the Vulgata with interpretations of the mythological significance and recommendations for appropriate translation that is meteorologically defensible.

* Thomas, Morley K. *Canadian Meteorological Milestones.* Cited above as item NB73.

For additional information on this topic see works cited in the biography sections on the following scientists:

Bjerknes, J.
Dove, H.W.
Ficker, H. von
Mendel, G.
Redfield, W.
Tannehill, I.R.
Vines, B.
Walker, G.T.

PD. Pollution and Air Chemistry

PD1 Bolin, B. "Atmospheric Chemistry and Environmental Pollution." *Meteorological Challenges: A History* (item M39), pp. 237-266.

A brief history dealing with atmospheric composition, with special emphasis on carbon dioxide, nitrogen and sulfur compounds in rain, and the various geochemical cycles. Many observations date back to the 19th century but the anthropogenic contribution and its influence has been principally scrutinized since 1950.

PD2 Brimblecombe, Peter. "Earliest Atmospheric Profile." *New Scientist* 81 (1977): 364-365.

Report on manuscript data of air analyses made by Henry Cavendish of samples taken to 3 km height by John Jeffries on a balloon flight above London on 30 November 1784.

* Conrads, L.A., and P.J. Jonker. "De regenmeter van Krecke wit 1849...." Cited above as item OA9.

For additional information on this topic see works cited in the biography sections on the following scientists:

Dalton, J.
Dobson, G.M.B.
Götz, F.W.P.
Koschmieder, H.
Mendeleev, D.I.

Q. ATMOSPHERIC PHYSICS

* Khrgian, A.Kh. *Ocherki razvitiia meteorologii*. Cited above as item M34.

* Körber, Hans-Günther. "Über Alexander von Humboldts Arbeiten zur Meteorologie und Klimatologie." Cited below as item T13.

For additional information on this topic see works cited in the biography sections on the following scientists:

De Luc, J.A.
Johnson, N.
Karman, T. von
Khanevskii, V.A.
Linke, F.
Lomonosov, M.V.
Mendeleev, D.I.
Mohorovičić, A.
Parry, C.H.
Prandtl, L.
Saussure, H.B. de
Schneider-Carius, K.

QA. Atmospheric Radiation

QA1 Ångström, Anders. "Survey of the Activities of the Radiation Commission of the International Meteorological Organization and of the International Union of Geodesy and Geophysics." *Tellus* I (1949): 65-71.

A comprehensive account of the two commissions, which set standards for actinometry throughout the world, with a list of reports and important references for atmospheric actinometric work.

QA2 Drummond, A.J. "Precision Radiometry and Its Significance in Atmospheric and Space Physics." *Advances in Geophysics* 14 (1970): 1-52.

A review which includes a survey of major developments since 1666, especially the work of K. Angstrom (1893-1903), C.G. Abbot and his colleagues at the Smithsonian (1911, 1932, 1954), and H.L. Callendar (1910).

QA3 Möller, Fritz. "Entwicklungslinien der meteorologischen Strahlungsforschung." *75 Jahre Meteorologisches Observatorium Potsdam, 1892-1967, Meteorologischer Dienst* (der Deutschen Demokratischen Republik) [n.d., 1969?], pp. 24-28.

An account of radiation research in the current century, especially instrumentation, theory, and aims, the latter now focusing on aerosols.

QA4 Möller, Fritz. "Radiation in the Atmosphere." *Meteorological Challenges: A History* (item M39), pp. 43-71.

Reviews the principal developments in atmospheric optics and radiation, including explanation of the blue sky and rainbow, the measurement of solar radiation, scattering and polarization of light, and the radiation balance of the earth.

* Mügge, Ratje, et al. *Meteorology and Physics of the Atmosphere*. Cited above as item M40.

For additional information on this topic see works cited in the biography sections on the following scientists:

Abbot, C.
Angström, A.
Arctowski, H.
Dorno, C.
Götz, F.W.P.
Gorczynski, W.
Lunelund, H.W.
Maurer, J.M.
Roche, E.A.
Somerville, M.
Süring, R.

QB. Atmospheric Electricity (Other Than Lightning)

QB1 Chauveau, A. *Introduction historique et bibliographique à l'étude de l'électricité atmospherique*. Paris: Gauthier-Villars, 1902. 70 pp.

"Extract des Annales du Bureau central météorologique de France, annel 1899, t.1., Memories." Discussions of the theories of Peltier (1835-44), F. Exner (1886-90), Edlund (1878-88), and others; bibliography arranged in chronological order, listing about 200 publications since 1750.

QB2 Chauveau, [Amyr] B[enjamin]. *Électricité atmospherique*. Première fascicule: Introduction historique. Paris: Doin, 1922. xi + 90 pp.

Discusses Franklin and Dalibard, J. de Romas. L.-G Lemonnier, J.-B. Beccaria, H.-B. de Saussure, A. Volta, P. Erman, A. Peltier, F. Dellmann, W. Thomson (Lord Kelvin), F. Exner, J. Elster and H. Geitel, G.C. Simpson, Luvini, Sohucke, Brillouin, P. Lenard, Edlund, Ebert, C.T.R. Wilson.

QB3 Cohen, I. Bernard. *Franklin and Newton: An Inquiry into Speculative Newtonian Experimental Science and Franklin's Work in Electricity as an Example Thereof*. Philadelphia: American Philosophical Society, 1956. xv + 657 pp.

Chapter 11: "The Reception of the Franklin Theory: Atmospheric Electricity and Lightning Rods."

QB4 Fedorov, Evgenii K. "'Slovo o iavleniiakh vozdushnykh ot elektricheskoi sily proiskhodiashchikh' Lomonosova i sovremenye predstavlenie ob atmosfernom elektrichestve" ["A Word About Atmospheric Phenomena of Electrical Origin" by Lomonosov and the Present Concept of Atmospheric Electricity]. *Izvestiya Akademiia Nauk SSSR Seriya Geografii i Geofisiki* 14 (1950): 25-36.

Advances the claim that Mikhail V. Lomonosov (1711-1765) anticipated modern theories of atmospheric electricity.

QB5 Heilbron, J.L. *Electricity in the 17th and 18th Centuries. A Study of Early Modern Physics*. Berkeley: University of California Press, 1979. xiv + 606 pp.

Chapter XIV, on Benjamin Franklin and the reception of his views in Europe, includes some information on atmospheric electricity.

QB6 Miller, Dayton Clarence. *Sparks, Lightning, Cosmic Rays. An Anecdotal History of Electricity*. New York: Macmillan, 1939. xvii + 192 pp.

"Christmas Week Lectures for Young People, 1937, The Franklin Institute." Mostly on the 18th century; the chapter on cosmic rays includes aurora, ionosphere, etc.

QB7 Mottelay, Paul Fleury. *Bibliographical History of Electricity and Magnetism*. London: Griffin, 1922. xx + 673 pp., plates. Reprinted, New York: Arno Press, 1975.

Includes summaries of several works on atmospheric electricity and terrestrial magnetism. The chronological sequence ends with Faraday's work in 1821.

For additional information on this topic see works cited in the biography sections on the following scientists:

Benndorf, H.
Hogg, A.R.
Simpson, G.

QC. Atmospheric Optics (Including Rainbows)

QC1 Boyer, Carl B. "William Gilbert on the Rainbow." *American Journal of Physics* 20 (1952): 416-422.

QC2 Boyer, Carl B. *The Rainbow from Myth to Mathematics*. New York: Yoseloff, 1959. 376 pp.

On the theories and observations of Alexander of Aphrodisias, Alhazen, Aristotle, Francis Bacon, Roger Bacon, René Descartes, Euclid, Johann Fleischer, William Gilbert, Edmund Halley, Christiaan Huygens, Johann Kepler, Marcus Marci, Franciscus Maurolycus, Wm. Hallowes Miller, Pieter van Musschenbroek, Isaac Newton, Olympiodorus, John Peckham, Giambattista della Porta, Richard Potter, Robert Grosseteste, Themo Judoii, Theodoric of Freiberg (Dietrich of Freiberg), Francesco Vimercati, Witelo, and Thomas Young.

QC3 Dietrich Von Frieberg. *Theodorcus Teutonicus de Vriberg. De Iride et Radialibus Impressionibus*. Edited by J. Wuerschmidt. Beitrage zur Geschichte der Philosophie des Mittelalters, Texte und untersuchungen, 12, Heft 5-6 (1914). ix + 205 pp.

With an introduction and detailed summary (in German) by the author.

QC4 Frisinger, H. Howard. "René Descartes: The Last of the Old and the First of the New Meteorologists." *Weather* 21 (1966): 443-446.

Although Descartes (1596-1650) approached meteorology as a natural philosopher he was far ahead of his times by urging wide-spread observations. He was the first to explain the rainbow phenomenon correctly.

QC5 Greenslade, Thomas B., Jr. "19th Century Textbook Illustrations, XLII. The Rainbow." *Physics Teacher* 20 (1982): 463.

Reproduces diagrams from books by Ganot and Amy Johnson.

* Hellmann, G., ed. *Neudrucke von Schriften und Karten über Meteorologie und Erdmagnetismus*. Cited above as item M27.

QC6 Middleton, W.E. Knowles. "Random Reflections on the History of Atmospheric Optics." *Journal of the Optical Society of America* 50 (1960): 97-100.

QC7 Middleton, W.E. Knowles. "The Early History of the Visibility Problem." *Applied Optics* 3 (1964): 599-602.

Discusses the contributions to atmospheric optics of J.J. d'Ortous de Mairan, P. Bouguer, J.H. Lambert, and L. Weber.

QC8 Spinoza, Benedictus de. *Algebraic Calculation of the Rainbow, 1687*. Facsimile of the original Dutch text, with an Introduction by G. ten Doesschate. Dutch Classics on History of Science, 5. Nieuwkoop: B. de Graaf, 1963. 26 + 20 pp.

Reprint of *Stelkonstige Reeckening van den Regenboog, dienende tot naedere samenknoping der natuurkunde met de wiskonsten*.

For additional information on this topic see works cited in the biography sections on the following scientists:

Arago, F.
Babinet, J.
Bergman, T.O.
Bouguer, P.
Descartes, R.
Dietrich von Freiberg
Exner, F.M.
Foster, H.
Koschmieder, H.
Mariotte, E.
Sharonov, V.V.

R. ATMOSPHERIC DYNAMICS AND THERMODYNAMICS

R1 Bôcher, Maxime. "The Meteorological Labors of Dove, Redfield and Espy." *American Meteorological Journal* 5 (1888): 1-13. Reprinted in the *Journal of the American Meteorological Society* 46 (1965): 448-455, with an annotation about Professor Bôcher of Harvard University.

This is an early appraisal of the contributions of three meteorologists to the theory of storms, their contrasting views and how they led to the evolution of more adequate ideas.

R2 Burstyn, Harold L. "The Deflecting Force of the Earth's Rotation from Galileo to Newton." *Annals of Science* 21 (1965): 47-80.

The theory of the effect of the earth's rotation on motions of the atmosphere and oceans is based on 17th-century research on the path of a body falling to the earth's surface. Thus it is related to the debate about whether the earth moves and whether the effects of its motion are observable. The author discusses the views of G.A. Borelli (1667) and the correspondence of Robert Hooke and Isaac Newton (1679-80) on this subject.

R3 Burstyn, Harold L. "The Deflecting Force and Coriolis." *Bulletin of the American Meteorological Society* 47 (1966): 887-891.

Discusses the development of the recognition of the role of the rotation of the earth on motions, including those in the atmosphere. Laplace (1778) should be credited with an early mathematical model rather than Coriolis (1835). In meteorology, however, Ferrel (1856) is the first to introduce the deflection into a coherent theory.

R4 Burstyn, H.L. "Early Explanations of the Role of the Earth's Rotation on the Circulation of the Atmosphere." *Isis* 57 (1966): 167-187.

A review of the evolution of a theory of wind and water motion that eventually linked the basic force of pressure differences with the deflecting influence of the earth's rotation. The role of George Hadley (1685-1744) and Colin Maclaurin (1698-1746) in this development is stressed.

R5 Burstyn, Harold L. "Theories of Winds and Ocean Currents from the Discoveries to the End of the 17th Century." *Terrae Incognita* 3 (1971): 7-31.

R6 Cannegieter, H.G. "Honderd jaar wet van Buys-Ballot." *Hemel en Dampkring* 56/3 (1958): 42-45.

Traces the history of the law relating pressure and wind first tentatively published by C.H.D. Buys-Ballot (1817-1890) in 1857 and then enunciated in final form in 1860.

R7 Carapiperis, Leon N. "The Etesian Winds; I. The Opinions of the Ancient Greeks on the Etesian Winds." *Ethnikon Asterospeion Hypomnemata*, Ser. II, *Meteorologica* 9 (1962) 17 pp.

R8 Clodman, J. "Small-Scale Motions." *Meteorological Challenges A History* (item M39), pp. 209-234.

Secondary circulations, such as mountain and valley breezes, land and sea breezes, and various forms of turbulence, although known from antiquity, have only in recent decades become amenable to physical analysis. The literature cited, with a few exceptions, covers principally post-World War II developments.

R9 Coutant, V., and V. Eichenlaub. "The *de Ventis* of Theophrastus: Its Contribution to the Theory of Winds." *Bulletin of the American Meteorological Society* 55 (1974): 1454-1462.

Theophrastus (372 B.C.-287 B.C.), in contrast to his teacher Aristotle, defined wind as moving air. He observed the regular land and sea breezes, commented on the speed of winds from various directions, and attributed advected temperatures to them. An interpretation of fact and fallacies in the book *de Ventis*.

R10 Cressman, G.P. "Dynamic Weather Prediction." *Meteorological Challenges: A History* (item M39), pp. 181-207.

Reviews the developments in numerical weather prediction since the end of World War II, the improvements in forecast reliability, and the problems of extended-range forecasts and gives the pertinent references.

R11 Dettwiller, J. "La Loi de Buys Ballot." *Bulletin d'Information Direction de la Météorologie*, no. 54 (January 1982): 36-42.

An illustrated article discussing the development of the theory of cyclones, particularly the recognition of the influence of the earth's rotation on wind circulation. Although it was known before him, the Dutch meteorologist C.H.D. Buys-Ballot (1817-1890) formulated in 1857 the rule of wind flow in low-pressure centers.

R12 Dettwiller, Jacques. *À Propos de la Loi de Buys Ballot.* MET MAR (Ministère des Transports, Météorologie Nationale), 2nd trimestre 1982, No. 115, pp. 46-53.

A brief history of development of views on wind circulation with reproductions of original illustrations from the 17th century onward, culminating in the 1857 pronouncement of the rule of circulation around low-pressure systems by H.D. Buys Ballot (1817-1890).

R13 Frisinger, H. Howard. "Mathematicians in the History of Meteorology: The Pressure-Height Problem." *Historia Mathematica* 1 (1974): 263-286.

R14 Grasnick, K.H. "Wärmehaushalts--Turbulenz--und Ozonforschung als Aufgabengebiete des Potsdamer Meteorologischen Observatoriums." *75 Jahre Meteorologisches Observatorium Potsdam, 1892-1967, Meteorologischer Dienst* (der Deutschen Demokratischen Republik) [n.d., 1969?], pp. 19-23.

A review of the important contributions of successive research workers at the Potsdam Observatory dealing with atmospheric heat balance, thermal convection, diffusion, and ozone.

* Hellmann, G., ed. *Neudrucke von Schriften und Karten über Meteorologie und Erdmagnetismus*. Cited above as item M27.

R15 Heyer, E. "Immanuel Kant über den Einfluss des Mondes auf die Witterung." *Zeitschrift für Meteorologie* 5 (1951): 285.

A summary of a 1794 paper by Kant (1724-1804), who thought that a lunar atmospheric tide could not be observed and hence a direct influence of the moon on the weather was unlikely.

R16 Jacobi, Max. "Immanuel Kant und die Lehre von den Winden." *Meteorologische Zeitschrift* 21 (1903): 419-421.

Kant explained wind as an equilibrium disturbance in the atmosphere brought about by unequal heating at the equator and the poles. He recognized the wind deflection on a rotating earth probably quite independently of the earlier work published in 1735 by Hadley (1685-1768) on the origin of trade winds.

R17 Jehn, K.H. *A Sea Breeze Bibliography: 1664-1972*. Austin: The University of Texas, Atmospheric Science Group Report No. 37 (NSF Grant GA 16167), 1973. 51 pp.

A catalog of investigators and publications on the sea breeze in two major sections, the first giving names of authors on the topic in chronological order, starting with F. Bacon (1664). The second is a listing of references in alphabetical order of authors.

R18 Jordan, J.L. "On Coriolis and the Deflective Force." *Bulletin of the American Meteorological Society* 47 (1966): 401-403.

An extensive review with many citations, telling how the eponym of G.G. Coriolis (1792-1843), based on a paper published in 1835, became attached to the deflecting force observed in wind systems. Although the process was known since G. Hadley (1685-1744) and had been mathematically modelled by William Ferrel (1817-1891) for meteorological systems in 1856, it was not until the 1940s that meteorologists began to refer to the deflecting process as "Coriolis Parameter."

R19 Khromov, Sergei Petrovich. "A fél évszázados szovjet meteorológia" [Fifty Years of Soviet Meteorology]. *Idöjaras* 72 (1968): 1-14.

A summary of research results by Soviet meteorologists in dynamic meteorology, radiation, general circulation, and climatic fluctuations in half a century.

R20 Kutzbach, Gisela. *The Thermal Theory of Cyclones: A History of Meteorological Thought in the Nineteenth Century*. Historical Monograph Series. Boston: American Meteorological Society, 1979. 255 pp.

A well-written, copiously annotated treatise on the gradual development of a rational explanation of rotating, propagating storm systems. It covers the period from about 1830 to the beginning of the 20th century when the Norwegian cyclone model began to dominate. It is based on a thorough coverage of the pertinent publications. An appendix gives brief but valuable biographical sketches of the principal contributors to the topic.

R21 Landsberg, H.E. "Note on 'Early Developments in the Theory of the Saturated Adiabatic Process.'" *Bulletin of the American Meteorological Society* 44 (1963): 511.

A comment on an article by J.E. McDonald on this topic (item R26) with some amendments and corrections.

R22 Landsberg, H.E. "Why Indeed Coriolis?" *Bulletin of the American Meteorological Society* 47 (1966): 887-889.

Traces the development of understanding of the deflection of moving bodies on the rotating earth, as applied to atmospheric wind system. Giving this deflecting principle the eponym Coriolis force is not justified by this engineer's complete lack of influence on the developments in atmospheric science. The main credit belongs to William Ferrel (1817-1891) and C.H.D. Buys Ballot (1817-1891).

R23 Ludlam, F.H. *The Cyclone Problem: A History of Models of the Cyclonic Storm*. Inaugural Lecture, 8 November 1966, Imperial College of Science and Technology, London.

Traces the development of theories of cyclonic storms in middle latitudes from H.W. Brandes (1820) to the Norwegian School (1919) and the first attempts at numerical weather prediction (ca. 1960). Well illustrated but devoid of literature citation.

R24 Ludlum, David M. "The Espy-Redfield Dispute." *Weatherwise* 22 (1969): 224-228.

An account of two conflicting hypotheses of wind circulation in cyclones, with brief biographies, illustrated by portraits of the two protagonists, James Pollard Espy (1785-1860) and William C. Redfield (1789-1857). The former advocated a convective, centripetal model, the latter a vortex model. The matter was debated both in papers (*American Journal of Science*, 1831-1834; *Transactions of the Geological Society of Pennsylvania*, 1835, initially) and meetings. The controversy lasted for two decades until the work of Dove (1803-1879) and Ferrel (1817-1891) indicated that the Redfield mechanism prevailed in most cases.

R25 Lugt, Hans J. *The Vortex Concept in the History of Science*. David Taylor Naval Ship Research and Development Center, Report CMD-04-76 (1976). 17 pp.

Traces the changing concepts of vortices in the atmosphere and fluids from antiquity to the present. The meteorological work of V. Bjerknes and C.G. Rossby is cited.

R26 McDonald, J.E. "Early Developments in the Theory of the Saturated Adiabatic Process." *Bulletin of the American Meteorological Society* 44 (1963): 203-211.

Gives a history of developments of the theory of the saturated adiabatic process, citing work by William Thomson (later Lord Kelvin) (1824-1907), Th. Reye (1838-1919), Julius Hann (1839-1921), and Heinrich Hertz (1857-1894), whose adiabatic diagram, the first in meteorology, is reproduced. See also item R21.

R27 Middleton, W.E. Knowles. "Carlo Rinaldini and the Discovery of Convection in Air." *Physis* 10 (1968): 299-305.

"It is shown that the process of convection in air was described in the 18th century ... from a meteorological standpoint. But it was recognized in the laboratory by Carlo Rinaldini in November, 1657."

R28 Middleton, W.E. Knowles. "Paolo del Buono on the Elasticity of Air." *Archive for History of Exact Sciences* 6 (1969): 1-28.

Publishes in its entirety a manuscript by Paolo del Buono (1625-1659) in the Bibliothèque Nationale, Paris, which contains "a clear description of a suggested experiment that is identical with the one that Boyle made in 1661 concerning the 'spring of the air.'" The present MS is dated September or October 1657. The MS contains also a description of the principle of the "diagonal barometer," an invention usually ascribed to Sir Samuel Morland about 1675.

* Mügge, Ratje, et al. *Meteorology and Physics of the Atmosphere*. Cited above as item M40.

R29 Neumann, J. "The Sea and Land Breezes in the Classical Greek Literature." *Bulletin of the American Meteorological Society* 54 (1973): 5-8.

An extensively annotated review indicating the great importance attached to winds in the classical Greek literature, the use of land and sea breezes in navigation, and the views of Aristotle and Theophrastus on their origin.

R30 Platzman, George W. "A Retrospective View of Richardson's Book on Weather Prediction." *Bulletin of the American Meteorological Society* 48 (1967): 514-550.

A modern review of Richardson's (1881-1949) classical treatise *Weather Prediction by Numerical Process* (Cambridge University Press, 1922; 236 pp.). This reports on an attempt to predict meteorological conditions by numerical solution of a model composed of differential equations representing atmospheric conditions. The experiment failed then because of inadequate observations and problems with the computations. The review contains not only contemporary opinions but much previously unpublished material from private archives. It also includes much material on the antecedents of Richardson in dynamic meteorology.

R31 Platzman, George W. "Richardson's Weather Prediction." *Bulletin of the American Meteorological Society* 49 (1968): 496-500.

A follow-up to the author's 1967 review with further correspondence of Richardson and other reminiscences sent to him by Richardson's colleagues and correspondents.

R32 Queney, Paul. "Rapport sur le dévéloppement de la météorologie dynamique aux États Unis pendant la période 1937-1947." *Annales de Géophysique* 5 (1949): 89-97.

Reviews the great advances made in dynamic meteorology in the United States in the interval when military exigencies dictated a proliferation of meteorological education and research at a number of universities, with C.G. Rossby (1898-1957) as a driving spirit. A list of 133 references is given.

R33 Siebert, Manfred. "Atmospheric Tides." *Advances in Geophysics* 7 (1961): 105-187.

Includes a review of the history of research since Laplace (1799) and Kelvin (1882). 108 references.

R34 Smagorinsky, J. "The General Circulation of the Atmosphere." *Meteorological Challenges: A History* (item M39), pp. 3-41.

Traces the development of understanding of the general atmospheric circulation from Halley (1686) and Hadley (1735) first to Ferrel (1856) and then to the present era, outlining the principal milestones, with a very good set of references.

R35 Stewart, R.W. "Atmospheric Boundary Layer." *Meteorological Challenges: A History* (item M39), pp. 267-281.

Traces the application of theories of fluid flow to the layer of air near the ground from Navier (1822) onward. Although the Swedish, German, and Russian contributors to the turbulent motions in the boundary layer are mentioned in the text, this rather well-written essay omits many of them from the references.

R36 Thompson, Philip Duncan. "The Mathematics of Meteorology." *Mathematics Today: Twelve Informal Essays*. Edited by L.A. Steen. New York: Springer-Verlag, 1978, pp. 127-152.

Historical sketch, 18th-20th centuries, including Bjerknes' Meteorological Manifesto (1904); work of Jule Charney and others on first successful numerical weather prediction using ENIAC (1948-50), and more recent numerical work.

R37 Weickmann, L. *Über Aerologische Diagrammpapiere*. 2 vols. Internationale Meteorologische Organisation, Internationale Aerologische Kommission. Berlin: Julius Springer, 1938. 123 pp. + 30 plates.

A comprehensive review of the development of thermodynamic graph papers to characterize the vertical structure of the atmosphere, starting from the first type proposed by Heinrich Hertz in 1884; with an extensive bibliography on theory, construction, and use of such diagrams. One volume of text and one volume of diagram types.

R38 Woolard, E.W. "Historical Note on the Deflecting Influence of the Rotation of the Earth." *Journal of the Franklin Institute* 233 (1942): 465-470.

On the contributions of W. Ferrel and others.

* Woolard, Edgar W. "L.F. Richardson on Weather Prediction by Numerical Process." Cited below as item S29.

For additional information on this topic see works cited in the biography sections on the following scientists:

Bjerknes, J.
Bjerknes, V.
Bonaventura, F.
Charney, J.

Dalton, J.
Ertel, H.
Exner, F.M.
Ferrel, W.
Guericke, O. v.
Halley, E.
Karman, T. von
Koschmieder, H.
Krick, I.
Margules, M.
Mariotte, E.
Mieghem, J. van
Mohn, H.
Rossby, C.-G.
Schmidt, C.A. Von
Shaw, W.N.
Sprung, A.F.W.
Taylor, G.I.
Wegener, A.
Weickmann, L.

S. FORECASTING AND SYNOPTIC METEOROLOGY

S1 [Abercromby, R.]. "Weather Forecast Center USA--1888-1969." *ESSA, U.S. Department of Commerce, Environmental Science Services Administration* 5, no. 1 (1970): 8-16.

An account of current chart and forecast products by the National Meteorological Center with an excerpt "Seas and Skies in Many Latitudes," from a book by the British meteorologist Ralph Abercromby (1842-1897), who visited the U.S. Weather Bureau, then part of the Signal Corps in 1888. He gives a lively account of personalities and procedures, including how forecasts were then produced.

* Cressman, G.P. "Dynamic Weather Prediction." Cited above as item R10.

S2 Dettwiller, J. "Le Verrier ou la naissance de la météorologie moderne." *Bulletin d'Information, Ministère des Transports, Direction de la Météorologie* 37 (1977). 22 pp.

Reviews the development of meteorology in France with particular emphasis of the merits of Le Verrier (1811-1877), who introduced synoptic weather maps and forecasts in France in 1855. It contains many other historical personalities in France and elsewhere with portraits and examples of early weather maps.

S3 Dettwiller, J. "En marge des prévisions--les lunaires." *Bulletin d'Information, Ministère des Transports, Direction de la Météorologie* 49 (October 1980): 29-40.

Commentary on astrological and lunar weather calendars, some dating back to the 10th century, with lunar effects still widely accepted in the 18th and 19th centuries. The title on the "margin of predictions" indicates that statistical studies do not support the hypothesis, a principal historical advocate of which was Guiseppe Toaldo (1719-1798). Interestingly illustrated but without references.

S4 Doane, Edith R. "Early Weather Forecasts Sent by U.S. Mail." *Weather* 25 (1970): 529-540.

Mail distribution of daily weather reports from the Signal Corps Weather Service started in December 1872. From 1876 onward Farmers' Bulletins were distributed. There are reproductions of some facsimile covers of mailed weather reports, which contained a synopsis of the past 24 hours and the "indications" for the future. The scheme continued for 35 years, even after the Weather Bureau had become a civilian service in 1890. It eventually folded in 1902 because of high costs for the free distribution. Thereafter, flag signals were substituted on railroad stations and post offices to give weather warnings.

S5 Douglas, C.K.M. "Reviews of Modern Meteorology--4. The Evolution of 20th Century Forecasting in the British Isles." *Quarterly Journal of the Royal Meteorological Society* 78 (1952): 1-21.

Traces the development of forecasting from purely empirical, through frontology, to three-dimensional aerology and attempts at quantitative reasoning.

S6 Friedman, Robert Marc. "Vilhelm Bjerknes and the Bergen School of Meteorology, 1918-1923: A Study of the Economic and Military Foundations for the Transformation of Atmospheric Science." Ph.D. dissertation, Johns Hopkins University, 1979.

This history of introduction of the new concepts of air masses, fronts, and cyclogenesis as well as the discovery of the polar front is based on archival studies of and interviews with the surviving members of the Bergen meteorological group (Jörgen Holmboe, Tor Bergeron). Based on the theoretical hydrodynamic studies of Vilhelm Bjerknes and spawned by World War I exigencies and lack of observations, the need for reliable weather forecasts for food production gave great impetus to the developments on which all of modern meteorology is based. These were strongly supported by the needs of aviation and chemical warfare. The strong Norwegian Fishery interests also gave motivation and support to these efforts. The correspondence of Vilhelm Bjerknes is frequently quoted. The contributions of other members of the team (Thor Hesselberg, E.G. Calwagen, Jacob Bjerknes, H. Solberg, Herbert Petzold) are stressed. The role of the International Convention on Aviation of 1919 and the 1919 Paris Meeting of the International Meteorological Committee for Meteorology is noted. The dissertation is stronger on terminology evolution than on development of physical concepts and it is completely devoid of mathematical formulations, which played a major role in the Norwegian meteorological contributions.

See also an article based on this dissertation, "Constitutin the Polar Front, 1919-1920." *Isis* 73 (1982): 343-362.

S7 Hay, R.F.M. "Some Landmarks in Meteorological Progress 1855-1955." *Marine Observer* 25 (1955): 11-16.

An account of the early difficulties in making storm forecasts for shipping and the gradual improvement in the current century by new technology of observations and vastly improved theory.

S8 Hellmann, Gustav. "Versuch einer Geschichte der Wettervorhersage im XVI. Jahrhundert." *Abhandlungen der Preussischen Akademie der Wissenschaften*, Jahrgang 1924, Physikalisch-Mathematische Klasse no. 1 (1924): 1-54.

A listing of long-range forecasts in calendars between 1500 and 1600, mostly based on astrological ideas.

* Hellmann, G., ed. *Neudrick von Schriften und Karten über Meteorologie und Erdmagnetismus*. Cited above as item M27.

S9 Jewell, Ralph. "The Bergen School of Meteorology--The Cradle of Modern Weather Forecasting." *Bulletin of the American Meteorological Society* 62 (1981): 824-850. Reprinted from *Research in Norway*, 1979, pp. 1-8.

Written by a historian of science, this article gives an account of the fundamental contributions to synoptic meteorology by Vilhelm Bjerknes, Jacob Bjerknes, Halvor Solberg, and Tor Bergeron from 1918 to 1920. It contains verbatim quotations from correspondence and facsimile reproductions of symbols and analyzed weather charts which became the hallmark of the Bergen (Norway) meteorological school.

* Khrgian, A.Kh. *Ocherki razvitiia meteorologii*. Cited above as item M34.

S10 Khromov, Sergei Petrovich. "Sto let nashey sluzhby pogody" [One Hundred Years of Weather Service]. *Meteorologiya i Gidrologiya* No. 10 (1972): 3-22.

A rather succinct history of synoptic meteorology in Russia since 1872 when the Central Physical Observatory in St. Petersburg began forecasting. Personalities and their contributions are given. The account ends with the beginning of World War II.

S11 Landsberg, H. "Storm of Balaklava and the Daily Weather Forecast." *The Scientific Monthly* 79 (1954): 347-352.

A brief history of early work in synoptic meteorology and weather forecasting and the impetus given by the devastating storm of Balaklava during the Crimean War, which stopped the campaign and sank many ships. It led directly to inauguration of weather services in France and England.

S12 McAllen, P.F. "Brief History of Weather Broadcasting." *Weather* 34 (1979): 436-441.

Reviews forecast dissemination by radio and television in England. On November 14, 1922, the BBC included the Meteorological Office forecast for the first time in its evening news service. Trials with presenting a weather chart on TV started in 1936. Now a wide range of weather information is available on both radio and television.

S13 Mook, Conrad P. *A Historical Note on the Development of Systematic Procedures for Short Range Weather Forecasting*. Washington, D.C.: U.S. Weather Bureau, Airport Station, 1949. 6 pp.

Between 1895 and 1906 new procedures for objective systems of weather forecasting were developed by Thomas Russell, Johan August Udden, and Louis Besson. Their charts are reproduced.

* Mügge, Ratje, et al. *Meteorology and Physics of the Atmosphere*. Cited above as item M40.

S14 Namias, J. "Long Range Weather Forecasting--History, Current Status and Outlook." *Bulletin of the American Meteorological Society* 49 (1968): 438-470.

Wexler Memorial Lecture with some reminiscences of Harry Wexler's (1911-1962) role in fostering research on long-range weather forecasting, followed by a chronology of prior work from 1883 onward with detailed references. This is followed by an account of the work of the author and his Weather Bureau collaborators in the preceding two decades.

S15 Namias, Jerome. "The Early Influence of the Bergen School on Synoptic Meteorology in the United States." *Pageoph* 119 (1980/81): 491-500.

Describes the resistance of U.S. Weather Bureau officials to introduction of the Norwegian weather analysis procedures in the 1920s and early 1930s. Presents the role of F.W. Reichelderfer (1895-), C.G. Rossby (1898-1957), and Namias himself in modernizing synoptic meteorology in the United States.

S16 Neumann, J., and D.A. Metaxas. "The Battle Between the Athenian and Peloponnesian Fleets, 429 B.C., and Thucydides' *Wind from the Gulf* (of Corinth)." *Meteorologische Rundschau* 32 (1979): 182-188.

Investigates the validity of a statement by Thucydides in his history of the Peloponnesian War that the commander of the Athenian fleet properly predicted and exploited a sea breeze in a battle in the Gulf of Patras. Modern meteorological data confirm the high frequency of such breezes during the time of year when that battle took place.

S17 Normand, Sir Charles. "Monsoon Seasonal Forecasting." *Quarterly Journal of the Royal Meteorological Society* 79 (1953): 463-473.

Traces the procedures used for predicting the intensity of the Indian monsoon since the first attempts in 1878.

S18 O'Neill, Thomas H.R. *Historical Survey of Statistical Weather Prediction*. U.S. Navy Weather Research Facility NWRF 41-1263-087 (1964). 21 pp.

Describes in chronological fashion the gradual development and various statistical techniques used in weather forecasting with a bibliography of 141 titles.

S19 Pagava, S.T. *Osnovy sinopticheskogo metoda dolgosrochnykh prognozov pogody maloi azblagovremennosti* [Basic Synoptic Method for Long-Range Weather Forecasts for Short Intervals]. Leningrad: Gidrometeorologicheskoe Izdatel'stvo, 1946. 76 pp.

The introduction of this monograph on extended range forecasting reviews the history of research on this topic in Russia starting in 1909. Also briefly reviews American and German work in the field.

S20 Pejml, Karel. "Johann Kepler (1571-1630) a česká meteorologie v jeho době" [Johann Kepler and the Meteorology of the Period in Bohemia]. *Meteorologické Zprávy* 24 (1971): 133-134.

Although Kepler's interest in meteorology centered on attempts to predict the weather from planetary constellations, he contributed to the explanation of origin of solid precipitation.

S21 Pelz, Jürgen. "Der 'Hundertjährige Kalendar.'" *Beilage zur Berliner Wetterkarte* 6/78, SO 2/78, 9 pp. Institut für Meteorologie, Freie Universität Berlin.

The Abbot Moritz Knauer (1612-1664) of Langheim, Bavaria, made seven years of weather observations, 1652-1658, and related them in the then prevalent astrological fashion to the position of sun, moon, Mercury, Venus, Mars, and Jupiter. From these observations he compiled a "perpetual" weather calendar. This was published posthumously in 1701 by a physician named Hellwig under the title "Immer-währender Curieuser Hausskalender," reprinted many times since as "Hundred Year Almanac."

* Philipps, H. "Historische Betrachtungen zur Meteorologie." Cited above as item M48.

S22 Philipps, H. "Rückschauende Betrachtungen zur Meteorologie." *Zeitschrift für Meteorologie* 11 (1957): 290-299.

Traces the history of weather forecasting and weather service from the 19th century onward and the development from empirical rules of forecasting to physico-dynamic procedures.

* Platzman, George W. "A Retrospective View of Richardson's Book on Weather Prediction." Cited above as item R30.

* Platzman, George W. "Richardson's Weather Prediction." Cited above as item R31.

S23 Reed, Richard J. "The Development and Status of Modern Weather Prediction." *Bulletin of the American Meteorological Society* 58 (1977): 390-400.

This memorial lecture in honor of Jacob Bjerknes (1897-1975) is an illustrated review of synoptic meteorology in three sections covering the empirical era from 1860 to 1920, the era of the Norwegian school, 1920-1950, and the post-1950 era, dominated by numerical models. The gradual improvement in forecasts is shown, as is the extension of forecasts to longer time intervals ahead. An extension of Bjerknes' work in the future is outlined. Literature references are very limited.

S24 Reuter, Heinz. "Die Entwicklung der Wettervorhersage in den vergangenen hundert Jahren." *Hundert Jahre Meteorologische Weltorganisation und die Entwicklung der Meteorologie in Österreich.* Zentralanstalt für Meteorologie und Geodynamik in Wien, Publication no. 207 (1975): 43-50.

An essay on the progress in scientific weather prediction, which started about 100 years ago with the universal introduction of synoptic weather maps, eventually leading to numerical prediction with the help of high-speed electronic computers. A large number of Austrians contributed both to theoretical and practical advances in weather prognosis.

S25 Reuter, Heinz. "Weather Prediction Past, Present, and Future." *Proceedings of the Symposium on Current Problems of Weather Prediction, Vienna, June 23-26, 1981.* Zentralanstalt für Meteorologie und Geodynamik, Publikation Nr. 253 (1981): ix-xii.

Reviews the development of weather forecasting from the pre-synoptic period to the present with emphasis on the concepts of the physico-mathematical solutions to the problem.

S26 Schlegel, Max. "Alte meteorologische Karten und Welt-Atlanten." *Berichte des Deutschen Wetterdienstes in the U.S. Zone* 38 (1952): 327-328.

A brief bibliographical note on early weather charts from 1686 to 1899 with illustrations.

S27 Schwerdtfeger, Werner. "Comments on Tor Bergeron's Contribution to Synoptic Meteorology." *Pageoph* 119 (1980/81): 501-509.

Bergeron (1891-1977) made important contributions to the development and acceptance of the Norwegian methods of weather analysis and prediction. Some antecedent research is cited and the changing practices in synoptic meteorology are discussed.

* Seilkopf, Heinrich. "Die meteorologische Tätigkeit der deutschen Seewarte für Seeschiffahrt und Seefischerei 1875-1945." Cited above as item NB64.

S28 Stagg, J.M. *Forecast for Overlord*. New York: W.W. Norton and Co., 1971. 128 pp.

The senior meteorological officer on General Eisenhower's staff gives an account of the forecast for D-day in 1944. The British-American organization of weather service is presented. A judgment is made on efficiency of the system and verification of the separate estimates by various participating meteorologists is made (with some bias). (Reviews by R.T.H. Collis and B.G. Holzman appeared in *Bulletin of the American Meteorological Society* 54 [1973]: 561-563.)

* Walker, J.M. "A Meteorological Curio." Cited above as item PA34.

S29 Woolard, Edgar W. "L.F. Richardson on Weather Prediction by Numerical Process." *Monthly Weather Review* 50 (1922): 72-74.

This is an early appraisal of a milestone in meteorological history (*Weather Prediction by Numerical Process*, Cambridge University Press, 1922). The review includes a good description of slow progress in forecasting and a clear recognition of the powerful potential of Richardson's approach.

For additional information on this topic see works cited in the biography sections on the following scientists:

Abbot, C.
Baur, F.
Bergeron, H.P.
Bjerknes, J.
Charney, J.
Chromov, S.P.
Exner, F.M.
Ficker, H. von
Fitzroy, R.
Guericke, O. v.
Kepler, J.
Kirwan, R.
Krick, I.
Loomis, E.
Montanari, G.
Petterssen, S.
Reichelderfer, F.W.
Richardson, L.F.
Scherhag, R.
Stagg, J.M.
Swoboda, G.J.H.
Truman, H.S.
Vangengeim, A.F.
Vines, B.

T. CLIMATOLOGY, BIOCLIMATOLOGY, AGRICULTURE, ANTHROPOGENIC INFLUENCES

T1 Alisow, B.P., O.A. Drosdow, and E. Rubinstein. "Kurzer geschichtlicher Abriss der Klimakunde." *Lehrbuch der Klimatologie*. Berlin: VEB Verlag der Wissenschaften, 1956, pp. 6-22 [German translation from the Russian: *Kurs Klimatologii*. Leningrad: Gidrometeorologicheskoe Izdatelstvo, 1952].

Presents a brief history of climatology in Russia, beginning with M.W. Lomonosov. The role of Peter the Great is stressed, and the major contributions of Voeikov are eulogized. Work in Russia after the October Revolution and organizational changes are discussed. A lengthy list of names of contributors to advances in the science is given, but no citations are included. One section contrasts climatic research in socialist and capitalist countries, which consists principally of an attack on Ellsworth Huntington and S.F. Markham, both geographers and advocates of climatic determinism in human affairs.

T2 Amelung, Walther. "In Memoriam Bernhard de Rudder." *International Journal of Biometeorology* 7 (1963): 111-112.

Biographical sketch of de Rudder, a physician who wrote extensively on problems of weather and seasons as disease factors and whose book *Grundriss einer Meteorobiologie des Menschen* (1931) appeared in three editions.

T3 Blanc, M.L., and L.P. Smith. "International Agricultural Meteorology." *Agricultural Meteorology* 1 (1964): 3-13.

T4 Brooks, Charles E.P. *Climate Through the Ages: A Study of the Climatic Factors and Their Variations*. Second revised edition. New York: Dover, 1970. 395 pp., illus.

* Brouilette, Benoit. "Quelques observations climatiques en Nouvelle-France au XVIIIe siècle." Cited above as item OD1.

T5 Burgos, Juan J. "Evolución de la meteorología agrícola." *Meteoros* 4 (1954): 108-111.

A brief account of developments in agricultural meteorology since R.A.F. de Réaumur (1683-1757).

T6 Cardy, Michael. "Discussion of the Theory of Climate in the *querelle des anciens et des modernes*." *Studies on Voltaire and the 18th Century* 163 (1976): 73-88.

T7 Court, Arnold. "Climatic Classification and Plant Geography in 1842." *Weather* 22 (1967): 276-282, 287-288.

Calls attention to the completely ignored effort of Richard Brinsley Hinds to devise a world classification of climate and distribution of vegetation. Author tries to explain why it received no attention.

T8 Flohn, Hermann. "Hippocrates und die heutige Meteorologie." *Meteorologische Rundschau* 1 (1948): 355-356.

A brief note calling attention to Hippocrates' (460-377 B.C.) work on the effects of weather and climate on disease.

* Gentilli, J. "A History of Meteorological and Climatological Studies in Australia." Cited above as item NB19.

T9 Gorham, Eville. "Robert Angus Smith, F.R.S., and 'Chemical Climatology.'" *Notes and Records of the Royal Society of London* 36 (1982): 267-272.

Smith (1817-1884) "appears to have been the first person to investigate the urban/industrial nature of 'acid rain'" starting in 1852.

T10 Hagarty, J.H. *History of Climatological Publications*. U.S. Weather Bureau, Key to Meteorological Records Documentation No. 4.1, Washington, D.C., 1958. 34 pp.

An annotated alphabetical list of Weather Bureau publications containing climatological information. Some of these are periodicals dating back to 1872.

T11 Hagarty, J.H. *History of Verification of Weather Records in the United States Weather Bureau*. Key to Meteorological Records Documentation No. 2.01. Washington, D.C.: Weather Bureau, U.S. Department of Commerce, 1962. 16 pp.

Quotes regulations issued for eliminating errors from the meteorological records taken by or for the U.S. Weather Bureau. Relates the development of checking centers to weather records processing centers and their ultimate consolidation (1962) into the National Weather Records Center (now National Climatic Center) at Asheville, N.C. The pamphlet also reports on the gradual adoption of machine processing of weather records.

T12 Hellmann, Gustav. *Entwicklungsgeschichte des klimatologischen Lehrbuches*. Berlin: Beiträge zur Geschichte der Meteorologie, Vol. 3, No. 11, Veröffentlichungen des Preussischen Meteorologischen Instituts No. 315, 1922.

A bibliographic review of textbooks on climatology.

* Hughes, Patrick. "A Century of Cooperation." Cited above as item NB32.

* Hughes, Patrick. "Weather and Crop Service: A Century of Cooperation." Cited above as item NB33.

* Khrgian, A.Kh. *Ocherki razvitiia meteorologii*. Cited above as item M34.

T13 Körber, Hans-Günther. "Über Alexander von Humboldts Arbeiten zur Meteorologie und Klimatologie." Akademie der Wissenschaften, *Alexander von Humboldt, 14.9.1769-6.5. 1859: Gedenkschrift zur 100. Wiederkehr seines Todestages*. Berlin: Akademie Verlag, 1959, pp. 291-335.

Amply annotated analysis of von Humboldt's contributions to climatology, especially horizontal and vertical temperature distribution, atmospheric physics, atmospheric composition, and organization of meteorological observations.

See also: Karl A. Sinnhuber, *Scottish Geographical Magazine* 75 (1959): 89-101.

T14 Landsberg, H.E. "Weather 'Normals' and Normal Weather." *Weekly Weather and Crop Bulletin*, 42, no. 5 (January 31, 1955): 7-8.

Traces the history of the changing use of the term "normal" in meteorology from the 18th century to the middle of the 20th century. The difference in usage in the field of medicine and the arbitrariness of correct use of the term are pointed out. Updated by item T18.

T15 Landsberg, H.E. *The Decennial United States Census of Climate 1960 and Its Antecedents*. Key to Meteorological Records Documentation No. 6.2, U.S. Department of Commerce, Weather Bureau, Washington, D.C., 1960. 23 pp.

Background information for the planned 1960 Census of Climate of the United States. Starting with Thomas Jefferson, Josiah Meigs, and Surgeon General James Tilton, weather observations were periodically summarized for climatological information. This paper documents these efforts and includes a facsimile letter of Jefferson and a tabulation of the number of stations, by states, included in the summaries.

T16 Landsberg, H.E. "Early Stages of Climatology in the United States." *Bulletin of the American Meteorological Society* 45 (1964): 268-274.

The main developments in organizing meteorological networks of observations in the United States in the 19th century are presented, and the first compilations and texts describing the U.S. climate are listed.

T17 Landsberg, H.E. "Roots of Modern Climatology." *Journal of the Washington Academy of Science* 54 (1964): 130-141.

Traces the history of climatology and its developments by contributions of scientists educated in various disciplines, including mathematics, physics, earth sciences, botany, and medicine. The principal milestones from the late 18th century to the middle of the 19th century are traced and the major formative publications are cited.

T18 Landsberg, H.E. "Weather 'Normals' and Normal Weather." *Environmental Data Service*, October 1972, U.S. Department of Commerce, National Oceanic and Atmospheric Administration, Washington, D.C., pp. 8-13.

An updated, illustrated version of item T14.

T19 Lawrence, E.N. "The Meteorology of Anglo-Saxon Month Names." *Weather* 26 (1971): 536-537.

Much annual climatology is contained in the old Anglo-Saxon month names, many of which refer to the state of the crops or weather, such as "stormy month" for March or "dry month" for June.

T20 Le Roy Ladurie, Emmanuel. *Times of Feast, Times of Famine: A History of Climate Since the Year 1000*. Translated by

Barbara Bray. Garden City, N.Y.: Doubleday, 1971. xvi + 426 pp.

T21 Ludlum, David M. "Thomas Jefferson and the American Climate." *Bulletin of the American Meteorological Society* 47 (1966): 974-975.

Outlines Jefferson's interest in climatology, which prompted him to make observations for 50 years. His views on climate were published posthumously in 1829.

* Mügge, Ratje, et al. *Meteorology and Physics of the Atmosphere*. Cited above as item M40.

* Munn, R.E. "Applied Meteorology and Environmental Utilization." Cited above as item M41.

T22 Munzar, Jan. "Alexander von Humboldt and His Isotherms." *Weather* 22 (1967): 360-363.

Shows a reproduction of Humboldt's (1769-1859) map of mean annual isotherms of the Northern Hemisphere, published in 1817. It is the first cartographical representation of climate and was based on the data collection of the Societas Meteorologica Palatina in the late 18th century.

T23 Nash, William P., and Dick M. Whiting. "The Periodic Summarization of United States Climate." *Environmental Data Service*, June 1972, U.S. Department of Commerce, National Oceanic and Atmospheric Administration, Washington, D.C., pp. 4-7.

Summarizations of weather records into so-called Normals in 1842 in the United States, based on Army Medical Department observations. In conformance with practices adopted by the World Meteorological Organization, such summaries are now prepared every 10 years for the preceding 30-year interval. The article lists the summaries proposed for the 1961-1970 interval.

* Oliver, J. "William Borlase's Contribution to Eighteenth-Century Meteorology and Climatology." Cited above as item M46.

T24 Ospovat, Dov. "Lyell's Theory of Climate." *Journal of the History of Biology* 10 (1977): 317-339.

On the ideas of the British geologist Charles Lyell.

T25 Pueyo, Guy. "Du Hamel du Monceau, précurseur des études climatiques et microclimatiques." *Actes, XIIe Congrès International d'Histoire des Sciences, Paris 1968* (pub. 1971) 12: 63-68.

T26 Reichel, Eberhard. "Entwicklungslinien der Klimatologie." *Naturwissenschaftliche Rundschau* 3 (1950): 440-446.

A summary of developments in climatological research over a 25-year period, covering such topics as climatic indices, air-mass climatology, microclimatology, and bioclimatology.

T27 Selianinov, Georgii Timofeevich. "Istoria sozdanya *Klimatov Zemnogo Shara* A.I. Voeikova i ikh znachenie v razvitii klimatologii" [History of A.I. Voeikov's work *The Climate of the Terrestrial Globe* and Its Significance in the Development of Climatology]. *Vsesoiuznoe Geograficheskoe Obshchestvo, SSSR Izvestiya* 80 (1948): 69-83.

Reviews the impact of Voeikov's (1842-1916) book on global climate on contemporary and subsequent climatic work.

T28 Shitara, Hiroshi. "Fifty Years of Climatology in Japan." *Japanese Progress in Climatology*. Tokyo, 1978, pp. 45-80.

A survey of climatological studies in Japan, with very few in existence from the 19th century. Before World War II emphasis was on classical climatological compilations, local studies, and climatic classification. In the post-war era a number of broad-scale studies were undertaken on methodology, microclimate, urban climate, agro-climate, local winds, climatic change, and related topics.

T29 Shitara, Hiroshi. "Fifty Years of Climatology in Japan." *Tohuku University Science Reports*, 7th series, Geography, 28 (1978): 395-430.

Reviews studies on regional climate of Japan, research on local wind fields, methods of analysis, and climatic fluctuations since Japanese work in the field became autochthonous.

* Steinhauser, Ferdinand. "Die Wissenschaftlichen Beiträge der Wiener Zentralanstalt...." Cited above as item NB69.

T30 Ward, R. de C. "Lorin Blodgett's *Climatology of the United States*: An Appreciation." *Monthly Weather Review* 42 (1914): 23-27.

Gives a thorough appraisal of the great textbook published in 1857 (J.B. Lippincott and Co., Philadelphia, 536 pp.) by Blodgett who, for a while, was climatologist for the Smithsonian Institution. It is an excellent summarization of all the early U.S. weather observations with charts and remarkably acute descriptions of climate.

T31 Woeikov (Voeikov), A. "Meteorology in Russia." *Report of the Secretary of the Smithsonian Institution for 1872*. Washington, D.C., 1873, pp. 267-298.

An account of the development of meteorology in Russia and its current status, with extensive tables on climate and the probable reasons for the observed conditions.

For additional information on this topic see works cited in the biography sections on the following scientists:

Baur, F.
Blodgett, L.
Brooks, C.E.P.
Brounov, P.I.
Conrad, V.A.
Cotte, L.
Dorno, C.
Dove, H.W.
Easton, C.
Fassig, O.L.
Fleming, J.
Hann, J.
Hellpach, W.
Hopkins, W.
Howard, L.
Humboldt, A. v.
Huntington, E.
Kassner, C.
Kendrew, W.G.
Knoch, K.
Koeppen, W.P.
Kratzer, A.
Kulik, M.S.
Lamarck, J. de
Lyell, C.
Maurer, J.M.
Meinardus, W.
Molga, M.
Pettersson, H.
Przhevalsky, N.M.
Thornthwaite, C.W.
Van der Stok, J.P.
Vangengeim, A.F.
Voeikov, A.I.

U. GEOMAGNETISM

U1 Allan, T.D. "A Review of Marine Geomagnetism." *Earth-Science Reviews* 5 (1969): 217-254.

"The publication in 1963 of the Vine-Matthews theory of symmetrical magnetic lineations about mid-ocean ridges caused a rapid reappraisal of the magnetometer as a geophysical tool. A review is given of the development of marine magnetic surveys from the first reported profile made with a towed magnetometer (Heezen et al., 1953) up to the present day. The results of magnetic surveys over continental margins and island arcs are reported followed by a review of the earlier work over mid-ocean ridges and rifts. The major part of the review traces the development of sea-floor spreading models" (Author's summary).

U2 Balmer, Heinz. "Beiträge zur Geschichte der Erkenntnis des Erdmagnetismus." *Gesnerus* 13 (1956): 65-81.

A brief survey from Petrus Peregrimus (1269) to Christopher Hansteen (1821).

U3 Bartels, Julius. "Mathematische Methoden der Geophysik." *Applied Mathematics*. FIAT Review of German Science, 1939-1946. Part V, pp. 89-99. Office of Military Government for Germany, Field Information Agencies Technical, British, French, U.S., 1948.

Review of work on application of mathematical statistics to problems in geophysics such as geomagnetism and atmospheric tides.

U4 Biermann, Kurt-R. "Der Brief Alexander von Humboldt an Wilhelm Weber von Ende 1831--ein bedeutendes Dokument zur Geschichte der Erforschung des Geomagnetismus." *Monatsbericht der Deutschen Akademie der Wissenschaften zu Berlin* 13 (1971): 234-242.

U5 Bromehead, C. "Alexander Neckam on the Compass-Needle." *Terrestrial Magnetism and Atmospheric Electricity* 50 (1945): 139-140.

Neckam (1157-1217) gave one of the earliest European accounts of the use of the directional property of a magnet in his *De Naturis Rerum* (1190). (The poet Guiot of Provence gave an account at about the same time.)

U6 Chapman, Sydney, and Julius Bartels. *Geomagnetism*. 2 vols. Oxford: Clarendon Press, 1940. xxviii + x + 1049 pp.

Chapter XXVI (pp. 898-937), "Historical Notes," recounts the major events up to 1850. Extensive bibliography (pp. 938-1007).

U7 Chapman, S[ydney]. "Charles Chree and His Work on Geomagnetism." *Proceedings of the Physical Society of London* 53 (1941): 629-634.

Part I of the first Charles Chree Address, delivered 25 July 1941. "After having gained distinction as an authority on the mathematical theory of elasticity, Charles Chree's interests were turned to the study of geomagnetism by his appointment, at the age of 33, as superintendent of the Kew Observatory. Though critical of current geophysical theories of the geomagnetic variations, and though he himself refrained from speculation on the physical causes of geomagnetic disturbance, his examination of the theoretical views of Maunder and Arrhenius, by the discussion of the magnetic data, led him to his finest achievements" (Author's abstract).

U8 Chapman, S. "Geomagnetic Time-Relationships." *Proceedings of the Physical Society of London* 53 (1941): 635-650.

Part II of the first Charles Chree Address (see above, item U7, for Part I).

"On the average, on days of notable geomagnetic disturbance, the sunspottedness is declining from a maximum value attained two or three days earlier; this is proved most directly by the *method of superposed epochs*, devised by Charles Chree for the study of geomagnetic time-relationships.... By the same method Chree gave the most convincing demonstration of the existence and interval of the 27-day recurrence tendency, which Maunder had previously demonstrated by means of a time-pattern of magnetic storms, and interpreted in terms of the solar rotation. The time-pattern method ... was developed further by Chree and [J.M.] Stagg and perfected by Bartels" (Author's abstract).

U9 Chapman, Sydney. "Edmund Halley and Geomagnetism." *Nature* 152 (1943): 231-237.

U10 Chapman, Sydney. "Alexander von Humboldt and Geomagnetic Science." *Archive for History of Exact Sciences* 2 (1962): 41-51. [A French translation was published in *Ciel et Terre* 75 (1959): 1-16.]

Humboldt's contributions are briefly reviewed in the context of the history of geomagnetism and current research.

U11 Cohen, I. Bernard. "The One Hundredth Anniversary of the Establishment of the Alexander Dallas Bache Magnetic Observatory." *Isis* 33 (1941): 336-337.

Includes an account of the career of Bache and of meetings held in Philadelphia to commemorate his work in geomagnetism.

U12 Cotter, Charles H. "The Royal Society and the Deviation of the Compass." *Notes and Records of the Royal Society of London* 31 (1977): 297-309.

On the correspondence between the Royal Society and the Board of Trade, initiated by Edward Sabine (P.R.S.) in 1865, concerning the use of magnetic compasses on board iron ships.

U13 Deb, S. "Evolution of Thoughts on the Genesis of Magnetic Ore-deposits from Early to Recent Days." *Science and Culture* 31 (1965): 605-610, 32 (1966): 9-14.

On the ideas of Agricola, Descartes, A.G. Werner, J. Hutton, Dolomieu, Elie de Beaumont, Daubree, J. Posepny, L. de Launay, J.E. Spurr, W. Lindgren, W.H. Goodchild, J.M. Cambell, G.W. Morey, H. Schneiderhöhn, J.S. Brown, C. Oftedahl.

U14 French, C.A. "Magnetic Work of the Dominion Observatory, Ottawa, Canada, 1907-32." *Terrestrial Magnetism and Atmospheric Electricity* 37 (1932): 335-342.

U15 Fukushima, N., ed. *Contributions to the Session on History of Geomagnetism and Aeronomy in the Pacific Area* (IAGA Commission IX), prepared for the Second General Scientific Assembly of the International Association of Geomagnetism and Aeronomy, Kyoto, September 1973. 24 pp.

Abstracts of the following papers: "Aspects of the History of Ionospheric Physics in the Asian-Pacific Area" by C.S. Gillmor; "A Summary of Geomagnetic Field Measurements Taken at Hong Kong ... from 1884-1973" by M. Neighbour, D.G. Rivers, and G.O. Walker; miscellaneous contributions of Wilfried Schröder on the history of geophysics and meteorology; magnetic observations in India, Burma, Ceylon, New Zealand, and Australia in the 19th and 20th centuries by B.N. Bhargava and A. Yacob; "Historical and Present Observations in Japan" by Naoshi Fukushima.

U16 Garland, G.D. "The Contributions of Carl Friedrich Gauss to Geomagnetism." *Historia Mathematica* 6 (1979): 5-29.

The contributions were in three areas: absolute measurement of the field, analysis in terms of spherical harmonics, and the organization and equipping of magnetic observatories. "Apparently Gauss spent little time in speculating on the origin of the field."

U17 Glen, William. "The First Potassium-Argon Geomagnetic Polarity Reversal Time Scale, A Premature Start by Martin G. Rutten." *Centaurus* 25 (1981): 222-238.

"In 1959 Rutten determined the magnetic polarities of some Italian rock units that had just been dated in the pioneering, young-rock, potassium-argon laboratory of Jack Evernden and Garniss Curtis at Berkeley. By combining their isotopic dates and his own polarity data, Rutten formulated the first (however crude) calendar of reversals of the earth's magnetic field. Though aware of the value of such a time scale to earth science, he did not follow up on his initial effort; likely, because he lacked training in both radiometry and paleomagnetism" (from Author's summary).

For further information on geomagnetic reversals and the role of this research in the development of plate tectonics, see the author's book *The Road to Jaramillo*, cited above as item HC17.

* Hall, D.H. *History of the Earth Sciences During the Scientific and Industrial Revolutions, with Special Emphasis on the Physical Geosciences*. Cited above as item A48.

U18 Harradon, H.D. "Some Early Contributions to the History of Geomagnetism I." *Terrestrial Magnetism and Atmospheric Electricity* 48 (1943): 3-17.

Contains the translation of a letter of Peter Peregrinus de Maricourt, indicating the existence of two unlike poles.

U19 Harradon, H.D. "Some Early Contributions to the History of Geomagnetism II and III." *Terrestrial Magnetism and Atmospheric Electricity* 48 (1943): 79-91.

Part II is an extract from the *Treatise on the Sphere and the Art of Navigation* by Francisco Falew (1535), giving methods of determining declination; Part III is an extract from the *Brief Compendium on the Sphere and Art of Navigating* by Martin Cortes (1551) on the magnetic needle and its variation.

U20 Harradon, H.D. "Some Early Contributions to the History of Geomagnetism--IV." *Terrestrial Magnetism and Atmospheric Electricity* 48 (1943): 127-130.

Translation (by S. Chapman) of a letter from Georg Hartmann to Duke Albrecht of Prussia, 4 March 1544, announcing the discovery of magnetic inclination, and the first determination of magnetic declination on land (probably made in 1510).

U21 Harradon, H.D. "Some Early Contributions to the History of Geomagnetism--VI." *Terrestrial Magnetism and Atmospheric Electricity* 48 (1943): 200-202.

Translation of a letter (1546) from Gerhard Mercator to the Biship of Arras, in which "we find for the first time the view expressed and substantiated that the earth has a magnetic pole." Mercator located the pole at longitude 168°W, latitude 79°N.

U22 Harradon, H.D. "Some Early Contributions to the History of Geomagnetism--VII." *Terrestrial Magnetism and Atmospheric Electricity* 49 (1944): 185-198.

Translation (by J. de Sampaio Ferray) of extracts on magnetic observations from log books of João de Castro, 1538-1539 and 1541. De Castro made numerous determinations of magnetic declination and discovered the magnetization of rocks. Moreover, "Hellman considered João de Castro as the most outstanding representative of scientific oceanography during the closing years of the age of discovery."

U23 Harradon, H.D. "Some Early Contributions to the History of Geomagnetism--VIII." *Terrestrial Magnetism and Atmospheric Electricity* 50 (1945): 63-68.

Extracts from the English version of the *Portuum Investigandorum Ratio* (*The Haven-Finding Art*, London,

1599, translated by E. Wright). "It contains the oldest list of values of the magnetic declination...."

U24 Hazard, D.L. *The Earth's Magnetism*. Washington, D.C.: U.S. Coast and Geodetic Survey, Serial No. 313, Special Publication, No. 117, 1925. 52 pp.

"Much of the historical material [pp. 2-26] has been taken from the 1902 publication" (*United States Magnetic Declination Tables* by L.A. Bauer). The writings of P. Fleury Mottelay, G. Hellmann, A. Wolkenhauer, P. Timoteo Bertelli, and W. van Bemmelen were also used. No references.

U25 Hazard, Daniel L. "Terrestrial Magnetism in the Twentieth Century." *Journal of the Washington Academy of Sciences* 15 (1925): 111-125.

Survey of contributions of L.A. Bauer; review of ideas about the cause of the earth's magnetism (Nippolat, 1921; Sutherland, 1900; etc.); observations of magnetic storms and auroras, correlation with sun spots.

* Heirtzler, James R. "This Week's Citation Classic.... Marine Magnetic Anomalies, Geomagnetic Field Reversals...." Cited above as item HC20.

* Hellmann, Gustav. *Repertorium der deutschen Meteorologie*. Cited above as item M23.

* Hellmann, Gustav. "Contribution to the Bibliography of Meteorology and Terrestrial Magnetism in the Fifteenth, Sixteenth, and Seventeenth Centuries." Cited above as item M24.

* Hellmann, G., ed. *Neudrucke von Schriften und Karten über Meteorologie und Erdmagnetismus*. Cited above as item M27.

U26 Hennig, R. "Ein Zusammenhang zwischen der Magnetberg-Fabel und der Kenntnis der Kompasses." *Archiv für Kulturgeschichte* 20 (1930): 350-369.

A discussion of ancient and medieval literature on magnetic mountains.

U27 Hennig, R. "Die Frühkenntnis der magnetischen Nordweisung." *Beiträge zur Geschichte der Technik und Industrie* 21 (1932): 25-42.

An historical survey of geomagnetic investigations, in antiquity and the middle ages. 161 references.

U28 Kellner, L. "Alexander von Humboldt and the Organization of International Collaboration in Geophysical Research." *Contemporary Physics* 1 (1959): 35-48.

On geomagnetic observations started in the 1830s.

U29 Knapp, D.G. "Origins of Geomagnetic Science." *Magnetism of the Earth*. [Edited by] A.K. Ludy and H.H. Howe. Washington, D.C.: U.S. Coast and Geodetic Survey, Serial 663, 1945, pp. 58-75.

Historical survey to the mid-19th century.

U30 Körber, H.G. "Über die Hillmannsche Sammlung von Sonnenuhren und Kompassen im Geomagnetischen Institut Potsdam." *Monatsberichte, der Deutschen Akademie der Wissenschaften, zu Berlin* 5 (1963): 641-650.

Description of a collection of 17th- and 18th-century sundials and compass.

U31 Körber, Hans-Gunther. "Zur Vorgeschichte der Grundung des Geomagnetischen Observatoriums Potsdam." *Jahrbuch 1963 des Adolf-Schmidt-Observatoriums für Erdmagetismus in Niemegk*, Berlin: Akademie-Verlag, 1965, pp. 126-133.

On W. Förster's activities in the 1870s.

U32 Konchev, S.K. "Iz istorii Kurskoi magnitnoi anomalii. Otkrytie i izuchenie Kurskoi magnitnoi anomalii v dorevolyutsionnoe vremya." *Voprosy istorii estestvoznaniia i tekhniki* 7 (1959): 67-74.

On the history of the Kursk magnetic anomaly, as studied from the 1880s to the 1950s.

U33 Malin, S.R.C. "British World Magnetic Charts." *Quarterly Journal of the Royal Astronomical Society* 10 (1969): 309-316.

The first chart was produced by E. Halley in 1702. Modern work was stimulated by a series of papers by S. Chapman in 1940-42.

* Mather, Kirtley F., ed. *Source Book in Geology, 1900-1950*. Cited above as item A65.

U34 McConnell, Anita. *Geomagnetic Instruments Before 1900: An Illustrated Account of Their Construction and Use.* London: Harriet Wynter, 1980. 75 pp.

The book includes 15 photographs of instruments in the Science Museum, London. In his review, R.P. Multhauf expressed surprise that so few seem to have survived in view of the intense interest in geomagnetism a century ago [*Isis* 73 (1982): 116-117].

U35 Michel, Henri. "Notes sur l'histoire de la boussole." *Mededelingen, Académie van Marine van Belgie/Communications, Académie de Marine de Belgique* 5 (1950): 161-171.

The compass in ancient China and medieval Europe.

U36 Mitchell, A. Crichton. "Chapters in the History of Terrestrial Magnetism. Chapter II--The Discovery of the of a Magnet in the Earth's Field and the Origin of the Nautical Compass." *Terrestrial Magnetism and Atmospheric Electricity* 37 (1932): 105-146.

Part I deals with the early discoveries in China and knowledge of the compass among the Arabs and Persians before A.D. 1250. Mitchell concludes that the use of the compass in Europe was independent of that in China.

U37 Mitchell, A. Crichton. "Chapters in the History of Terrestrial Magnetism. Chapter II--The Discovery of the Magnetic Declination." *Terrestrial Magnetism and Atmospheric Electricity* 42 (1937): 241-280.

A comprehensive survey, including several quotations from original sources (with English translations) and 152 references.

U38 Mitchell, A. Crichton. "Chapters in the History of Terrestrial Magnetism. Chapter III. The Discovery of the Magnetic Inclination." *Terrestrial Magnetism and Atmospheric Electricity* 44 (1939): 77-80.

The earliest known observations of inclination was by G. Hartmann (1544). R. Norman (1581) measured the dip with better accuracy.

U39 Mitchell, A. Crichton. "Suggestions for Reprinting Articles of Historic Interest in Geomagnetism." *Terrestrial Magnetism and Atmospheric Electricity* 52 (1947): 232.

Calls attention to the historical writings of Fr. Timoteo Bertelli and Fr.E. Schlund.

* Mottelay, Paul Fleury. *Bibliographical History of Electricity and Magnetism*. Cited above as item QB7.

U40 Pushkin, N.F. "The One-Hundred-and-Tenth Anniversary (1938) of the Kazan Magnetic Observatory." *Terrestrial Magnetism and Atmospheric Electricity* 50 (1945): 233-235.

Translated by H.D. Harradon from *Information Book on Terrestrial Magnetism and Electricity* (USSR), 5, Pt. 1, pp. 62-64 (1940).

U41 Ronan, Colin A. "Edmond Halley and Early Geophysics." *Geophysical Journal of the Royal Astronomical Society* 15 (1968): 241-248.

Edmond (not Edmund) Halley (1656-1742) carried out original work on geomagnetism, on the aurora, the behavior and cause of the trade winds, and the age of the earth. He proposed a core-shell model for the earth to account for magnetic variations. His ideas about the antiquity of the earth, based on salinity of the oceans, incited religious opposition and may have delayed his appointment to a professorship at Oxford.

* Runcorn, S.K., ed. *Earth Sciences*. Cited above as item A86.

U42 Schuck, Albert. *Der Kompass*. 3 vols. Hamburg: Selbst-Verlag des Verfassers, 1911-1917. 30 + 58 + 34 pp., 88 plates.

Primarily a collection of illustrations, many reproduced in color, with notes on sources. Volume II includes a 57 (double-column) page essay on the early history of the compass, and a discussion of terms such as magnet, calamita, and bussole.

U43 Smith, Peter J., and Joseph Needham. "Magnetic Declination in Mediaeval China." *Nature, Lond.* 214 (1967): 1213-1214.

New evidence is presented for the priority of discovery of I-Hsing; observations in China and Japan from 720 to 1829 are surveyed.

U44 Srochynskii, R. "Predstavleniia o magnetizme v prevnostie." *Voprosy Istorii Estestvoznaniia i tekhniki* 60, no. 3 (1978): 84-86.

Ancient ideas about magnetism. Linguistic analysis of the word "magnet" suggests where the property of the magnet to attract iron was first discovered and shows that in some countries (especially France) information about the magnet was probably first obtained from China. Moreover, the English name of "magnet" allows one to date its first use in Europe as a compass.

* [Tai, Nieu-Tsu]. "Ancient Records in Modern Chinese Science." Cited below as item V59.

U45 Takeuchi, H., S. Uyeda, and H. Kanamori. *Debate About the Earth. Approach to Geophysics Through Analysis of Continental Drift*. San Francisco: Freeman, Cooper & Co., 1967; revised edition, 1970. 281 pp.

A detailed account of the debate about continental drift and the revivial of the theory in the 1960s, emphasizing the role of geomagnetic evidence.

U46 Wasserfall, K.F. "Hansteen's Magnetic Instrument." *Terrestrial Magnetism and Atmospheric Electricity* 42 (1937): 45-47.

Description of the instrument used by Hansteen from 1843 to 1866 at the Oslo Observatory.

For additional information on this topic see works cited in the biography sections on the following scientists:

Bache, A.D.
Bartels, J.
Bellinsgauzen, F.F.
Carlheim-Gyllensköld, V.
Chapman, S.
Duperrey, L.
Ellis, W.
Elsasser, W.M.
Evans, F.J.O.
Foster, H.
Gauss, C.F.
Halley, E.
Hassler, F.R.
Knott, C.G.
Lamb, H.
Lazarev, P.P.
Lisboa, J.D.
Matuyama, M.
Norman, R.
Ogg, A.
Ricco, A.
Ross, J.C.
Sabine, E.
Schuster, A.
Stewart, B.
Van der Stok, J.P.

V. UPPER ATMOSPHERE, IONOSPHERE, MAGNETOSPHERE, AURORA

V1 Akasofu, S.-I., S. Chapman, and A.B. Meinel. "The Aurora." *Handbuch der Physik* 49, no. 1 (1966): 1-158.

Historical review of theories, pp. 124-129.

V2 Akasofu, Syun-Ichi. "The Aurora and the Magnetosphere." *Planetary and Space Science* 22 (1974): 885-923.

This Sydney Chapman Memorial Lecture gives a chronology of early work on solar-terrestrial relations and later developments leading to an explanation of auroral phenomena. Includes an extensive list of references.

V3 Appleton, E.V. "The Ionosphere." *Occasional Notes of the Royal Astronomical Society* no. 3 (1939): 33-41.

Includes historical remarks.

V4 [Appleton, E.V.]. "Collections Completed, October 1981-March 1982." *Progress Report No. 17, Contemporary Scientific Archives Centre, Oxford* (1982): 3-4.

Reports that CSAC Catalogue no. 82/6/81, 172 pp., has been deposited in the University Library, Edinburgh. "The collection, as finally dispatched, is a substantial one, though concerned almost entirely with Appleton's scientific work in ionospheric physics." There is a brief note on the papers of Sir Edward Bullard under "collections in progress."

V5 Brekke, Asgeir, and Alv Egeland. *Nordlyset. Fra mytologi til romforskning*. Oslo: Grondahl & Son Forlag A.S., 1979. 143 pp.

Well-illustrated history of auroral research, including contributions of C. Hansteen, H.C. Oersted. A.J. Angström, K.S. Lemström, E. Edlund, S. Tromholt, A. Paulsen, F.C.M. Størmer, L. Vegard, K.O.B. Birkeland; survey of modern magnetospheric research

V6 Brekke, Asgeir. "Ancient Norwegian Literature in Relation to the Auroral Oval." *Exploration in the Polar Upper Atmosphere*. Edited by C.S. Deehr and J.A. Holtet. Boston: Reidel, 1980, pp. 431-442.

V7 Briggs, J.M., Jr. "Aurora and Enlightenment. Eighteenth Century Explanations of the Aurora Borealis." *Isis* 58 (1967): 491-503.

Discusses the reports by E. Halley, J.J. Dortous de Mairan, J.P. Maraldi, and L. Euler. In their attempts to explain the phenomena "all the proper ingredients were mentioned: terrestrial magnetism, electricity, the glowing of particles of the atmosphere, even the role of the sun in triggering the aurora."

V8 Carpenter, D.L. "This Week's Citation Classic. Carpenter, D.L. Whistler Studies of the Plasmapause in the Magnetosphere. 1. Temporal Variations in the Position of the Knee, and Some Evidence on Plasma Motions near the Knee. *J. Geophys. Res.* 71: 693-709, 1966." *Current Contents, Physical, Chemical & Earth Sciences* 20, no. 5 (February 4, 1980): 16.

The author recalls the circumstances of writing this paper, which has been cited over 295 times since 1966.

V9 Chamberlain, Joseph W. "Theories of the Aurora." *Advances in Geophysics* 4 (1958): 109-215.

Includes a detailed review of the theory of C. Stormer (1907, 1911-12), and discusses the later work of W.H. Bennett and E.O. Hulburt (1954), S. Chapman and V.C.A. Ferraro (1931-33, 1940), and others. 113 references.

V10 Chapman, Sydney. "History of the Aurora and Airglow." *Aurora and Airglow*. Edited by B.M. McCormac. Proceedings of the NATO Advanced Study Institute, held at the University of Keele, Staffordshire, England, August 15-26, 1966. New York: Reinhold Pub. Corp., 1967, pp. 15-28.

Brief survey with bibliography of original sources.

V11 Chapman, Sydney. "Historical Introduction to Aurora and Magnetic Storms." *Annales de Geophysique* 24 (1968): 497-505.

"The uncertainties as to the height of the aurora and the misconceptions as to its spectrum, at the time when [K.] Birkeland began his work in this field, are indicated;

also the mistaken views then current about a connection between the aurora and cirrus clouds.... Birkeland, though greatly interested in the aurora, and active in promoting auroral observations, contributed thereto mainly indirectly, through the stimulus he gave and the auroral interest he aroused in [C.] Störmer, [O.] Krogness, [L.] Vegard and their successors. His electron beam theories of auroras and magnetic storms were untenable, but he was a leading pioneer in magnetic storm studies at a time when they were much neglected. Through the magnetic records he obtained by his arctic expeditions, which he coordinated with data from many other observations, he contributed greatly to our understanding of magnetic storm morphology, and was the first discoverer of the auroral electrojet."

V12 Chapman, Sydney. "Auroral Science, 1600-1965, Toward Its Golden Age?" *Atmospheric Emissions: Proceedings of the* [NATO Advanced Study] *Institute*. Edited by B.M. McCormac and Anders Omholt. New York: Van Nostrand Reinhold Co., 1969, pp. 11-25.

Review of early auroral theories, discovery of the influence of the earth's magnetic field, magnetic storm hypothesis, the vastly expanded information gathered during the International Geophysical Year 1957/58; but even better understanding lies still ahead.

V13 Chapman, Sydney. "Auroral Physics." *Annual Review of Astronomy and Astrophysics* 8 (1970): 61-83.

Includes historical review of studies of the aurora since Halley's first published description in 1716.

V14 Cowling, T.G. "Vincent Ferraro as a Pioneer of Hydromagnetics." *Quarterly Journal of the Royal Astronomical Society* 16 (1975): 136-144.

Discusses the history of the Chapman-Ferraro theory of geomagnetic storms (1930), with comments on the work of H. Alfvén.

V15 Curtis, S.A. "The Use of the German V-2 in U.S. for Upper Atmosphere Research." *Journal of the British Interplanetary Society* 32 (1979): 442-448.

Project Hermes "which originally was heavily oriented toward developing missile technology took a strong scientific slant." But this article is more concerned with hardware than scientific results.

V16 Dobson, G.M.B. "Forty Years' Research on Atmospheric Ozone at Oxford: A History." *Applied Optics* 7 (1968): 387-405.

"The development of research on atmospheric ozone at Oxford is traced from the year 1922, when one single instrument was used there to make measurements of the total ozone, to 1906, when some hundred instruments were distributed all over the world. In recent years an important advance has been made by the measurement of the vertical distribution of the ozone in the atmosphere. A digression, covering the war years, describes the measurement of the amount of water vapor in the stratosphere which, however, proved to have a bearing on the effect of the ozone on the temperature of the upper atmosphere" (Author's abstract).

The cover of this issue has a photograph of participants in the Paris Ozone Conference of 1929.

V17 Eather, Robert H. *Majestic Lights. The Aurora in Science, History, and the Arts*. Washington, D.C.: American Geophysical Union, 1980. ix + 323 pp.

Comprehensive history of auroral research with many illustrations.

V18 Egeland, A., Ø. Holter, and A. Omholt, eds. *Cosmical Geophysics*. Oslo: Universitetsforlaget, 1973. 360 pp.

The "historic preamble" by the editors, pp. 11-17, discusses aurora, ionospheric research, terrestrial magnetism, and solar radiation, with a chronology for the period 1600-1900.

V19 Farley, Donald T. "This Week's Citation Classic: Farley, D.T., Jr. A Plasma Instability Resulting in Field-Aligned Irregularities in the Ionosphere. *J. Geophys. Res.* 68: 6083-97, 1963." *Current Contents, Physical, Chemical & Earth Sciences* 21, no. 51 (December 21, 1981): 22.

Farley recalls the circumstances of his original research leading to the writing of this paper, which has been cited over 265 times since 1963.

V20 Farley, T.A. "The Growth of Our Knowledge of the Earth's Radiation Belt." *Reviews of Geophysics* 1 (1963): 1-34.

V21 Feldstein, Yasha I. "This Week's Citation Classic: Feldstein, Y.I. & Starkov, G.V. Dynamics of Auroral Belt and Polar Geomagnetic Disturbances. *Planet. Space Sci.* 15: 209-29, 1967." *Current Contents, Physical, Chemical & Earth Sciences* 20, no. 19 (May 12, 1980): 14.

The author recalls the circumstances of writing this paper and discusses its impact on the field; the paper has been cited over 120 times since 1967.

V22 Gardiner, G.W. "Origin of the Term 'Ionosphere.'" *Nature* 224 (1969): 1096.

The term was proposed by R.A. Watson-Watt in 1926.

V23 Gillmor, C.S. "The History of the Term 'Ionosphere.'" *Nature* 262 (1976): 347-348.

The term was suggested by R.A. Watson-Watt and E.V. Appleton in separate letters in November 1926, and in print by Watson-Watt in 1929.

V24 Gillmor, C.S. "The Early History of Upper Atmospheric Physics Research in Antarctica." *Upper Atmosphere Research in Antarctica*. Edited by L. Lanzerotti and C. Park. Antarctic Research Series, vol. 29. Washington, D.C.: American Geophysical Union, 1978, pp. 236-262.

"This chapter discusses the historical circumstances of the first application of several specialties and new techniques of upper atmospheric physics research to Antarctica; radio propagation studies, Kennelly-Heaviside echo soundings, VLF phenomena and whistlers, cosmic ray studies, and certain details concerning early attempts at photographic recording of the aurorae." The first ionospheric echo-sounding measurements were taken by Malcolm P. Hansen in 1929 and 1930. John N. Dyer began studies of VLF phenomena in 1934. Cosmic ray research was started by Erwin H. Bramhall and Arthur A. Zuhn, in 1933.

V25 Gillmor, C. Stewart. "Threshold to Space: Early Studies of the Ionosphere." *Space Science Comes of Age. Perspectives in the History of the Space Sciences*. Edited by Paul A. Hanle and Von Del Chamberlain. Washington, D.C.: National Air and Space Museum/Smithsonian Institution Press, 1981, pp. 101-104.

"This paper examines early concepts of the form or profile of electron density above the earth in the ionosphere and the effects of radio measuring instruments upon those concepts." Evidence for an ionized upper layer came from the work of L.F. Fuller and Lee de Forest in

1912-14, and later of Edward Appleton, J.A. Ratcliffe, Merle Tuve and Gregory Breit. In 1931 Sydney Chapman presented a theoretical model to account for the formation of an ionized layer. Whether postwar research with rockets and satellites simply confirmed earlier ideas or completely changed them is a matter of controversy and viewpoint.

V26 Gillmor, C. Stewart. "Wilhelm Altar, Edward V. Appleton, and the Magneto-Ionic Theory." *Proceedings of the American Philosophical Society* 126 (1982): 395-440.

"I will discuss the genesis of the magneto-ionic theory including previously unknown aspects of its history, based upon archival research and oral interviews made over the last 10 years, and discuss especially the contributions of the Austrian Wilhelm Altar. Until 1976 it was virtually unknown that Altar collaborated with Appleton on magneto-ionic theory in 1925-26...." The appendix includes a translation of Altar's MS, "Wave Propagation in Ionized Gases Under the Influence of a Magnetic Field," written in 1925-26.

V27 Gillmor, C. Stewart, and Douglas Gran. "Research in Ionospheric Physics." *Proceedings of the XIIIth International Congress of the History of Science, Moscow, 1971* (pub. 1974) 6: 160-164.

Preliminary report on a statistical study of papers published after 1920.

V28 Gillmor, C.S., and C.J. Terman. "Communication Modes of Geophysics: The Case of Ionospheric Physics." *Eos* 54 (1973): 900-908.

Statistical analysis of published articles, and results of a survey in which respondents assessed the usefulness of journals, meetings, and correspondence.

V29 Günther, S. "Wann enstand die erste wissenschaftliche Theorie des Polarlichtes?" *Mitteilungen zur Geschichte der Medizin, der Naturwissenschaften und der Technik* 14 (1915): 231-240.

"The first to offer a scientific theory of the aurora borealis was the Norwegian Jens Kristian Spidberg (1684-1762). Of course this was a purely optical theory."--G. Sarton, *Isis* Critical Bibliography No. 8, p. 341.

V30 Harrison, D.N. "British Radiosonde: Its Debt to Kew." *Meteorological Magazine* 98 (1969): 186-190.

Traces the development and testing of the British radiosonde since 1937. Design of the latest of these upper-air sounding devices is given.

V31 Healy, T.R. "Development of Meteorological Exploration in the Upper Air Over Australia." *Weather* 27 (1972): 440-447.

After a review of early surface observations in Australia, emphasis is on determinations of upper winds by cloud observations with nephoscopes since 1895, by free balloons since 1916, by kites and pilot balloons since 1921, and by airplane observations since 1937.

* Hellmann, G. "Die Meteorologie in den deutschen Flugschriften und Flugblättern des XVI. Jahrhunderts...." Cited above as item M26.

V32 Kennel, Charles F. "This Week's Citation Classic: Kennel, C.F. & Petschek, H.E. Limit on Stably Trapped Particle Fluxes." *J. Geophys. Res.* 71: 1-28, 1966." *Current Contents, Physical, Chemical & Earth Sciences* 20, no. 30 (July 28, 1980): 20.

The author recalls the circumstances of writing this paper, which has been cited more than 510 times since 1966.

V33 Kosibowa, Stefania. "Historia badań oblaków mesoferycznych" [Historical Review of Mesospheric Clouds]. *Prezcgląd Geofysiczny* 16 (1971): 341-345.

A historical review of observations of noctilucent clouds in both hemispheres since 1885.

V34 Link, Frantisek. "On the History of the Aurora Borealis." *Vistas in Astronomy* 9 (1967): 297-306.

A brief summary of ancient and medieval reports.

V35 Maeda, Ken-ichi. "Ionospheric Research." *Science in Japan*. Edited by Arthur H. Livermore. Washington, D.C.: AAAS, 1965, pp. 199-238.

Survey of Japanese research since 1931.

V36 Murgatroyd, R.J. "Upper Atmosphere Meteorology." *Meteorological Challenges: A History* (item M39), pp. 127-156.

A brief review of recent work on the structure of the upper atmosphere, with a more historical account of views on its composition and its motion. References are nearly all to work of the three decades preceding publication of article.

V37 Newell, Homer E. *Beyond the Atmosphere. Early Years of Space Science*. Washington, D.C.: National Aeronautics and Space Administration (document number NASA SP-4211), 1980. xviii + 497.

A history of scientific research made possible or significantly aided by rockets, satellites, and space probes.

V38 Oosterkamp, W.J., and E.M. Bruins. "Discovery of Radiation Belts." *Physics Today* 20, no. 2 (February 1967): 9-13.

Oosterkamp points out in a brief letter that E.M. Bruins implied the existence of a belt of charged particles at a distance of 2-4 earth radii in 1935. A long letter from Bruins recounts the background of his discovery, and notes that he sent 400 copies of his thesis (with a long English summary) to other scientists and libraries all over the world, so there is no excuse for its having been ignored.

V39 Petrie, William. *Keoeeit--The Story of the Aurora Borealis*. New York: Macmillan, 1963. 134 pp.

Includes brief historical section.

V40 Ratcliffe, J.A., et al. "A Discussion of the Early Days of Ionospheric Research." *Philosophical Transactions of the Royal Society of London* 280 (1975): 3-55. Special issue.

Contents: J.A. Ratcliffe, "The Early Ionosphere Investigations of Appleton and His Colleagues," pp. 3-9; A.H. Waynick, "The Early History of Ionospheric Investigations in the U.S.," pp. 11-25; W. Dieminger, "Trends in Early Ionospheric Research in Germany," pp. 27-34; Sir F. White, "Early Work in Australia, New Zealand, and at the Halley Steward Laboratory, London," pp. 35-46; W.J.G. Benyon, "URSI and the Early History of the Ionsphere," pp. 47-55.

Proceedings of a meeting, 5-6 December 1974. The speakers discussed the work of E.V. Appleton, M.A.F. Farnett, C.G. Darwin, A.E. Kennelly, E.O. Hulburt, A.H. Taylor, H. Lassen, A.L. Green, G.H. Munro, F.W.G. White, M.V. Wilkes, and others.

V41 Rossi, V. "On the Development of Experimental Aerology in Finland." *Vaisala News* No. 58 (1973): 3-36.

A history of the Finnish meteorological research which started in 1840, with special emphasis on kite, balloon, and radiosonde exploration of the upper atmosphere.

V42 Schmauss, August. "Die Entdeckung der Stratosphäre im Jahr 1902." *Meteorologische Rundschau* 5 (1952): 161-163.

In May of 1902, Richard Assmann (1845-1918) and Leon Philippe Teisserenc de Bort (1855-1913) almost simultaneously discovered the existence of the stratosphere.

V43 Schröder, Wilfried. "Zur Vorgeschichte der Erforschung der leuchtenden Nachtwolken." *Beiträge zur Geophysik* 80 (1971): 371-374.

O. Jesse pioneered observations of noctilucent clouds in the interval 1885-1901.

V44 Schröder, W. "Wilhelm Foerster und sein Einfluss auf die Erforschung der leuchtenden Nachtwolken." *Zeitschrift für Meteorologie* 23 (1973): 302-305.

An account of Foerster's (1832-1921) interest in noctilucent clouds and in observations of anomalous twilight phenomena as a result of the Krakatoa eruption in 1883. His accounts of efforts to make systematic observations of these phenomena are cited from the reports of the Berlin astronomical observatory, of which he was director. His ardent support of O. Jesse (1838-1901) in noctilucent cloud observations is cited.

V45 Schröder, W. "Das Wirken von Otto Jesse fur die Erforschung der Hochatmosphäre." *Archive for History of Exact Sciences* 11 (1973): 267-272.

Summary of the research of Jesse (1838-1901) on the upper atmosphere, especially on noctilucent clouds.

V46 Schröder, W. "Zur Erforschung der leuchtenden Nachtwolken in den Jahren 1885-1901." *Zeitschrift für Meteorologie* 23 (1973): 296-301.

An account of early work on observation and attempts at interpretation of noctilucent clouds in the interval 1885-1901, with special reference to the merit of O. Jesse in exploring these rare events.

V47 Schroeder, W. *Entwicklungsphasen der Erforschung der leuchtenden Nachtwolken* (Eine Darstellung unter besonderer Beruecksichtigung der Beitraege der Berliner Sternwarte der Akademie der Wissenschaften). Berlin: Akademie-Verlag, 1975. vi + 64 pp.

Surveys the history of research on noctilucent clouds at the Berlin Observatory, starting in 1885, including the contributions of O. Jesse and W. Foerster.

V48 Schröder, W. "Aspekte einer Geschichte der Polarlichtforschung." *Acta Geodaetica, Geophysica et Montanistica, Academia Hungarica* 11 (1976): 137-142.

Cites important sources for a history of research on auroras.

V49 Schröder, W. "Otto Jesse--Wegbereiter der Erforschung der Hochatmosphäre." *Meteorologische Rundschau* 29 (1976): 94.

A brief note on the pioneering efforts of Jesse (1838-1901) to make precise measurements of noctilucent clouds. The article contains first publication of a letter by Jesse.

V50 Schröder, Wilfried. "Some Aspects of History of Auroral Research." *EOS Transactions of the American Geophysical Union* 60 (1979): 1035-1036.

Auroras were probably known in antiquity, although the first description did not occur until 1561 when Conrad Gessner (1516-1565) published an account. In the middle of the 18th century the relation to magnetic fluctuations was noted. In the 19th century the relation to sunspots was established and charts of equal frequency of observations were published.

V51 Schröder, Wilfried. "Zur Geschichte der Polarlichtforschung." *Physikalische Blätter* 35 (1979): 160-164.

A review of auroral observations from antiquity to the 20th century. Principal emphasis is on the contributions of Conrad Gessner (1516-1565), Mairan, L. Euler, M.W. Lomonosov, W. Boller, R. Wolf, H. Fritz, and C. Störmer (1874-1957).

V52 Schroeder, Wilfried. "Friedrich S. Archenhold und des Berliner Programm zur Erforschung der Hochatmosphaere." *NTM* 17, no. 1 (1980): 49-60.

V53 Schroeder, W. "Alfred Wegener und die Physik der Hochatmosphaere." *Astronomische Nachrichten* 302 (1981): 197-201.

"Extracts from the papers of A.W. relating to noctilucent clouds, aurora (including the so-called geocoronium) and other upper atmospheric phenomena are presented here so as to give an idea of his contribution to the subject made during the years of 1900-1930" (Author's summary).

V54 Schröder, Wilfried. "Hermann Fritz, Wegbereiter der Polarlichtforschung." *Vierteljahrsschrift der Naturforschenden Gesellschaft in Zürich* 126 (1981): 199-204.

The contributions of Hermann Fritz (1830-1893) to aurora research are appraised. Fritz taught at the Zurich Institute of Technology.

V55 Silverman, S.M. "Franklin's Theory of the Aurora." *Journal of the Franklin Institute* 290 (1970): 177-178.

Franklin proposed "a theory in which a meridional circulation of the atmosphere transported warm, moist, electrified air from equatorial to polar regions, where it condensed as snow and, because of the poor conducting properties of the polar ice cover, was discharged through the vacuum of the upper atmosphere...." Followed by a survey article by Silverman, "The Polar Cap as a Distinct Geophysical Entity," pp. 179-196.

V56 Singer, S.F. "Research in the Upper Atmosphere with Sounding Rockets and Earth Satellite Vehicles." *Journal of the British Interplanetary Society* 11 (1952): 61-73.

Cites the early history of upper atmospheric explorations in the United States, starting in 1945 with the use of V-2, Aerobee, and Viking rockets; makes an early plea for the use of satellites.

V57 Siscoe, George L. "An Historical Footnote on the Origin of 'Aurora Borealis.'" *EOS* 59 (1978): 994-997.

The metaphor "aurora borealis" (northern dawn) did not originate with P. Gassendi, who first published it in 1649, but appeared in 1619 in a work at least partially written by Galileo (*Discourse on the Comets*).

* Siscoe, George L. "Solar-Terrestrial Relations: Stone Age to Space Age." Cited below as item W6.

* Special Committee for the International Geophysical Year. *The Histories of the International Polar Years and the Inception and Development of the International Geophysical Year*. Cited above as item C24.

V58 Stothers, Richard. "Ancient Aurora." *Isis* 70 (1979): 85-95.

A survey of Greek and Roman reports. Analysis of data in the interval 223 B.C. to 91 B.C. shows a cycle of 11.5 years.

V59 [Tai, Nieu-Tsu]. "Ancient Records in Modern Chinese Science." *New Scientist* 69 (1976): 233.

Brief note on an article in *Kexue Tongbao* (no. 10, 1975, p. 1975) which reports the use of aurora observations going back to the 1st century B.C. to infer shifts in the geomagnetic pole.

For additional information on this topic see works cited in the biography sections on the following scientists:

Appleton, E.V.
Bergman, T.O.
Carlheim-Gyllensköld, V.
Chapman, S.
Dalton, J.
Dobson, G.M.B.
Franklin, B.
Glaisher, J.
Gutenberg, B.
Heaviside, O.
Mitra, S.K.
Molchanov, P.A.
Rayleigh, R.J.S.
Scherhag, R.
Schmidt, A.
Schmidt, C.A. Von
Stewart, B.
Störmer, F.C.M.
Swinden, J.H. Van

W. LUNAR-SOLAR-TERRESTRIAL RELATIONS

W1 Felber, H.-J. "Kant's Beitrag zur Frage der Verzögerung der Erdrotation." *Die Sterne* 50 (1974): 82-90.

W2 Fick. "Historical Notice in Reference to the Retardation of the Earth's Velocity of Rotation." *Philosophical Magazine*, series 4, 31 (1866): 322.

Calls attention to I. Kant's 1754 work on this subject, preceding that of J.R. Mayer (1848).

W3 Fleming, John Ambrose. "The Sun and the Earth's Magnetic Field." *Annual Report of the Board of Regents of the Smithsonian Institution for the Year Ended June 30, 1942*. Washington, D.C.: U.S. Government Printing Office, 1943, pp. 173-208.

Semi-historical review.

W4 Meadows, A.J., and J.E. Kennedy. "The Origin of Solar-Terrestrial Studies." *Vistas in Astronomy* 25 (1982): 419-426.

The development and contemporary impact of 19th-century solar-terrestrial studies are examined under the headings: solar variations; magnetic variations; auroral variations; meteorological variations.

W5 Medunin, A.E."Izuchenie v Rossii deformatsii Zemli, vyzvannykh lunno-solnechnym pritiazheniem (1892-1920)" [The Study in Russia of the Deformation of the Earth Caused by Lunar and Solar Attraction (1892-1920)]. *Trudy, Institut istorii estestvoznaniia i tekniki, Akademiia Nauk SSSR* 43 (1961): 151-164.

On the work of G.V. Levitskii (1852-1917), A.Ya. Orlov (1880-1954), and L.S. Leibenzon (1879-1951).

W6 Siscoe, George L. "Solar-Terrestrial Relations: Stone Age to Space Age." *Technology Review* 78, no. 3 (January 1976): 26-37.

Survey of reports on auroral displays, paleomagnetism, and implications for archeology.

W7 Weiss, J.E., and N.O. Weiss. "Andrew Marvell and the Maunder Minimum." *Quarterly Journal of the Royal Astronomical Society* 20 (1979): 115-118.

Evidence for disappearance of sunspots in late 17th century.

For additional information on this topic see works cited in the biography sections on the following scientists:

Bartels, J.
Chapman, S.
Childrey, J.
Ellis, W.
Hopkins, W.
Laplace, P.
Mendel, G.
Ricco, A.
Rodés, L.

NAME INDEX

SUBJECT INDEX